Crystallization Process Systems

Crystallization Process Systems

A.G. Jones

*Professor of Chemical Engineering,
Department of Chemical Engineering,
UCL (University College London),
London, UK*

OXFORD AMSTERDAM BOSTON LONDON NEW YORK PARIS
SAN DIEGO SAN FRANCISCO SINGAPORE SYDNEY TOKYO

Butterworth-Heinemann
An imprint of Elsevier Science
Linacre House, Jordan Hill, Oxford OX2 8DP
225 Wildwood Avenue, Woburn, MA 01801-2041

First published 2002

Copyright © 2002, A.G. Jones. All rights reserved

The right of A.G. Jones to be identified as the author of this work has been asserted
in accordance with the Copyright, Designs and Patents Act 1988

All rights reserved. No part of this publication
may be reproduced in any material form (including
photocopying or storing in any medium by electronic
means and whether or not transiently or incidentally
to some other use of this publication) without the
written permission of the copyright holder except in
accordance with the provisions of the Copyright,
Designs and Patents Act 1988 or under the terms of a
licence issued by the Copyright Licensing Agency Ltd,
90 Tottenham Court Road, London, England W1T 4LP.
Applications for the copyright holder's written permission
to reproduce any part of this publication should be addressed
to the publishers

British Library Cataloguing in Publication Data
A catalogue record for this book is available from the British Library

Library of Congress Cataloguing in Publication Data
A catalogue record for this book is available from the Library of Congress

ISBN 0 7506 5520 8

For information on all Butterworth-Heinemann publications visit our website at www.bh.com

Typeset by Integra Software Services Pvt. Ltd., Pondicherry, India
www.integra-india.com
Printed and bound in Great Britain by Biddles Ltd
www.biddles.co.uk

FOR EVERY TITLE THAT WE PUBLISH, BUTTERWORTH-HEINEMANN
WILL PAY FOR BTCV TO PLANT AND CARE FOR A TREE.

Contents

Preface — xii
Nomenclature and abbreviations — xv

1 Particulate crystal characteristics — 1
Crystal characteristics — 1
   Crystal structure — 1
   Crystal symmetry — 2
   Crystal systems — 3
   Crystal habit — 4
   Polymorphs and enantiomorphs — 4
   Crystal defects — 6
   Particle characteristics — 7
Single particles — 7
   Particle size — 7
   Equivalent dimensions — 9
   Particle shape — 9
   Shape factors — 9
   Fractals — 12
Particle size distributions — 12
   Frequency histogram — 12
   Mass fractions — 12
   Mean particle size — 14
   Mean sizes of distributions — 14
   Specific surface (S_p) — 16
   Coefficient of variation (CV) — 16
Packing characteristics — 17
   Voidage and internal surface area — 17
   Effect of size distributions — 18
Techniques for particle sizing and characterization — 18
   Zone sensing — 19
   Laser diffraction — 19
   The sieve test — 19
   Mesh number (N) — 20
   Sieve test method — 21
   Data presentation and analysis from sizing tests — 21

2 Fluid–particle transport processes — 26
Particle–fluid hydrodynamics — 26
Sedimentation — 27
   Drag on particles — 28
   Force balance — 29

Laminar flow: Stokes law ($Re_p < 0.2$)	29
Particle Reynolds number	30
Terminal velocity in Stokes' flow ($Re_p < 0.2$)	30
Turbulent flow	30
Slurries	31
Hindered settling	31
Hindered settling rates	32
Hindered drag coefficients	33
Polydisperse suspensions	34
Sedimentation of slurries	35
Typical settling rates	36
Mass flux	37
Fluid flow through porous media	37
Laminar flow	38
Compressible beds	40
Empirical methods	41
Turbulent flow	41
Friction factors	41
Suspension of settling particles by agitation	43
Energy dissipation	44
Conservation equations	45
Computational fluid dynamics (CFD)	47
Simulation of the fluid dynamics in agitated vessels	48
Mixing models	49
Macromixing and micromixing	49
Mesomixing and micromixing	50
Phenomenological or mechanistic models	50
Physical models	51
The population balance	52
Numerical methods	56
3 Crystallization principles and techniques	**58**
Phase equilibria	58
Solubility	58
Crystallization and precipitation modes	61
Supersaturation	62
Types of crystallizer	63
Unstirred vessels	63
Agitated vessels	63
Draft-tube and baffled (DTB)	64
Swenson forced circulation (FC)	65
Oslo fluidized-bed	65
The mixed suspension, mixed product removal (MSMPR) crystallizer	65
Crystallizer modelling	66
Crystal yield	66
Mass balance	66

Prediction and analysis of CSD	67
Continuous MSMPR operation	68
Population balance	68
MSMPR crystal size distribution (CSD)	69
Crystal number, area and mass	71
Mass modal size	73
Mean sizes of the CSD	74
Mass mean size	75
Mass median size	75
Coefficient of variation	75
Anomalous crystal growth	75
Precipitation	76
Phase transformation	77
Microscopic analysis	78

4 Solid–liquid separation processes — 80

Settling tanks	80
Thickeners	82
Filters and filtration	87
Methods of filtration	88
Types of filter	88
Filter press	88
Rotary vacuum filter	88
Belt filter	90
Theory of filtration	90
Modified Darcy equation	90
Modes of operation	92
Effect of the filter medium	95
Simplified filtration equation	96
Leaf test	97
Optimum cycle time during batch filtration	101
Maximizing filter throughput	102
Pre-coat and bodyfeed	103
Continuous rotary vacuum filtration	103
Operating principle	103
Submergence fraction	104
Operating equations	104
Effect of drum speed	105
Centrifuges	106
Classification of centrifuge types	106
Theory of centrifugation	107
Solid bowl centrifuge	107
Thin layer solid bowl centrifuge	109
Centrifugal filtration	112
Hydrocyclones	114
Crystal washing	116

	Crystal drying	116
	Equilibrium moisture content	116
	Drying time	120
	Types of drier	120
5	**Crystal formation and breakage**	**123**
	Crystallization and precipitation kinetics	123
	Nucleation	124
	Primary nucleation	125
	Homogeneous nucleation	125
	Heterogeneous nucleation	125
	Crystal growth	125
	Diffusion-reaction model	127
	Surface integration models	128
	Hydrodynamic determinants of crystal growth rates	130
	Induction periods	131
	Case 1: $t_n \gg t_g$	132
	Case 2: $t_g \gg t_n$	133
	Case 3: $t_n \cong t_g$	133
	Interfacial tension	135
	Determination of crystal growth and nucleation kinetics	135
	Batch methods	135
	Continuous methods	136
	Particle breakage processes	136
	Nature of comminution	137
	Single particle breakage	137
	Typical comminution process	138
	Energy for size reduction	139
	Prediction of product particle size distribution	141
	Population balance	141
	Hydrodynamic determinants of attrition	142
	Impact attrition	143
	Turbulent attrition	143
	Secondary nucleation	146
	MSMPR crystallizer studies	147
	Mechanisms of secondary nucleation	148
	Secondary nucleation kinetics	148
	Scale-up of secondary nucleation kinetics	150
	Growth of secondary nuclei, small crystals and attrition fragments	151
	Effect of supersaturation on secondary nucleation	151
6	**Crystal agglomeration and disruption**	**155**
	Agglomerate particle form	155
	Agglomerative crystallization	155
	Primary agglomeration	157
	Secondary agglomeration	160

Colloid and surface science	161
Particle forces	162
Surface charge	162
Particle interactions	163
Steric interaction	163
Adsorption	164
Stability	164
Population balance	166
Analytical approximations	168
Agglomeration and disruption during precipitation	171
Collection efficiency	173
Agglomerate strength	174
Determination of precipitation kinetics	175
Parameter estimation	177
Growth and nucleation	177
Agglomeration kernel	178
Smoluchowski kernel (Smoluchowski, 1916)	178
Thompson kernel (Thompson, 1968)	179
Invariant agglomeration kernel	179
Disruption kernel	179
Experimental method	180
Crystal growth rates	181
Nucleation rates	182
Disruption kernel	184
Agglomeration kernel	185
Troubleshooting agglomeration	187
7 Crystallizer design and performance	**190**
Design and performance of batch crystallizers	190
Techniques	190
Equipment	191
Batch crystallizer design and operation	192
Vessel sizing	192
Operating policy	192
Performance measures	192
Heat and mass balances	192
Prediction of CSD	193
Population balance	194
Control of supersaturation	195
Programmed cooling crystallization	197
Optimal operation	197
Programmed cooling crystallizer	199
Fines destruction during batch crystallization	201
Programmed precipitation	201
Design and performance of continuous crystallizers	203
MSMPR interactions	203

Supersaturation	204
Design equations for MSMPR crystallizers	205
Effect of residence time on mean crystal size	207
Crystal size-dependent product removal crystallizers	210
Fines destruction	210
Classified product crystal withdrawal	212
Crystallizer dynamics	212

8 Crystallizer mixing and scale-up 215

Liquid–liquid crystallization systems 215

Mixing model: Segregated Feed Model (SFM)	216
SFM applied to continuous precipitation	217
Determination of mixing times using Computational Fluid Dynamics (CFD)	220
Model solutions	221
Model validation	221
Continuous mode of operation	221
Volume mean size L_{43}	223
Semibatch mode of operation	223
Effect of feed point position	226
Scale-up criteria	227
SFM methodology	228
Probability Density Function (PDF) methods	229

Gas–liquid crystal precipitation 231

Industrial gas–liquid precipitation reactions	232
Types of reactor	234
Role of gas–liquid mass transfer	236
Gas–liquid reactor modelling	250
Reactor design	255

Microcrystallizers 257

9 Design of crystallization process systems 261

Crystallization process systems design 261

Crystallization process equipment	262
Crystallization design data	263

Crystallization process models 264

Crystal size distribution	264
Median crystal size	264
Crystallizer volume	265
Solids hold-up	265
Solid–liquid separation	265
Particle characteristics	266
Crystallization kinetics	267
Crystal slurry filtrability	268
Improving solid/liquid separability	269
Effect of downstream processing on product characteristics	270

Crystallization process synthesis	271
Costs	271
Crystallization process optimization	272
Multi-component systems	275
Biotechnological systems	276
Crystallization process synthesis procedure	277
Crystallization process systems simulation	277
Machine learning methodology	278
Stage-wise crystallization process synthesis	279
Network synthesis methodology	281
Problem formulation for crystallizer networks	282
Crystallization process instrumentation and control	287
Batch crystallizers	287
Continuous crystallizers	289
Physical model based control	291
Black box modelling for dynamics	293
Crystallization in practice	295
Scale up	295
Crystal encrustation	296
Crystal caking	296
Sustainability	297
References	299
Author index	329
Subject index	335

Preface

Crystallization from solution is a core technology in major sectors of the chemical process and allied industries. Crystals are produced in varying sizes ranging from as small as a few tens of nanometers to several millimetres or more, both as discrete particles and as structured agglomerates. Well-established examples include bulk and fine chemicals and their intermediates, such as common salt, sodium carbonate, zeolite catalysts and absorbents, ceramic and polyester pre-cursors, detergents, fertilizers, foodstuffs, pharmaceuticals and pigments. Applications that are more recent include crystalline materials and substances for electronics devices, healthcare products, and a wide variety of speciality applications. Thus, the tonnage and variety of particulate crystal products worldwide is enormous, amounting to about half the output of the modern chemical industry. The economic value, social benefit and technical sophistication of crystal products and processes are ever increasing, particularly in the newer high added value sectors of global markets. This places yet greater demands on the knowledge, skill and ingenuity of the scientist and engineer to form novel materials of the required product characteristics and to devise viable process engineering schemes for their manufacture.

Particulate crystallization processes often require subsequent solid–liquid separation. Thus, the unit operation of crystallization is normally only part of a wider processing system. These systems should preferably be designed and optimized as a whole – problems detected in one part of the plant (poor filtration say) may in fact arise in another (inadequate crystallizer control). Attention to the latter rather than the former can result in a simpler, cheaper and more robust solution. Similarly, the scale of crystallizer operation can have a large effect on crystal product characteristics and hence their subsequent separation requirements. Previously a largely empirical art, the design of process systems for manufacturing particulate crystals has now begun to be put on a rational basis and the more complex precipitation processes whereby crystallization follows fast chemical reactions have been analysed more deeply. This progress has been aided by the growing power of the population balance and kinetic models, computational fluid dynamics, and mixing theory. This not only increases understanding of existing processes but also enhances the possibility of innovative product and process designs, and speedier times to market. Several large gaps in knowledge remain to be filled, however, thereby providing opportunities for further research. This perspective gives the reason for writing the book, and provides its theme.

Crystallization Process Systems brings together essential aspects of the concepts, information and techniques for the design, operation and scale up of particulate crystallization processes as integrated crystal formation and solid–liquid separation systems. The focus of the book, however, is on crystallization; only dealing with related unit operations as far as is necessary. It is therefore not intended to be comprehensive but is designed to be complementary to existing texts on the unit operations of crystallization, solid–liquid separation and allied techniques, and important sources of further detailed information are given for the interested reader. The work is presented initially at an introductory level together with examples while later providing a window into the details of more advanced and research topics. Particular attention is paid both to the fundamental mechanisms and the formulation of computer aided mathematical methods, whilst emphasizing throughout the continuing need for careful yet efficient practical experimentation to collect basic data; the latter being essential in order to discriminate between competing theories, to inform and validate process models, and to discover the unexpected.

The book is mainly based on undergraduate BEng and MEng, and graduate MSc lecture courses that I have given at UCL since 1980 on Particulate Systems and Crystallization, respectively, together with those on a continuing professional development (CPD) course on Industrial Crystallization. To these is added the results of several research projects. Consequently, the book is aimed equally at students, researchers and practitioners in industry, particularly chemical engineers, process chemists and materials scientists.

The book is divided into 9 chapters. For the guidance of lecturers and students, Chapters 1–4 largely comprise the earlier undergraduate taught topics together with some enhanced material; with Chapters 4–8 mainly covering the more advanced and research topics. Chapter 9 should be of particular help to those undertaking crystallization process design projects. Thus, Chapter 1 provides the definition of the basic characteristics that are common to all particulate crystals, notably their molecular structure, particle size, size distribution and shape. This is followed in Chapter 2 by the relevant transport processes of chemical engineering as applied to crystallization, namely the hydrodynamic factors affecting crystallizers and solid–liquid separation equipment, viz. the motion of single particles, slurries and suspensions in both agitated and quiescent vessels, and flow through porous media, respectively. Then follows a more advanced description of mixing models, the unified population balance approach to the analysis and prediction of particle size distributions and its coupling with fluid flow. Having provided the theoretical basis, these chapters naturally lead on to an introduction to the fundamentals and techniques of crystallization and related precipitation processes *per se* in Chapter 3. As mentioned above, the next process stage is normally solid–liquid separation and drying. Thus, simple procedures for the design and operation of gravity settlers, thickeners, filters, centrifuges and driers are considered relatively briefly in Chapter 4, this time in the light of the underlying solid–liquid transport theory covered in Chapter 2. Particle formation processes occurring within crystallizers, viz. nucleation, crystal growth, agglomeration and breakage, together with methods for their determination, are then considered in greater depth in Chapters 5 and 6. Then in Chapter 7, the design and performance of well-mixed batch and continuous crystallizers is considered with the population balance from Chapter 2 underlying their theoretical analysis. Chapter 8 extends these considerations to crystallization systems where imperfect fluid–particle-mixing is an important factor, and is illustrated in detail in respect of the scale up of precipitation processes. Finally, in Chapter 9 the design of crystallization process systems is considered as a whole, bringing together the essential features of the analysis of particle formation and solid–liquid separation processes described earlier with the logic of process synthesis, optimization and control, again illustrated by some examples.

I am indebted to John Mullin for his encouragement during the writing of this book. Many other academic and industrial colleagues are also worthy of thanks, together with visitors and research students in the UCL crystallization group over the last two decades. Too numerous to mention individually, they each contributed much for which I am most grateful. I hope that each will see some of their many contributions reflected with due acknowledgement in the text. Thanks are also due to present and former students for their patience in working through the examples and giving feedback on courses. Finally, but most importantly, my special thanks are due to Judith, Robert and Stephen for their continual support during the preparation of the manuscript.

Alan Jones
University College London
October 2001

Nomenclature and abbreviations

a_i	coefficients
A	area, m^2
a_L	specific surface area based on liquid volume, m^{-1}
a_R	specific surface area based on reactor volume, m^{-1}
B^0	nucleation rate, #/m^3 s
B_a	birth rate of particles due to aggregation, #/m^3 s
B_d	birth rate of particles due to disruption, #/m^3 s
B_I	birth rate in region I, #/m^3 s
B_{II}	birth rate in region II, #/m^3 s
B_L	source function, #/m^3 s
B_n	birth rate due to nucleation, #/m^3 s
c	solution concentration, kg anhydrous solute/kg solvent
c^*	equilibrium saturation concentration, kg anhydrous substance/kg solvent
C	cooling rate (dθ/dt), K s^{-1}
C_p	specific heat of solution, kJ kg^{-1}
Cr	contraction factor at the inlet of riser
Δc	concentration driving force ($c - c^*$), kg anhydrous substance/kg solvent^{-1}
d	diameter, m
D_a	death rate of particles due to aggregation, #/m^3 s
d_B	bubble diameter, m
D_d	death rate of particles due to disruption, #/m^3 s
D_d	rate of particle death due to disruption, #/m^3 s
d_L	droplet diameter, m
D_M	molecular diffusivity, m^2 s^{-1}
D_p	particle diffusivity, m^2 s^{-1}
D_a	death rate of particles due to aggregation, #/m^3 s^{-1}
D_d	death rate of particles due to disruption, #/m^3 s^{-1}
E	energy of contact, kJ
E_L	longitudinal dispersion coefficient of liquid, m^2 s^{-1}
$\dot{E}_t$	rate of energy transfer to crystals by collision, kJ s^{-1}
f	mixture fraction for the non-reacting fluid $= c_A^0/c_{A0}$
$f(\alpha)$	fraction of crystals exiting CFR at point α
f_{ij}	fraction of crystals exiting discretized CFR at point α_{ij}
f_s	surface shape factor (area $= f_s L^2$)
f_v	volume shape factor (volume $= f_s L^3$)
f	collision frequency, exchange rate between zones I and II
F	crystal bed permeability, m^{-2}
F	flow rate of gas, m^3 s^{-1}
F	fraction of solvent evaporated, kg solvent evaporated/kg original solvent
G	linear growth rate, m s^{-1}
G'	mean velocity gradient, fluid shear rate, s^{-1}

g	gravitational acceleration, m s^{-2}
H	column height, m
Ha	Hatta number
h_c	enthalpy of crystallization, kJ kg^{-1}
h_{lg}	enthalpy of vaporization, kJ kg^{-1}
J_n	primary nucleation rate, #/m^3 s
J_s	secondary nucleation rate, #/m^3 s
k	Boltzmann constant, 1.38054×10^{-22} J K^{-1}
K	Kozeny coefficient
k	reaction rate constant, m^3 mol^{-1} s^{-1}
K_a	aggregation constant
k_b	number nucleation rate coefficient
K_d	disruption constant
k_g	crystal growth rate constant
k_i	kinetic rate coefficients
k_L	mass transfer coefficient, m s^{-1}
k_n	nucleation rate constant
k_{sp}	solubility product
k'_g	mass growth rate coefficient
k_m	mass nucleation rate coefficient
K_N	rate coefficient
K_R	rate coefficient
L	characteristic dimension, size co-ordinate, crystal size, m
L_0	size of nuclei, m
L_{so}	initial size of seed crystals, m
L_p	size of product crystals, m
M_c	mass of crystals, kg
M_l	mass of liquor in crystallizer, kg
M_g	mass of solvent evaporated, kg
M_h	mass solvent in crystallizer, kg
M_{anh}	molecular mass of anhydrous substance, kg kmol^{-1}
M_{hyd}	molecular mass of hydrated substance, kg kmol^{-1}
M_p	product crystal mass, kg
M_{so}	initial mass of seed crystals, kg
M_T	suspension, or magma, density, kg m^{-3}
$n(L,t)$	population density distribution function, #/m^4
n^0	nuclei population density (m^{-1} m^{-3})
N	total number of particles, #/m^{-3}
N_0	initial total number of particles, #/m^{-3}
N_a	agitator speed, Hz
N_{so}	initial number of seed crystals, #/kg solvent
n_C	number of crystals, #/m^{-3}
n_I	number density of crystals in region I, #/m^4
n_{II}	number density of crystals in region II, #/m^4
P	crystal production rate, kg s^{-1}
P	degree of agglomeration
p	population density of particles, #/m^4

P_0	power number
p_b	pressure at bottom, Pa
P_I	power input, W
P_N	power number
p_t	pressure at top, Pa
$q(\alpha)$	fraction of feed entering the system at point α
q_{ij}	fraction of feed entering discretized CFR at point α_{ij}
Q	volumetric throughput, m^3 s^{-1}
Q	heat input, kJ
Q_0	pumping (flow) number
Q_m	mol flow rate, mol s^{-1}
r	rate of chemical reaction, mol/m^3/s
R	universal gas constant, 8.315 J mol^{-1} K^{-1}
R	ratio of molecular masses (M_{hyd}/M_{anh})
Re	Reynolds number
S	relative supersaturation
S	co-ordinate of crystal number
Sc	Schmidt number
Sh	Sherwood number
S_p	bed specific surface area, m^{-1}
t	time, s
T	temperature, K
u, v	particle volume, m^3
u_{pi}	local instantaneous value of particle velocity
V	volume, m^3
x	distance from gas–liquid interface, m
X	solids moisture content
Y	dimensionless crystal size, $(L_p - L_{so})/L_{so}$
Y	gas humidity
Y	Young's modulus of elasticity, N m^{-2}
Z	dimensionless time, t/τ

Superscripts

b	'order' of nucleation
c	shear rate dependence of maximum agglomerate size
g	'order' of growth
i	relative kinetic order ($=b/g$)
j	magma density dependence of nucleation rate
$*$	equilibrium value

Subscripts

c	crystal
f	final, at time τ
g	growth
I	instantaneous
i	summation index
j	summation index

l	liquor (solution)
mm	mass mean
n	nuclei
0	initial, or zeroth
s	speed
t	time, t
wm	weight mean

Greek

α	time along the length of a network
β_a	aggregation kernel, $m^3 s^{-1}$
β_d	disruption kernel, $m^3 s^{-1}$
$\delta(L-L)$	Dirac delta function of crystal size, m
ε	voidage fraction
ε	specific power input, $W m^{-3}$
$\dot{\gamma}$	shear rate, #/s
μ	fluid viscosity, Pa s
μ_j	jth moment of size distribution
$\phi(f)$	probability density function for the statistics of fluid elements
ν	kinematic viscosity, $m^2 s^{-1}$
ρ	fluid density, $kg m^{-3}$
ρ_c	crystal density, $kg m^{-3}$
ρ_L	liquid density, $kg m^{-3}$
ρ_s	particle density, $kg m^{-3}$
σ	Poisson's ratio, absolute or relative supersaturation
τ	mean residence time in vessel, s
ϕ	coefficient
θ	solution temperature, K
τ	batch time, mean residence time in vessel, s
ψ	coefficient
η_T	impeller target efficiency

Others

$\langle \rangle$	ensemble average
–	average or mean value
#	number of particles or crystals

Abbreviations

ATR	Attenuated Total Reflection
CSD	Crystal Size Distribution
CV	Coefficient of Variation
FTIR	Fourier Transform Infra Red
HSE	Health, safety and environment
MFB	Micro Force Balance
MSMPR	Mixed Suspension, Mixed Product Removal
PDF	Probability Density Function
PSD	Particle Size Distribution

1 Particulate crystal characteristics

Crystals are familiar to everyone, common examples being salt and sugar. Less common but more alluring are diamonds and other gemstones. More prosaic are the innumerable crystals manufactured in the bulk, fine chemical and pharmaceutical industries, in both primary products, secondary formulations and their intermediates. As illustrated by the crystal products in Table 1.1, their range is immense and includes some highly sophisticated materials. Similarly, worldwide production rates and value are ever increasing.

Crystallization is an important separation process that purifies fluids by forming solids. Crystallization is also a particle formation process by which molecules in solution or vapour are transformed into a solid phase of regular lattice structure, which is reflected on the external faces. Crystallization may be further described as a self-assembly molecular building process. Crystallographic and molecular factors are thus very important in affecting the shape (habit), purity and structure of crystals, as considered in detail by, for example, Mullin (2001) and Myerson (1999). In this chapter the internal crystal structure and external particle characteristics of size and shape are considered, which are important indicators of product quality and can affect downstream processing, such as solid–liquid separation markedly. Larger particles separate out from fluids more quickly than fines and are less prone to dust formation whilst smaller particles dissolve more rapidly.

Crystal characteristics

Crystal structure

The ideal solid crystal comprises a rigid lattice of ions, atoms or molecules, the location of which are characteristic of the substance. The regularity of the internal structure of this solid body results in the crystal having a characteristic shape; smooth surfaces or faces develop as a crystal grows, and the planes of these faces are parallel to atomic planes in the lattice.

A particular feature of crystals is that the angles between two corresponding faces of a given substance are constant reflecting their internal lattice structure

Table 1.1 *Some industrially important particulate crystals*

Bulk chemicals	Catalysts	Ceramic precursors
Detergents	Dyestuffs	Electronic materials
Fertilizers	Fibre intermediates	Fine chemicals
Foodstuffs	Mineral ores	Pharmaceuticals
Polymers	Proteins	Speciality chemicals

(see McKie and McKie, 1986). The relative size of the individual faces can vary, however, changing the overall growth shape, or habit, but corresponding angles remain constant. Thus, in practice, two crystals rarely look exactly alike due to the different conditions experienced during their growth histories (Mullin, 2001).

Crystal symmetry

Many (but not all) of the geometric shapes that appear in the crystalline state are readily recognized as being to some degree symmetrical, and this fact can be used as a means of crystal classification. The three simple elements of symmetry that can be considered are

1. symmetry about a point (a *centre* of symmetry)
2. symmetry about a line (an *axis* of symmetry)
3. symmetry about a plane (a *plane* of symmetry).

The cube for example, has 1 centre of symmetry, 13 axes of symmetry and 9 planes of symmetry (Figure 1.1). An octahedron also has 23 elements of symmetry and is therefore crystallographically related to the cube.

Some crystals exhibit a fourth element known as 'compound, or alternating' symmetry about a 'rotation–reflection axis' or 'axis of rotary inversion'.

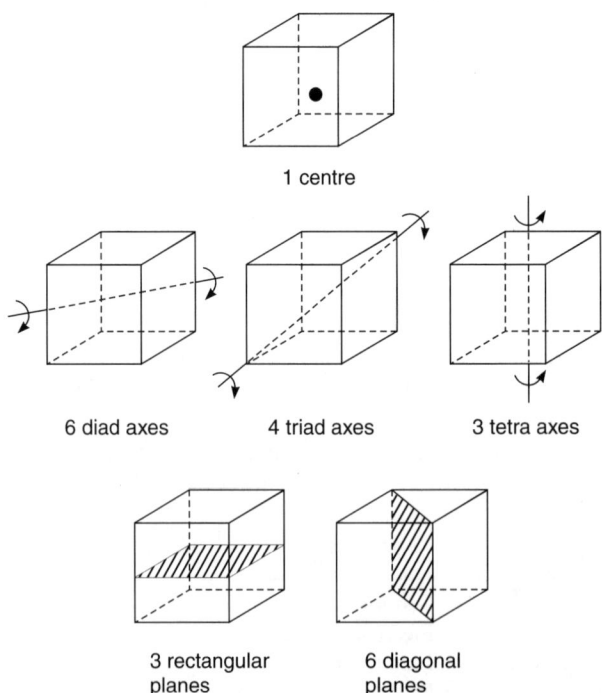

Figure 1.1 *Symmetry in the cube*

Crystal systems

There are only 32 possible combinations of the above-mentioned elements of symmetry (including the asymmetric state), and these are called the 32 classes or point groups. For convenience these 32 classes are grouped into 7 systems characterized by the angles between their $\underline{x}$, $\underline{y}$ and $\underline{z}$ axes. Crystals of course, can exhibit combination forms of the crystal systems.

Miller indices are used to numerically define the shape of crystals in terms of their faces. All the faces of a crystal can be described and numbered in terms of their axial intercepts (usually three though sometimes four are required). If, for example, three crystallographic axes have been decided upon, a plane that is inclined to all three axes is chosen as the standard or parametral plane. The intercepts $\underline{X}$, $\underline{Y}$, $\underline{Z}$ of this plane on the axes $\underline{x}$, $\underline{y}$, and $\underline{z}$ are called parameters $\underline{a}$, $\underline{b}$ and $\underline{c}$. The ratios of the parameters $\underline{a{:}b}$ and $\underline{b{:}c}$ are called the axial ratios, and by convention the values of the parameters are reduced so that the value of $\underline{b}$ is unity.

The Miller indices, $\underline{h}$, $\underline{k}$ and $\underline{l}$ of the crystal faces or planes are then defined by

$$\underline{h} = \frac{a}{X}, \quad \underline{k} = \frac{b}{Y} \quad \text{and} \quad \underline{l} = \frac{c}{Z} \tag{1.1}$$

For the parametral plane, the axial intercepts $\underline{X}$, $\underline{Y}$, and $\underline{Z}$ are the parameters $\underline{a}$, $\underline{b}$ and $\underline{c}$, so that indices, $\underline{h}$, $\underline{k}$ and $\underline{l}$ are $\underline{a/a}$, $\underline{b/b}$ and $\underline{c/c}$, i.e. 1, 1 and 1, usually written (111). The indices for the other faces of the crystal are calculated from the values of their respective intercepts $\underline{X}$, $\underline{Y}$, and $\underline{Z}$, and these intercepts can always be represented by $\underline{ma}$, $\underline{nb}$ and $\underline{pc}$, where m, n and p whole numbers or infinity.

For the cube (Figure 1.2) since no face is included none can be chosen as the parametral plane (111). The intercepts $\underline{Y}$ and $\underline{Z}$ of face $\underline{A}$ on the axes y and z are at infinity, so the Miller indices $\underline{h}$, $\underline{k}$ and $\underline{l}$ for this face will be $\underline{a/a}$, $\underline{b/\infty}$ and $\underline{c/\infty}$, or (100). Similarly, faces $\underline{B}$ and $\underline{C}$ are designated (010) and (001), respectively. Several characteristic crystal forms of some common substances are given in Mullin (2001).

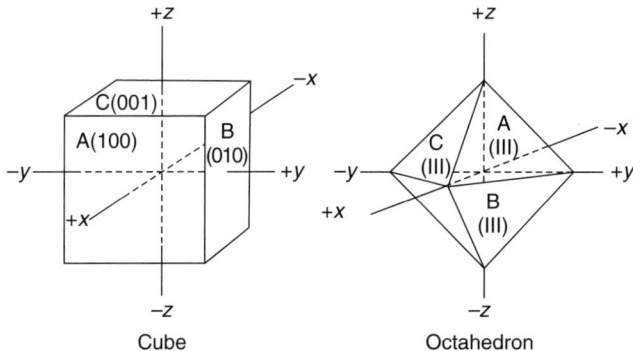

Figure 1.2 Miller indices: cube and octahedron

Crystal habit

Crystal habit is the shape a crystal adopts according to the proportions occupied by each crystal face. Following the heterogeneous equilibrium ideas of Gibbs (1948) applied to crystals by Curie (1885), Wulff (1901) suggested that the equilibrium shape of a crystal is related to the free energies of the faces and suggested that the crystal faces would grow at rates proportional to their respective free energies. Bravais (1866), Friedel (1907) and Donnay and Harker (1937) initially investigated the relationship between the external structure of a crystal and its internal molecular arrangement, the studies of whom are collectively known as the BFDH theory. They offered a quick approach for identifying the most likely crystallographic forms that dominate a crystal habit. Later, the periodic bond chain (PBC) theory due to Hartman and Perdok (1955) quantified crystal morphology in terms of the interaction energy between crystallizing units. X-ray crystallography data provide the necessary asymmetric co-ordinates of the unit cell for each crystal. These data are then used to predict the crystal morphology. Molecular modelling now offers a sophisticated method for predicting inorganic and molecular crystal shapes (see Docherty *et al.*, 1988; 1991; Coombes *et al.*, 1996, 1997; Price and Wibly, 1997; Willock *et al.*, 1995; Myerson, 1999; Potter *et al.*, 1999 and Lommerse *et al.*, 2000).

Many substances can, of course, exhibit more than one form, or habit. It is also worth noting that although crystals can be classified according to the seven crystal systems, the relative sizes of the faces of a particular crystal can vary considerably depending on their relative crystal growth rates (Chapter 5). In general very fast growing faces have little or no effect on the overall crystal habit, which is governed by the slowest growing faces.

Crystal habit can be markedly affected by the presence of the solvent, or by impurities, which are often used deliberatively as additives industrially. Changes to the structure of the interface between crystal and solvent may occur differentially across the faces, which then grow at different rates. Similarly, impostor molecules may differentially be incorporated into the growing crystal lattice whilst offering an incompatible functionality and stereochemistry to impede further growth of specific faces, which then dominate the crystal habit. Davey and Garside (2000) and Lahav and Leiserowitz (2001), respectively, provide extensive accounts of the effect of impurities and solvents on crystal growth and morphology.

Polymorphs and enantiomorphs

Isomorphous substances crystallize in almost identical form whereas a polymorph is a substance capable of crystallizing into different, but chemically similar, forms. For example, calcium carbonate can appear as calcite (trigonal–rhombohedral), aragonite (orthorhombic) and vaterite (hexagonal), whilst it is well-known that carbon can occur in two forms of very differing values, graphite (hexagonal) and diamond (regular), respectively. Interestingly, calcium carbonate can also precipitate as an amorphous, non-crystalline, form under certain circumstances.

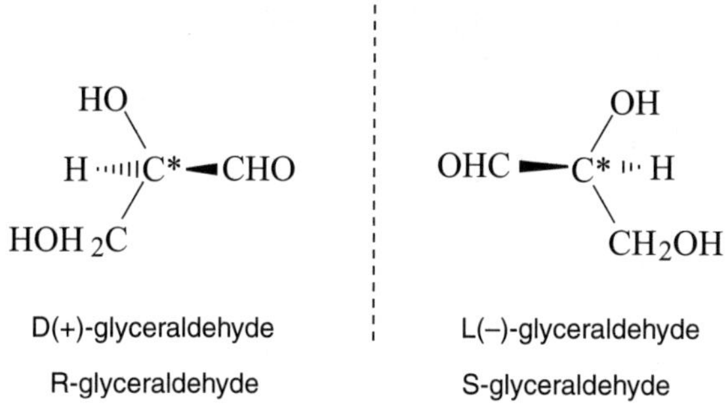

```
        HO                              OH
          \                              /
    H ····||C*◄─CHO         OHC─►C* ··· H
          /                              \
      HOH₂C                            CH₂OH

   D(+)-glyceraldehyde         L(–)-glyceraldehyde
    R-glyceraldehyde            S-glyceraldehyde
```

Figure 1.3 *Enantiomer pair of glyceraldehyde*

Enantiomers, or optical isomers, exist in pairs and represent an important and common case of stereoisomerism. Stereoisomerism is exhibited by molecules which are made up of the same atoms and functional groups, but which are arranged spatially, or configured, in a different way. Enantiomer pairs are non-superimposable mirror images of each other, as illustrated by the simple example of glyceraldehyde in Figure 1.3. Here, four different groups are arranged tetrahedrally around the asterisked carbon atom to give the two mirror-image configurations shown.

Crystallization continues to be the most widely used method of separating or resolving enantiomers (optical resolutions). The manufacture of chemicals and pharmaceuticals as purified optical isomers, or enantiomers, has taken on a pivotal importance in the pharmaceutical, agricultural and fine chemicals industries over the past 15–20 years. Crystallization has been and continues to be the most widely used method of separating or resolving enantiomers (optical resolutions), and is particularly well suited to separations at large scale in manufacturing processes (Jacques *et al.*, 1981; Roth *et al.*, 1988; Wood, 1997; Cains, 1999).

Enantiomer pairs possess identical physical properties, but their biological activities and effects can be markedly different. A dramatic example of this was the thalidomide problem in the 1960s, where the drug was administered as a mixture of the two enantiomer forms. One form mediated the desired effect as a remedy for morning sickness; the other form was severely teratogenic. More recently, the isolation and purification of single enantiomer products has become an important component of manufacture. In 1992, for example, the US FDA promulgated a set of 'proper development guidelines' for drug actives, which requires that best efforts be made to isolate and purify enantiomeric products. Generic pharmaceuticals that are resolved by crystallization include amoxycillin, ampicillin, captopril, cefalexin, cefadroxil, chloramphenicol, ethambutol, α-methyldopa and timolol.

Crystallization methods are widely used for the separation, or resolution, of enantiomer pairs. Enantiomer mixtures may essentially crystallize in two different ways. In around 8 per cent of cases, each enantiomer crystallizes separately, giving rise to a mechanical mixture of crystals of the two forms, known as a conglomerate. Conglomerates may usually be separated by physical methods alone. In the remaining 92 per cent of cases, the two enantiomeric forms co-crystallize in equal amounts within the unit cell of the crystal, giving rise to a racemic crystal structure. Where such racemic crystals form, separation cannot be achieved by physical means alone, and is usually brought about by reacting the enantiomer pair with a single-enantiomer resolving agent, to form a pair of diastereomer adducts. Diastereomers are stereoisomers for which the mirror–image relationship does not exist, and which often exhibit significant differences in their physical properties that enable their separation by crystallization.

By far the commonest adduction method is via an acid-base reaction, in which an enantiomer pair containing an acid function (e.g. a carboxylic acid) is reacted with a base resolving agent, or vice versa.

$$(\pm)A^-H^+ + (-)B \rightarrow [(-)A^- \cdot BH^+] + [(+)A^- \cdot BH^+]$$
$$\text{'p' – salt} \qquad \text{'n' – salt}$$

At present, trial-and-error procedures of experimental screening are usually employed. Leusen *et al.* (1993) hypothesized that separability correlates with the difference in the internal or lattice energy of the two diastereomer adducts and the extent to which molecular modelling methods may be applied to estimate such energy differences.

Crystal defects

Crystalline solids at temperatures above absolute zero are never perfect in that all lattice sites are occupied in a completely regular manner. Imperfections exist. Formation of such sites is endothermic; a small quantity of energy is required. Point defects occur, of which the principal two are a *vacant* lattice site, and an *interstitial* atom that occupies a volume between a group of atoms on normal sites that affects crystal purity (see Hull and Bacon, 2001, for a detailed exposition).

The other major defects in crystalline solids occupy much more of the volume in the lattice. They are known as line defects. There are two types viz. *edge dislocations* and *screw dislocations* (Figure 1.4). Line defects play an important role in determining crystal growth and secondary nucleation process (Chapter 5).

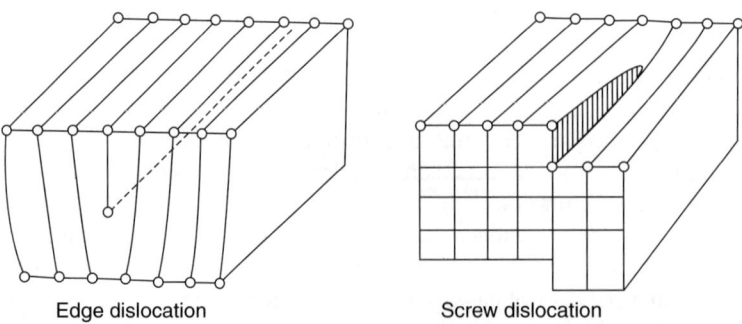

Figure 1.4 *Structures of edge and screw dislocations*

Particle characteristics

A particularly interesting feature of industrial crystallization systems is the relatively wide range of particle sizes encountered. Particle sizes range over several orders of magnitude from the sub micron (nanometers) to several millimetres or more, i.e. from 'colloidal' to 'coarse'. Such particles comprise a large part of the world on a human scale and a great source of industrially generated wealth.

Purely physical laws mainly control the behaviour of very large particles. Further down the particle size range, however, specific surface area, i.e. surface area per unit mass, increases rapidly. Chemical effects then become important, as in the nucleation and growth of crystals. Thus, a study of particulate systems within this size range of interest has become very much within the ambit of chemical engineering, physical chemistry and materials science.

Solid particles have a distinct form, which can strongly affect their appearance, product quality and processing behaviour. Thus, in addition to chemical composition, particulate solids have to be additionally characterized by particle size and shape. Furthermore, particles can be generated at any point within the process. For example, nucleation occurs within a crystallization process and large particles are broken down to numerous smaller ones in a comminution process or within a drier.

Thus, a common feature of crystallization process systems is the formation of particle populations with a range of characteristics. Real particles, however, are rarely exactly the same form as each other and irregular particles have many individual dimensions – they are in fact 'multi-dimensional', together with several angles and faces giving rise to the corresponding problem of mathematical dimensionality (Kaye, 1986). Therefore, some simplification of the problem is in order. In order to analyse such particulate populations it is first necessary to define what is meant by 'particle size', 'size distribution' and 'particle shape' (or 'habit' in crystallization parlance). Conventionally, this starts with a consideration of the simplest single particles and trying to relate real particles to them.

For a more detailed exposition, the reader is referred to specialist texts such as those on particle technology (Beddow, 1980; Rhodes, 1990; Rumpf, 1990), theory particulate of processes (Randolph and Larson, 1988) and particle characterization (Allen, 1996; Coulson and Richardson, 1991).

Single particles

Particle size

The dimensions of particles are expressed in terms of *Length*; with units: $1 \text{ m} \equiv 10^3 \text{ mm} \equiv 10^6 \text{ μm} \equiv 10^9 \text{ nm}$. The micrometer (micron) is probably the commonest unit used in practice. The dimensions of individual particles of various shapes will now be considered.

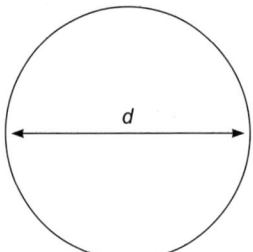

Figure 1.5 *The sphere*

(i) Sphere

The sphere (Figure 1.5) is unique. It is the only particle whose form can be totally described by a single dimension: its diameter d: Volume = $\pi d^3/6$; Area = πd^2.

Because of the uniqueness and simplicity of the sphere, the characteristics of non-spherical particles (which most real ones are) are often related back in some way to the size of an 'equivalent' sphere which has some shared characteristic, such as the same volume or surface area.

(ii) Cube

The cube (Figure 1.6) and other regular geometric shapes have more than one dimension.

In the cube there is a choice for the characteristic dimension. Firstly, the side length = L could be selected whence the surface area = $6L^2$ and volume = L^3. Alternatively, the cube 'diameter' could be chosen, i.e. the distance between diagonally opposed corners = $\sqrt{3}L$.

(iii) Irregular particles

Irregular particles have a unique maximum characteristic chord length (Figure 1.7) and this, or a mean value is sometimes used as a characteristic dimension.

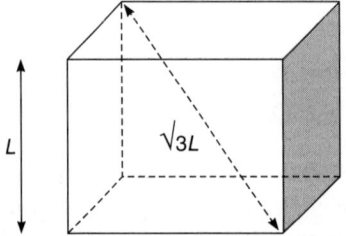

Figure 1.6 *The cube*

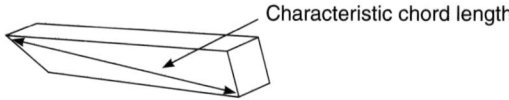

Figure 1.7 *Irregular particle*

Normally, however, irregular particles are characterized by some form of 'equivalent dimension'.

Equivalent dimensions

Any 'size' quoted for real particles normally depends on the method of measurement and is usually expressed in terms of an *equivalent dimension* of which there are two main types:

1. *Equivalent diameter* (d): This is frequently used and is usually related to the sphere.
2. *Characteristic length* (L): This is most frequently determined by sieving.

Particle shape

Particle shape can be described qualitatively using standards terms such as *granular, flaky, needle*, etc. (BS 2955). Quantitative description, however, requires use of 'shape factors' that enables a particle's surface area and volume to be calculated on knowing its 'size'.

Shape factors

Shape factors relate the characteristic dimension of a particle to its surface area or volume respectively. Any characteristic dimension can be chosen, but the corresponding shape factor 'belongs' to the characteristic dimension selected. It should not be used with any other.

(i) Surface shape factor (f_s)

The surface shape factor f_s relates the external surface area A of a particle to its characteristic dimension L:

$$A = f_s L^2 \tag{1.2}$$

For a sphere of characteristic diameter $L = d$, $f_s = \pi$. For all other particles $f_s > \pi$. For example, for a cube of characteristic side length L then $f_s = 6$.

(ii) Volume shape factor (f_v)

The volume shape factor f_v relates the particle solid volume V to its characteristic dimension L.

$$V = f_v L^3 \tag{1.3}$$

For a sphere: $f_v = \pi/6$. For all other particles $f_v > \pi/6$. For example, for a cube of characteristic side length L then $f_v = 1.0$.

Three other commonly used shape factors are:

(iii) Specific surface shape factor (F)

The specific surface shape factor F is simply the ratio of surface to volume shape factors respectively considered above, i.e.

$$F = \frac{f_s}{f_v} \tag{1.4}$$

For both a sphere of diameter d and cube of characteristic side length L, $F = 6$. Note the ambiguity of using F alone to characterize particle shape factors.

(iv) Sphericity (ψ)

The sphericity ψ quantifies the 'deviation' of the particle from spherical

$$\psi = \text{surface area of sphere/surface area of particle (of equal volumes)} \tag{1.5}$$

For the sphere $\psi = 1$, for all other particle shapes $\psi < 1$

(v) Specific surface (S_p)

The specific surface S_p defines the surface area per unit volume (or mass) of particle

$$S_p = \frac{A}{V} \tag{1.6}$$

Therefore

$$F = \frac{f_s}{f_v} = S_p L \tag{1.7}$$

Note: all the above definitions relate only to external surfaces; internal area, due to porosity, requires further definition.

Example

A rectangular parallelepiped has sides $200\,\mu m \times 100\,\mu m \times 20\,\mu m$. Determine:
(1) The characteristic dimension, by:
    (i) projected area diameter, d_a
    (ii) equivalent surface diameter, d_s, based on the sphere
    (iii) equivalent volume diameter, d_v, based on the sphere
    (iv) sieving, L_s.
(2) From the 'sieve size' determine:
    (a) the surface shape factor
    (b) the volume shape factor
    (c) the specific surface shape factor
    (d) the sphericity, from first principles.

Solution

1. (i) Projected area, i.e. in the 'plane of greatest stability' = $200 \times 100 = 2 \times 10^4 \, \mu m^2$

   X-sectional area of sphere = $\dfrac{\pi d_a^2}{4}$

   Therefore $d_a = \sqrt{\dfrac{2 \times 10^4 \times 4}{\pi}} = 160 \, \mu m$

   (ii) Surface area

   $= 2 \times 200 \times 100 + 2 \times 200 \times 20 + 2 \times 100 \times 20 = \underline{5.2 \times 10^4} \, \mu m^2$

   Therefore, Equivalent surface area of sphere = $\pi d_s^2 = 5.2 \times 10^4$

   $\Rightarrow d_s = \sqrt{\dfrac{5.2 \times 10^4}{\pi}} = 129 \, \mu m$

   (iii) Volume = $200 \times 100 \times 100 = 4 \times 10^5 \, \mu m^3$
   Therefore

   $d_v = \sqrt[3]{\dfrac{4 \times 10^5 \times 6}{\pi}} = 91 \, \mu m$

   (iv) $L = 100 \, \mu m$ (Assuming sieve size = second largest dimension)

2. (a) Surface shape factor

   $f_s = \dfrac{A}{L^2} = \dfrac{5.2 \times 10^4}{1 \times 10^4} = 5.2$

   (b) Volume shape factor

   $f_v = \dfrac{V}{L^3} = \dfrac{4 \times 10^5}{1 \times 10^6} = 0.4$

   (c) Overall shape factor

   $F = \dfrac{f_s}{f_v} = \dfrac{5.2}{0.4} = 13$

   (d) Sphericity = surface area of sphere/surface area of particle (of equal volume)
   Based on sieve size

   $L_s = 100 \, \mu m$

   Area = $f_s \times (10^2)^2 = 5.2 \times 10^4$

   Volume = $f_v \times (10^2)^3 = 0.4 \times 10^6$

   Diameter of sphere of same volume = $\left(\dfrac{0.4 \times 10^6 \times 6}{\pi}\right)^{\frac{1}{3}}$

   Area = $\pi d^2 = \pi \left(\dfrac{0.4 \times 10^6 \times 6}{\pi}\right)^{\frac{2}{3}}$

Therefore

$$\psi = \frac{\pi \left(\frac{0.4 \times 10^6 \times 6}{\pi}\right)^{\frac{2}{3}}}{5.2 \times 10^4} = 0.5$$

Or, based on equivalent volume diameter:
Sphere of equal volume, diameter $d_v = 91$ μm (as 1 (iii))
Therefore

$$\psi = \frac{\pi \times 91^2}{5.2 \times 10^4} = 0.5$$

Or, based on the formula

$$\psi = 4.84 \times \frac{f_v^{\frac{2}{3}}}{f_s} = 4.84 \times \frac{0.4^{\frac{2}{3}}}{5.2} = 0.5$$

Thus, the sphericity of the particle is 0.5, or 50 per cent.

Fractals

Unlike a perfectly smooth regular object, an adequate representation of typical large irregular particles such as agglomerates defies simple description (Kaye, 1986). Any attempt, for example, to specify the outline or surface area depends upon the resolution, or step length, of the characterizing probe used, e.g. sieve aperture, wavelength of light, and molecular dimensions of adsorbed gas or liquid. The concept of fractals (*fractus*, Latin – broken, fragmented) is a (relatively) recent mathematical technique developed by Mandlebroot (1977, 1982) to quantify such ruggedness of form. It has application in the use of image analysis to characterize particle shape (see Clark, 1986; Simons, 1996) and simulate processes for the growth of complex shapes (Sander, 1986).

Particle size distributions

As mentioned above, industrial particles are rarely 'monodisperse' – of equal size (although there are many cases where this is striven for), but exhibit 'polydispersity' – they occupy a size range.

Frequency histogram

Bulk particulates of various sizes can be analysed to display their particle size distribution and a 'frequency histogram' plotted, as shown in Figure 1.8.

Mass fractions

Sequentially adding up each individual segment on the histogram gives the *Cumulative mass fraction, m(d)*

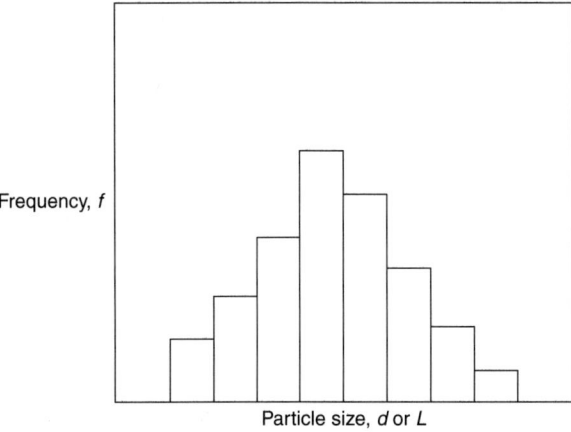

Figure 1.8 *Frequency histogram*

$$m(d) = \frac{M(d)}{M_T} = \frac{\text{mass to size } d}{\text{total mass}} \qquad (1.8)$$

Differentiating the cumulative mass fraction gives the *Differential mass fraction*

$$x = \frac{dm}{dd} \qquad (1.9)$$

$$\int_0^\infty x \, dd = 1 \qquad (1.10)$$

Alternatively, in terms of size increments, let

$$x_i = \frac{\text{mass in size at interval } i}{\text{total mass of all particles}}$$

n_i = number of particles × volume of particles × density of particles

Then the mass fraction x_i is

$$x_i = n_i f_{vi} \rho_{si} d_i^3 \qquad (1.11)$$

Or the number of particles in mass fraction x_i is

$$n_i = \left(\frac{1}{\rho_s f_v}\right) \frac{x_i}{d_i^3} \qquad (1.12)$$

Note that in both equations 1.10 and 1.11 constant particle shape and density are assumed.

The differential and cumulative size distributions are clearly related, as shown in Figure 1.9. Differentiating the cumulative distribution restores the original histogram but in a smoother form. Two important properties can be defined, the modal and median sizes.

The modal size is at the point of maximum frequency on the differential distribution. The median size is at the 50 per cent point on the cumulative plot. Note that the modal size of a distribution is not necessarily equal to the median.

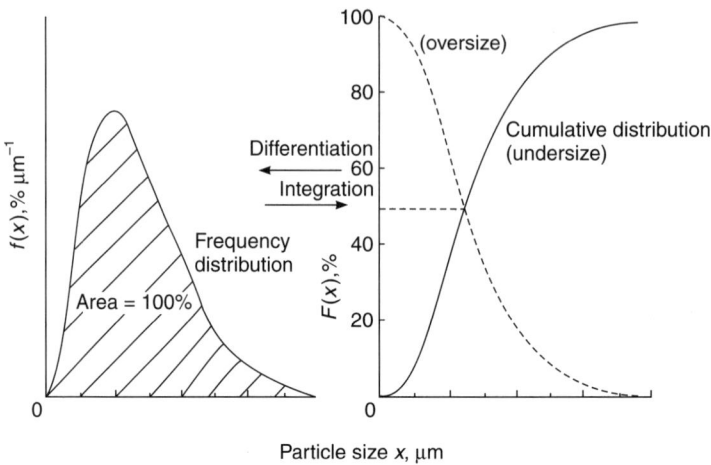

Figure 1.9 *Relationship between cumulative and differential mass fractions*

Mean particle size

The complete mathematical definition of a particle size distribution is often cumbersome and it is more convenient to use one or two single numbers representing say the *mean* and *spread* of the distribution. The *mean particle size* thus enables a distribution to be represented by a single dimension while its *standard deviation* indicates its spread about the mean. There are two classes of means:

(i) Property mean size

   Examples   weight mean size
              surface: volume mean size

Property mean sizes are derived by calculation from the actual distribution using a defined weighting for each size interval.

(ii) Mean property size

   Example   mean surface size

Mean property sizes are based on a fictitious equivalent ideal mono-disperse distribution containing the same of the same number of particles as the real one.

Note: In general, property mean size is not equal to mean property size, (i) property mean size is the more commonly used definition.

Mean sizes of distributions

(i) Weight (or volume) mean size (d_{wm})

This is the mean abscissa of a graph of cumulative mass fraction versus size. Note $d_{wm} \neq d_{median}$.

$$d_{wm} = \frac{\sum_i \text{mass} \times \text{size}}{\sum_i \text{mass}} \tag{1.13}$$

$$d_{wm} = \frac{\sum_i x_i d_i}{\sum_i x_i} = \sum_i x_i d_i \quad \text{since} \quad \sum_i x_i = 1, \text{ by definition} \tag{1.14}$$

or, in terms of particle numbers

$$d_{wm} = \frac{\sum_i n_i d_i^4}{\sum_i n_i d_i^3} \tag{1.15}$$

(ii) Mean weight (or volume) size (d_{mw})

This is the size of each particle of an artificial mono-disperse powder such that the same number of particles to equal the mass of the real powder.

$$f_v d_{mw}^3 \sum_i n_i = f_v \sum_i (n_i d_i^3) \tag{1.16}$$

$$\text{mean} \qquad \text{actual sample}$$

i.e.

$$d_{wm} = \sqrt[3]{\frac{\sum_i (n_i d_i^3)}{\sum_i n_i}} \tag{1.17}$$

Substituting for n_i

$$d_{mw} = \sqrt[3]{\frac{1}{\sum_i \left(\frac{x_i}{d_i^3}\right)}} \tag{1.18}$$

Note: In general $d_{mw} \neq d_{wm}$

(iii) Surface mean size (d_{sm})

This is the mean abscissa of a graph of cumulative area fraction versus size. Otherwise known as the *Surface: Volume Mean (or Sauter Mean)*

$$d_{sm} = \frac{\sum \text{Volume}}{\sum \text{Surface}} = \frac{\sum_i d_i S_i}{\sum_i S_i} = \frac{\sum_i n_i d_i^3}{\sum_i n_i d_i^2} \tag{1.19}$$

noting that f_s cancels.
Substituting for n_i

$$d_{sm} = \frac{\sum_i x_i}{\sum_i \frac{x_i}{d_i}} = \frac{1}{\sum_i \frac{x_i}{d_i}} \tag{1.20}$$

noting that ρ_s and f_v cancel.

The *Sauter mean* is often used in sieve analysis, and in situations in which the surface area is important, e.g. mass transfer calculations, etc.

(iv) Mean surface size (d_{ms})

This is the size at which each particle of a mono-disperse representation would have for the same number of particles to have equal surface area as the real powder.

$$\sum_i n_i f_s d_{ms}^2 = \sum_i (n_i f_s d_i^2) \tag{1.21}$$

surface area of uniform particles — surface area of actual powder

i.e.

$$d_{ms} = \sqrt{\frac{\sum_i (n_i d_i^2)}{\sum_i n_i}} \tag{1.22}$$

Substituting for n_i

$$d_{ms} = \sqrt{\frac{\sum_i \frac{x_i}{d_i}}{\sum_i \left(\frac{x_i}{d_i^3}\right)}} \tag{1.23}$$

The quantity d_{ms} is often used in adsorption, crushing, and light diffusion applications.

Again, note that in general $d_{ms} \neq d_{sm}$. Similar means based on length and number may also be defined.

Specific surface (S_p)

The specific surface is the surface area of a powder per unit mass (or volume). It can be calculated from the Sauter Mean size

$$S_p = \frac{\text{Surface Area}}{\text{Unit Mass (or Volume)}} \tag{1.24}$$

$$S_p = \frac{f_s d_{sm}^2}{\rho_s f_v d_{sm}^3} = \frac{F}{\rho_s d_{sm}} \tag{1.25}$$

which has units of m²/kg, or m²/m³ (the latter omitting ρ_s)

Coefficient of variation (CV)

The coefficient of variation is a statistical measure of the 'spread' of the distribution about the mean. In effect, it is a second term to quantify a distribution in terms of its shape since many distributions may have a common mean size (in some respect) but differ considerably in the spread about that mean.

$$\text{CV} = \frac{\text{Standard deviation}}{\text{Mean size}} \tag{1.26}$$

i.e.

$$\text{CV} = \frac{\sigma}{\text{d}} \tag{1.27}$$

where by definition of variance σ

$$\sigma^2 = \int_0^\infty (d - \bar{d})^2 f(d) \mathrm{d}d \tag{1.28}$$

and d is the characteristic size and the bar denotes the mean.

The magnitude of CV is normally calculated numerically, although graphical methods are also possible. Exceptionally, CV can be evaluated analytically for a defined distribution function (see Randolph and Larson, 1988). The gamma distribution is especially important in crystallization, and is considered in more detail in Chapter 3.

Packing characteristics

The bulk properties of particulate material are often strongly affected by the packing characteristics of the powder, as illustrated in Table 1.2. Efforts to relate packing characteristics to particle characteristics, notably voidage that is important in cake filtration for example (see Chapter 2), have been made by several authors (see Gray, 1968; Scott, 1968a; Bierwagen and Saunders, 1974; Wakeman, 1975; Ouchyama and Tanaka, 1981; Yu and Standish, 1987).

Voidage and internal surface area

The interstitial space existing between a mass of particles can also be important e.g. in retaining filtrate liquor which may lead to occlusions if overgrown, and hence the particulate 'voidage' or 'void fraction', ε, is defined, as follows.

$$\varepsilon = \frac{\text{Volume of particulate voids}}{(\text{Volume of voids} + \text{volume of solids, i.e. total bed volume})}$$

Particulate voidage is generally found to be in the range $0.48 < \varepsilon < 0.26$.

The voidage of irregular particles has to be determined empirically at present (Cumberland and Crawford, 1987). For random packing, e.g. particulate material poured into container and then tapped, voidage is generally in the range $= 0.40$ (loose) $< \varepsilon < 0.36$ (dense).

Table 1.2 *Relationship between primary and bulk particulate properties*

Primary particles characteristics	Packing characteristics	Bulk properties
Size, shape distribution	Voidage, bulk density, contacts, etc.	Fluid flow, particle flow, caking, etc.

18 Crystallization Process Systems

For dense random packing

$$(1 - \varepsilon) \cong 2/\pi = 0.6366 \tag{1.34}$$

an expression derived empirically due to Scott (1968a).

Agglomerated fine particles such as catalysts and precipitates may exhibit internal surface areas orders of magnitude greater than that available on the exterior surface. Again, if agglomerate particle size is used as the characteristic dimension, the calculated area is that of the external surface. If the internal surface area is to be calculated then additional knowledge of the constituent crystal size is required, e.g. via microscopy or adsorption.

Effect of size distributions

Generally, the small particles in a distribution may fit in between larger ones with the overall voidage passing through a minimum as the fines fraction increases. Wider size distributions give greater packing density (i.e. lower ε). Thus 'fines' are particularly important in determining bulk properties of particulates, e.g. filtrability in solid–liquid separation and evaporative mass transfer during drying.

Generalized methods for calculating random packing fractions and coordination number of size distributions have been proposed, but are very complex and usually apply to spheres having particular size distributions. Unfortunately, the products of industrial crystallizers frequently contain a wide variety of both shape and size distribution not amenable to conventional theories, leading to error in voidage estimation. Often therefore, voidage is simply a measured quantity for a particular industrial material.

Techniques for particle sizing and characterization

Numerous methods exist for determining crystal particle characteristics including those listed in Table 1.3. See for example, Allen (1990) for a detailed survey and description of each technique. Three commonly used methods are illustrated below whilst microscopy is illustrated in Chapter 3 (Figure 3.11).

Table 1.3 *Some particle sizing techniques*

Technique	Approximate size range (μm)	'Size' measured
Sieving	50–5000	aperture
Microscope	1–1000	projected area
Electron-microscope	0.001.5	projected area
Sedimentation	3–60	hydrodynamic
Gas adsorption	0.001–10	surface area
Zone sensing (Coutler®)	1–100	volume
Laser light-scattering	1–1800	mean projected area

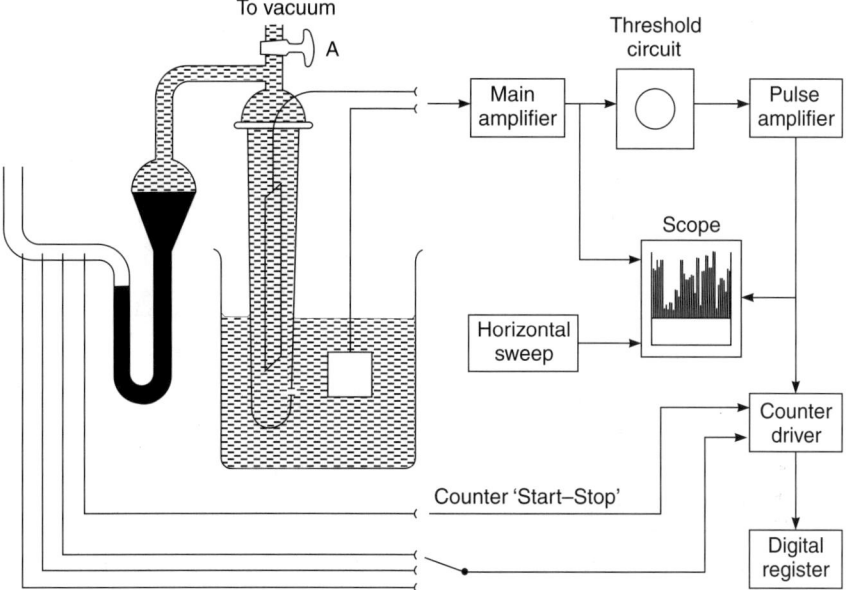

Figure 1.10 *The Coulter principle*®

Zone sensing

Particles are suspended in an electrolyte and are drawn through a small orifice containing an electrode at either side (Figure 1.10).

The principle of this zone sensing (Coulter Counter®) technique is as follows:

$$\left.\begin{array}{ll} \text{Voltage} \propto \text{particle volume} \\ \text{Pulse frequency} \propto \text{particle number} \end{array}\right\} \Rightarrow \text{Particle size distribution}$$

This technique measures size distribution on a volume and number basis.

Laser diffraction

Particle sizing via laser light scattering has become very popular due to its simplicity of operation (Figure 1.11).

The principle of this technique is based on the fact that small particles scatter light further than do larger ones. Therefore, though the mathematics is complex and normally incorporated within vendors' software, by determining the light intensity at positions in the focal plane, the size distribution can be inferred.

The sieve test

The sieve test is probably still the most commonly used sizing technique both to characterize particulate distributions and separate them into fractions on the

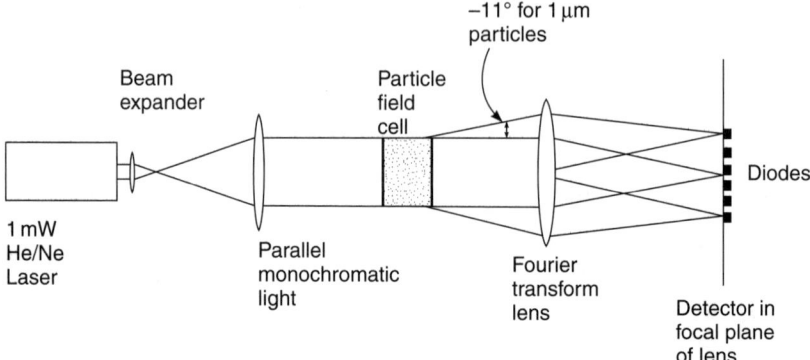

Figure 1.11 *Laser light scattering*

plant. In fact, test sieves are made to great precision and are subject to various national and international standards (ISO 565, 1990). The size range covered is approximately 50–5000 μm. Although one of the oldest techniques, the basic principles used in the analysis of the data are, however, common to many modern electronic techniques in which a distribution is separated in various size 'fractions', 'channels' or 'classes'. The sieve may be perforated or wire woven, as illustrated in Figure 1.12.

Mesh number (N)

The mesh number is the number of apertures per unit length of sieve.

$$N = \frac{1}{(L+W)} \tag{1.29}$$

Note, since N depends on the wire width as well as aperture size; aperture size itself is therefore the better quantity to quote.

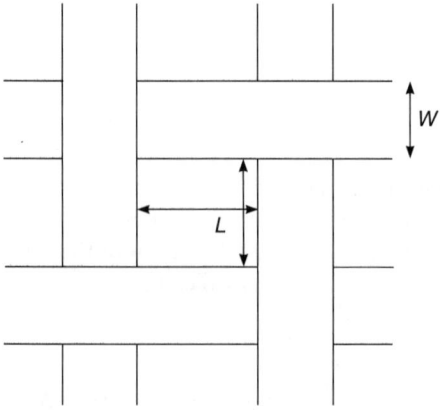

Figure 1.12 *Sieve mesh: L, aperture size; W, wire width*

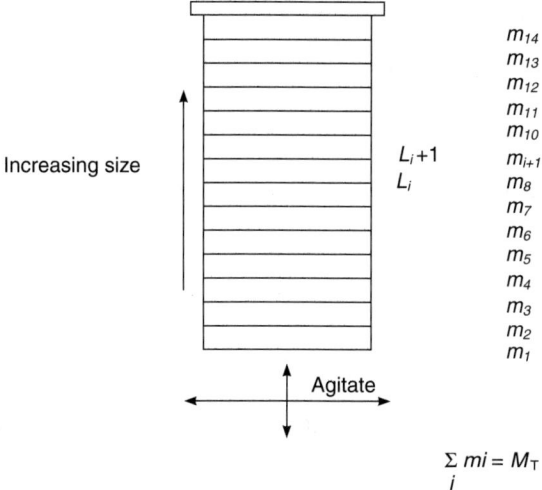

Figure 1.13 *Sieve test method for particle size analysis*

Sieve test method

In the *Sieve test*, a small, representative, sample of the material to be analysed is placed on the top pan of a nest of sieves stacked in ascending order of mesh size, Figure 1.13. The sieve nest is then gently agitated such that the particles pass through their 'just largest' sieve aperture and are retained on the one immediately below without undue breakage. Thus the very finest sized particles pass through the nest and are retained in the pan at the base with all other size fractions ranged above them. The mass of particles on each sieve and on the pan is then determined by weighing each sieve or pan and the statistics of the distribution calculated.

The resulting sieve analysis data are normally expressed initially as either

1. frequency polygon
2. frequency histogram
3. cumulative undersize (or oversize) distribution.

Data presentation and analysis from sizing tests

Sizing tests produce data in the form of the mass of powder retained between successive size intervals. Normally the 'size' of the particles in the fraction is considered to be either the *lower* screen aperture, L_l, on which the powder is retained, or the (arithmetic) mean of the lower and higher screen apertures, $(L_l + L_h)/2$.
Thus

$$\bar{L} = \frac{L_l + L_h}{2} = L_l + \frac{\Delta L}{2} \qquad (1.30)$$

where $\Delta L = L_h - L_l$. $\qquad (1.31)$

The *weight fraction*, x, is defined by

$$x_i = \frac{m_i}{\sum m_i} \quad (1.32)$$

where $\sum m_i$ is the total mass sieved (note that x is often expressed as a percentage).

The cumulative undersize fraction, ϕ = the total fraction smaller than size $\bar{L}$ (or L_1)

$$\phi = \sum_0^{\bar{L} \text{ or } L_l} x = \int_0^L x \, dL \quad \text{(for continuous curves)} \quad (1.33)$$

Use of arithmetic graph paper tends to bunch at the lower end (due to the $\sqrt{2}$ expansion of the sieve size scale). This can be overcome by use of semi-log or log–log graph paper. Also note that cumulative graphs have the effect of smoothing out the data. This can be a useful effect in the presentation of data! Care should be exercised, however, to prevent important features of the distribution being obscured in analytical work.

Example

A quantity of sodium bicarbonate crystals has the following sieve analysis

Sieve aperture (μm)	Mass retained (g)
500	0.00
355	4.75
250	4.75
180	28.50
125	47.50
90	3.80
63	3.80
pan	1.90

Microscopic examination of the crystals indicates $f_s = 4.0$, $f_V = 0.5$. Density = 2160 kg/m^3.

Determine

(a) the median size on a mass basis
(b) the weight mean size
(c) the surface mean size
(d) the linear mean size
(e) the mean size on a number basis, i.e. population mean size
(f) the specific surface area.

Solution

The first step is to reconstruct the data table to include derived mass fraction data and sums and quotients required for the calculation of mean. This is conveniently achieved by means of a spreadsheet.

Particulate crystal characteristics 23

Noting $\sum_i m = 95\,g$ enables the x_i values to be calculated from equation 1.32 and the ϕ value from equation 1.33.

Sieve size (μm)	Mean size (μm)	Mass fraction	Cumulative fraction u/s	L x	x/L	x/L*2	x/L*3
500			1				
	428	0.05		21.4	0.000117	2.729E-07	6.377E-10
355			0.95				
	303	0.05		15.15	0.000165	5.446E-07	1.797E-09
250			0.9				
	215	0.3		64.5	0.001395	6.490E-06	3.019E-08
180			0.6				
	153	0.5		76.5	0.003268	2.136E-05	1.396E-07
125			0.1				
	108	0.04		4.32	0.000370	3.429E-06	3.175E-08
90			0.06				
	77	0.04		3.08	0.000519	6.747E-06	8.762E-08
63			0.02				
	32	0.02		0.64	0.000625	1.953E-05	6.104E-07
pan			0				
b		1		185.59	0.006460	5.837E-05	9.019E-07
c					154.798462		
d						110.6659883	
e							64.72000825

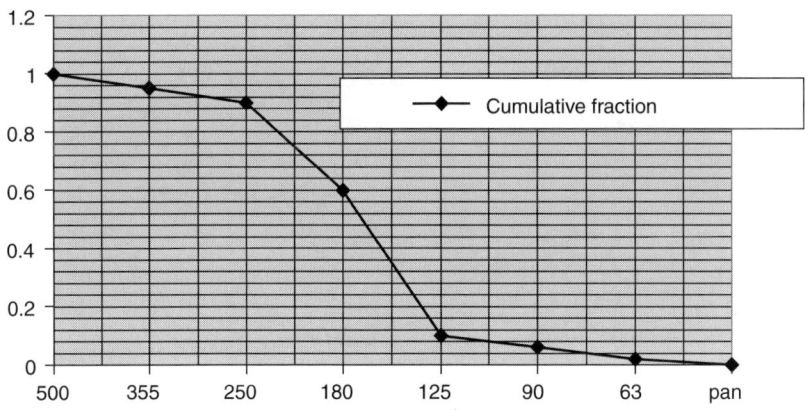

(a) From the graph of cumulative undersize fraction against aperture size, by inspection

$d_{50} = 170\,\mu m$

This means that 50 per cent of the material, by weight, lies above 170 μm, 50 per cent below.

(b) Noting that

$$\text{Weight mean size, } L_{wm} = \frac{\sum_i x_i d_i}{\sum_i x_i}$$

hence

$$L_{wm} = 186 \, \mu m$$

(c) Noting that

Surface mean size (Sauter mean), $L_{sm} = \dfrac{1}{\sum\limits_i \dfrac{x_i}{d_i}}$

hence

$$L_{sm} = 155 \, \mu m$$

(d) Length mean size
Similarly

$$L_{lm} = \frac{\sum\limits_i \dfrac{x_i}{L_i}}{\sum\limits_i \dfrac{x_i}{L_i^2}}$$

hence

$$L_{lm} = \frac{6.47 \times 10^{-3}}{5.80 \times 10^{-3}} = 112 \, \mu m$$

(e) Population mean size (L_{pm})
By similar arguments as for the weight, area and length means, the population mean (i.e. number mean) is defined as

$$L_{pm} = \frac{\sum\limits_i n_i L_i}{\sum\limits_i n_i}$$

where

$$n_i = \frac{x_i}{\rho_s f_v L_i^3}$$

$$L_{pm} = \frac{\sum\limits_i \dfrac{x_i L_i}{\rho_s f_v L_i^3}}{\sum\limits_i \dfrac{x_i}{\rho_s f_v L_i^3}} = \frac{\sum\limits_i \dfrac{x_i}{L_i^2}}{\sum\limits_i \dfrac{x_i}{L_i^3}}$$

hence

$$L_{pm} = \frac{5.80 \times 10^{-5}}{9.02 \times 10^{-7}} = 65 \, \mu m$$

(f) Noting that the specific surface is given by

$$S_p = \frac{F}{\rho_s L_{sm}} = \frac{F}{\rho_s} \sum\limits_i \frac{x_i}{L_i}$$

$$F = \frac{f_s}{f_v} = \frac{4.0}{0.5} = 8$$

$$S_p = \frac{8}{2160 \times 155 \times 10^{-6}} = 2.39 \times 10^{-5} \times 10^6 = 24\,\text{m}^2/\text{kg}$$

or, on a solids volume basis

$$S'_p = 2160 \times 24 = 51840\,\text{m}^2/\text{m}^3$$

Thus, the 'mean size' varies from 65 to 186 µm for the same sample. It is thus most important to carefully define which mean is intended. Clearly, the mean size on a number basis is much smaller than that a weight basis (since mass αL^3). These calculations show that in this example most mass lies between 125–250 µm but most crystals lie between 0–90 µm.

Summary

Crystals are ubiquitous. They vary enormously in form, size and shape, partly reflecting their internal structure, partly their growth history. Particle size and shape are quantified by use of characteristic dimensions and shape factors that, in combination, permit calculation of important properties such as particle volume (mass) and surface area. Relating them to the shape of ideal particles, e.g. the sphere often approximates real particles. Similarly, the 'size' of a mass of particles can be expressed in terms of a characteristic mean and spread. The voidage of a mass of particles is influenced by both these quantities. It will be shown in subsequent chapters that these particle characteristics can have a determining effect on both their processing behaviour and properties in application. They are therefore very important for the process engineer or scientist to measure, predict and control in a particulate crystallization process system.

2 Fluid–particle transport processes

The formation and subsequent solid–liquid separation of particulate crystals from solution commonly involves alternate periods of suspension and sedimentation during which they experience relative fluid–particle motion. Similarly, solid matter may change phase from liquid to solid or vice versa. New particles may be generated or existing ones lost e.g. in crystallizers or mills. They may be separated from fluids by flow through vessels e.g. settlers, thickeners or filters. Thus the solid phase is subject to the physical laws of change: continuity (for conservation of both mass and particle) and flow (solid through liquids and liquid through particle arrays etc.). These are known collectively as transport processes.

In this chapter, the transport processes relating to particle conservation and flow are considered. It starts with a brief introduction to *fluid–particle hydrodynamics* that describes the motion of crystals suspended in liquors (Chapter 3) and also enables solid–liquid separation equipment to be sized (Chapter 4). This is followed by the *momentum and population balances* respectively, which describe the complex flows and mixing within crystallizers and, together with particulate crystal formation processes (Chapters 5 and 6), enable particle size distributions from crystallizers to be analysed and predicted (Chapters 7 and 8).

Further details can be found in texts concerning fluid mixing and particle suspension (Šterbáček and Tausk, 1965; Holland and Chapman, 1966; Oldshue, 1983; Uhl and Gray, 1986; Allen, 1990; Coulson and Richardson, 1991; Harnby *et al.*, 1992 and Gibilaro, 2001), the theory particulate of processes (Randolph and Larson, 1988; Ramkrishna, 2000) and turbulent flows (Pope, 2000).

Particle–fluid hydrodynamics

There are numerous applications of particle–fluid flow in crystallization process systems including:

- sedimentation and elutriation
- thickening of slurries
- centrifugation
- filtration.

This section reviews with the theory of the following cases in turn, starting from the simplest i.e. single particle case, in the hope that understanding this will facilitate dealing the more complex particulate slurries and arrays.

1. single particles in fluids
2. particles in slurries
3. fluid flow through porous media, etc.

It is quickly evident, however, that it is necessary to blend theory with experiment to achieve the engineering objectives of predicting fluid–particle flows. Fortunately, there are several semi-empirical techniques available to do so (see Di Felice, 1995 for a review). Firstly, however, it is useful to define some more terms that will be used frequently.

Voidage (ε)

ε = void (fluid) volume/(total volume i.e. fluid + particles)

$0 < \varepsilon < 1 \ [\text{m}^3/\text{m}^3]$ (2.1)

Specific surface of particles (S_p)

S_p = Surface area of particles/volume of particles

$= \dfrac{F}{d_{sm}} \ [\text{m}^2/\text{m}^3]$ (2.2)

Specific surface of bed (S_B)

S_B = Surface area of particles/total volume of bed

$= (1 - \varepsilon) S_p \ [\text{m}^2/\text{m}^3]$ (2.3)

Superficial velocity (u)

u = Flowrate/cross-sectional area of bed (i.e. of both particles and voids)

$= \dfrac{Q}{A} \ [\text{m}^3/\text{m}^2\text{s, i.e. m/s}]$ (2.4)

Note that u is effectively a 'velocity flux' (volumetric flowrate per unit cross-sectional area).

Interstitial velocity (v)

v = Flowrate/cross-sectional of voids

$= \dfrac{Q}{\varepsilon A} = \dfrac{u}{\varepsilon} \ [\text{m}^3/\text{m}^2\text{s, i.e. m/s}]$ (2.5)

Note that v is therefore the 'true' linear velocity of the fluid.

Sedimentation

An ideal particle moves relative to the fluid with velocity, v, dependent on just four quantities viz. fluid viscosity and density μ and ρ respectively (i.e. fluid properties), and particle solid density and characteristic size ρ_s and d respectively (i.e. particle characteristics), subject to the forces of buoyancy and friction, as illustrated in Figure 2.1.

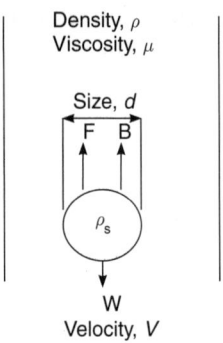

Figure 2.1 *Particle–fluid interactions*

A relationship between these four variables is required in order to predict particle velocity in a variety of circumstances. In doing so, it is noted that as a particle moves through a fluid it experiences drag and vice versa as the fluid molecules move across and around the surface of the particle. There is thus a *fluid–particle interaction* due to interfacial surface drag.

Wallis (1969) has presented the general one-dimensional momentum equation for the particle phase of a fluid–particle continuum as

$$\rho_s \left(\frac{\partial v}{\partial t} + v \frac{\partial v}{\partial z} \right) = B + F - \frac{\partial p}{\partial z} \quad (2.6)$$

where, on the right-hand side, b represents all the body forces, $-\partial p/\partial z$ is the pressure gradient force, which, in general, should include contributions from both the fluid and the particle pressures, and F accounts for all the surface forces acting on the solid phase that are not included in the pressure gradient term. In application, certain simplifications can be made. Thus particle contact forces are usually neglected whilst uniform and steady flows are often assumed.

Drag on particles

By definition, the drag force per unit area on a single particle at infinite dilution is related to the kinetic energy of the fluid by the expression

$$\frac{F}{A} = c_{D\infty} \frac{1}{2} \rho v^2 \quad (2.7)$$

where F = force, A = projected area, $c_{D\infty}$ = single particle 'infinite dilution' drag coefficient, ρ = fluid density and v = relative velocity.

For the sphere (the 'ideal' particle)

$$A = \frac{\pi d^2}{4} \quad (2.8)$$

Therefore

$$F = c_{D\infty} \frac{\pi d^2}{4} \frac{\rho v^2}{2} \quad (2.9)$$

Force balance

To determine the equation of motion, a force balance on the particle is applied to the particle. From Newton's second law of motion, the rate of change of particle momentum, $\dot{M}$, is equal to the net force acting.

$$W - B - F = \dot{M} \tag{2.10}$$

i.e.

$$mg - m'g - F = m\frac{dv}{dt} \tag{2.11}$$

where m = mass of particle, m' = mass of fluid displaced (buoyancy), F = drag force and v = velocity.

The *terminal velocity* occurs when the particle acceleration is zero i.e. $v = v_t$; $dv/dt = 0$. In other words

accelerative force (gravity−buoyancy) = resistive force (drag) (2.12)

or, in symbols

$$\frac{\pi}{6}d^3(\rho_s - \rho)g = c_{D\infty}\frac{\pi d^2}{4}\frac{\rho v_t^2}{2} \tag{2.13}$$

where ρ_s = solid density.
Thus

$$c_{D\infty} = \frac{4d(\rho_s - \rho)g}{3\rho v_t^2} \tag{2.14}$$

$$v_t = \left(\frac{4d(\rho_s - \rho)g}{3\rho c_{D\infty}}\right)^{\frac{1}{2}} \tag{2.15}$$

Equation 2.15 is thus an expression for the terminal velocity v_t in terms of particle and fluid properties and the drag coefficient.

But in order to predict v_t, knowledge of $c_{D\infty}$ is required as a function of the fluid properties and particle characteristic dimension, viz. the particle Reynolds number. This is straightforward for particles experiencing laminar flow for which there exists an analytical solution. In turbulent conditions, however, the flow is much more complex and analytical solutions are not available. Fortunately, in these cases resort can be made to semi-empirical (or semi-theoretical) formulae or charts that correlate reported experimental data over a wide range of conditions.

Laminar flow: Stokes law ($Re_p < 0.2$)

For laminar flow – flow in which the layers of fluid are stratified across which there is no mixing apart from that due to molecular diffusion – Stokes Law (Stokes, 1851) applies. Firstly, however, it is necessary to define an index of the flow to indicate whether it is laminar or turbulent. This is done through the dimensionless group known as the Reynolds number familiar in fluid flow but applied here to particles moving within fluids.

Particle Reynolds number

The particle Reynolds number is defined by

$$\text{Re}_p = \frac{\rho v d}{\mu} \tag{2.16}$$

where d = particle dimension [m] and μ = dynamic viscosity [Ns/m^2].

For a critical value of the Reynolds number $\text{Re}_p < 0.2$ Stokes showed analytically that the net force acting on a sphere is given by

$$F = 3\pi\mu dv \tag{2.17}$$

i.e.

$$c_{D\infty} = \frac{3\pi dv}{\frac{\pi d^2}{4}\frac{\rho v^2}{2}} = \frac{24}{\text{Re}_p} \tag{2.18}$$

Thus for a given value of particle Reynolds number, Re_p, the particle velocity, v, can be calculated. This has a special value when the accelerative and drag forces are in balance, known as the terminal settling velocity, v_t. It may be evaluated as follows.

Terminal velocity in Stokes' flow ($\text{Re}_p < 0.2$)

Force balance

$$\frac{\pi}{6}d^3(\rho_s - \rho)g = 3\pi\mu dv_t = F \tag{2.19}$$

$$v_t = \frac{d^2 g(\rho_s - \rho)}{18\mu}; \quad c_{D\infty} = \frac{24}{\text{Re}_p} \tag{2.20}$$

or, rearranging in terms of the *Stokes Diameter*, d_{st}

$$d_{st} = \left(\frac{18\mu v_t}{g(\rho_s - \rho)}\right)^{\frac{1}{2}} \tag{2.21}$$

Thus the Stokes diameter of any particle is that of an equivalent sphere having same terminal settling velocity and is a useful additional particle characteristic for particulate systems involving fluid motion.

Turbulent flow

As the fluid velocity past the sphere increases, the laminar layers begin to break-up into circulating packets known as eddies – a complex chaotic motion that is more difficult to analyse than the simple laminar flow case. Thus for At

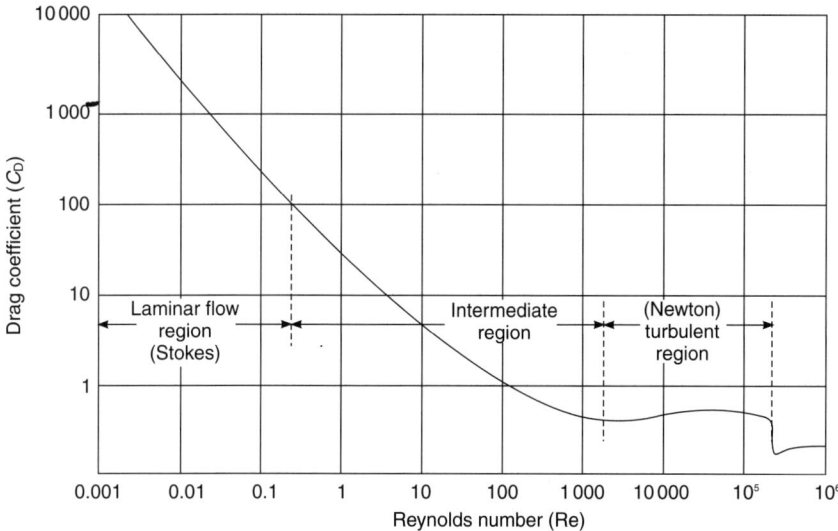

Figure 2.2 *Drag coefficients for the sphere as a function of particle Reynolds number*

higher Re (>0.2) empirical formulae for the drag coefficient are normally employed to determine the particle velocity (Figure 2.2).

Intermediate region $0.2 < \text{Re}_p < 1000$

$$c_{D\infty} \approx \frac{10}{\text{Re}_p^{0.5}} \tag{2.22}$$

Newton's Law region $1000 < \text{Re}_p < 2 \times 10^5$

$$c_{D\infty} \approx 0.44 \tag{2.23}$$

Fully turbulent region $2 \times 10^5 < \text{Re}_p$

$$c_{D\infty} \approx 0.1 \tag{2.24}$$

An approximate correlation due to Dallavalle (1948)

$$c_{D\infty} \left(0.63 + \frac{4.8}{\text{Re}^{0.5}}\right)^2 \tag{2.25}$$

is said to be applicable over the full practical range.

Slurries

Hindered settling

At high solids contents, settling occurs in a multi-particle system at a slower rate than single particle case. This is known as 'hindered settling'. There are several potential causes

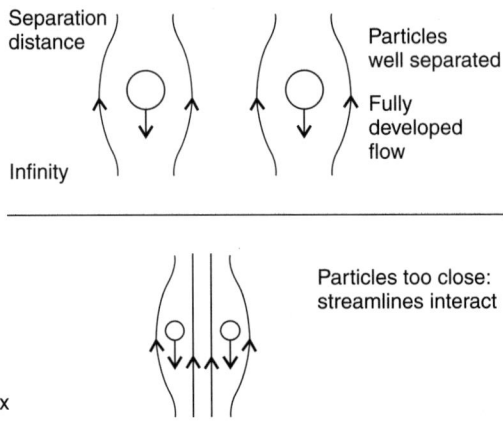

Figure 2.3 *Interaction of fluid streamlines around particles*

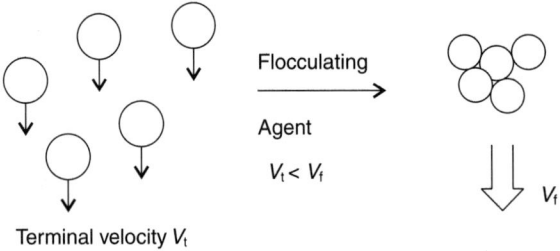

Figure 2.4 *Particle flocculation*

1. Interaction with streamlines. Streamlines can interact restricting fluid flow as shown in Figure 2.3.
2. Fluid displacement upwards due to settling particles. Thus, the *net* downward velocity of the particle is reduced.
3. Fines can increase the apparent viscosity and density resulting in reduced settling rates.

Incidentally, *enhanced* settling rates can be obtained, however, by use of flocculating agents which increase the effective particle size, as indicated in Figure 2.4.

Hindered settling rates

Monodisperse particles

The following semi-empirical equation relates the (hindered) settling velocity of a slurry of particles to the settling velocity of a single particle, known as the Richardson and Zaki (1954) (RZ) equation. The RZ equation is also used for liquid fluidization whereby particles are supported by an up-flow of fluid.

$$v_c = v_t \varepsilon^n \qquad (2.26)$$

where v_c = settling velocity at concentration, c, (equivalent to voidage ε), v_t = terminal settling velocity, ε = voidage and n = expansion index.

As liquid flowrate increases beyond the minimum required to suspend the particles, the bed expands. Unfortunately, however, the 'expansion index' n is not constant over all flow regimes but is a function of the flow. Thus

$$n = f\left(\text{Re}_p, \frac{d}{d_t}\right)$$
$$4.6 + \frac{20d}{d_t} \text{ (laminar)} > n > 2.4 \text{ (fully turbulent)} \tag{2.27}$$

where Re_p = particle Reynolds number and d_t = tube diameter
Empirically, the dependence of n on Re_p is given by

$$\frac{4.7 - n}{n - 2.35} = 0.175 \text{Re}_t^{0.75} \tag{2.28}$$

due to Rowe (1987).

Hindered drag coefficients

Force balance

An alternative method for predicting slurry settling rates is based on the drag coefficient approach (already used above for single particles) has been reviewed by Di Felice (1994). Again, starting from the force balance and noting that the buoyancy is due to the suspension (i.e. particles and fluid), not just the fluid

$$F = V(\rho_s - \rho_c)g \tag{2.29}$$

where ρ_s = solid density and ρ_c = slurry density.
Thus, by definition

$$F = \varepsilon V(\rho_s - \rho)g \tag{2.30}$$

$$F = \frac{1}{2} c_{D\varepsilon} A \rho v^2 \tag{2.31}$$

Putting $u = v\varepsilon$ then

$$u = \left[\frac{4\varepsilon^3 d(\rho_s - \rho)g}{3\rho c_{D\varepsilon}}\right]^{\frac{1}{2}} \tag{2.32}$$

Equation 2.32 thus reduces to the velocity of a single particle for $\varepsilon \to 1$. For the general case, however, an expression for the hindered drag coefficient, $c_{D\varepsilon}$, is needed. This can be obtained as follows.

For a single particle (at any velocity) by definition

$$F_{D\infty}(u) = c_D \frac{\pi d^2}{4} \frac{\rho u^2}{2} \tag{2.33}$$

For each particle in a suspension the force exerted is that of a particle at infinite dilution corrected by some, as yet unknown, function of voidage, $g(\varepsilon)$, thus

$$F_D = g(\varepsilon) F_{D\infty} = g(\varepsilon) c_{D\infty} \frac{\rho u^2}{2} \frac{\pi d^2}{4} \tag{2.34}$$

hence

$$c_{D\varepsilon} = g(\varepsilon)c_{D\infty} \tag{2.35}$$

Now, from the force balance

$$F_D = F_{Dt}\varepsilon = c_{Dt}\varepsilon \frac{\rho v_t^2}{2} \frac{\pi d^2}{4} \tag{2.36}$$

so that the voidage function $g(\varepsilon)$ becomes

$$g(\varepsilon) = \frac{F_D}{F_{D\infty}} = \frac{c_{Dt}}{c_{D\infty}}\left(\frac{u}{v_t}\right)^{-2}\varepsilon \tag{2.37}$$

At low Re_p (laminar shear flow)

$$\frac{c_{Dt}}{c_{D\infty}} = \frac{\frac{24}{Re_{pt}}}{\frac{24}{Re_p}} = \frac{u}{v_t} \Rightarrow g(\varepsilon) = \varepsilon^{1-n} \tag{2.38}$$

At high Re_p (turbulent inertial flow)

$$\frac{c_{Dt}}{c_{D\infty}} = 1 \Rightarrow g(\varepsilon) = \varepsilon^{1-2n} \tag{2.39}$$

Interestingly, it turns out that despite the differing functions of the expansion index, n, the correction factor, $g(\varepsilon)$, is approximately constant at the extremes of the flow regimes due to the particular values of the expansion index. The value of n is not known entirely theoretically but can be obtained from the Richardson and Zaki equation. Thus
At low Re_p ($n = 4.6$)

$$1 - n = 1 - 4.6 = 3.6 \tag{2.40}$$

and at high Re_p ($n = 2.4$)

$$1 - 2n = 1 - 2 \times 2.4 = 3.8 \tag{2.41}$$

Thus an approximate interpolated expression for $g(\varepsilon)$ can be used as

$$c_{D\varepsilon} = c_{D\infty}\varepsilon^{-3.7} \tag{2.42}$$

and is depicted in Figure 2.5.

Note that the abscissa in Figure 2.4 starts at a value of ~ 0.4, which corresponds to the voidage of a randomly packed bed. Equation 2.44 is valid for the extremes of flow regimes but strictly requires correction for the intermediate case (Khan and Richardson, 1990; Di Felice, 1994).

Polydisperse suspensions

Several authors have sought to provide equations for particles varying in shape and density (see Patwardhan and ChiTien, 1985; Davis and Gecol, 1994). Whilst these have been successfully tested on bidisperse systems, more work

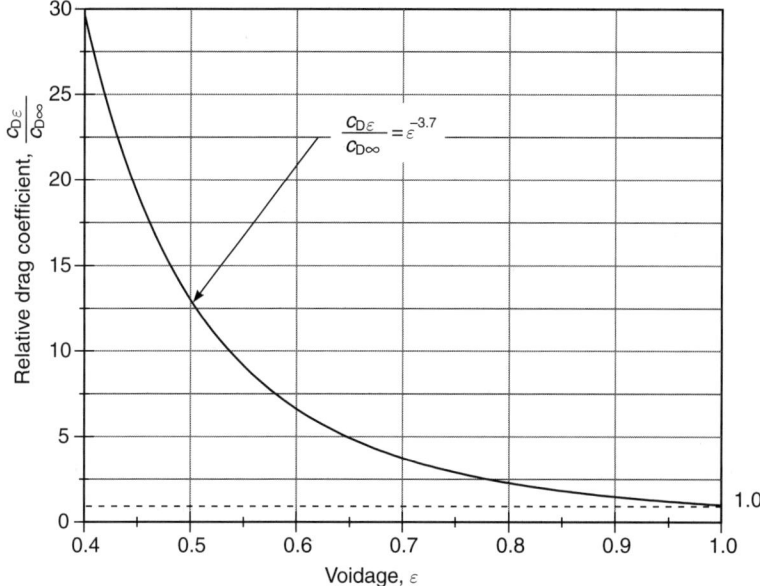

Figure 2.5 *Hindered drag coefficients for slurries (after Foscolo et al., 1983)*

is required for truly polydisperse systems such as those encountered in crystallizers. In these cases laboratory tests prove useful, as follows.

Sedimentation of slurries

In the 'Jar Test', an initially uniform, homogeneous suspension of particles slurries settles out under the force of gravity into distinct zones, as illustrated in Figure 2.6.

Starting from zone A of uniform concentration the slurry settles quickly at first (Figures 2.6 and 2.7) creating a clarified zone at the top (D) and increasing the concentration of the zone immediately below (B) with sediment at the base (C). As the solids content of the concentrating zone (B) increases settling occurs more slowly as hindered settling occurs (Figure 2.3). At the same time, sediment builds up from the base, merges with the concentrating zone and, given sufficient time, finally compacts.

Settled volumes are typically as follows
If

v_A = actual volume of particles = 1 say

Then

v_B = bulk volume of particles = 1.67
v_s = settled volume of particles = 10
v_f = final volume of particles = 8

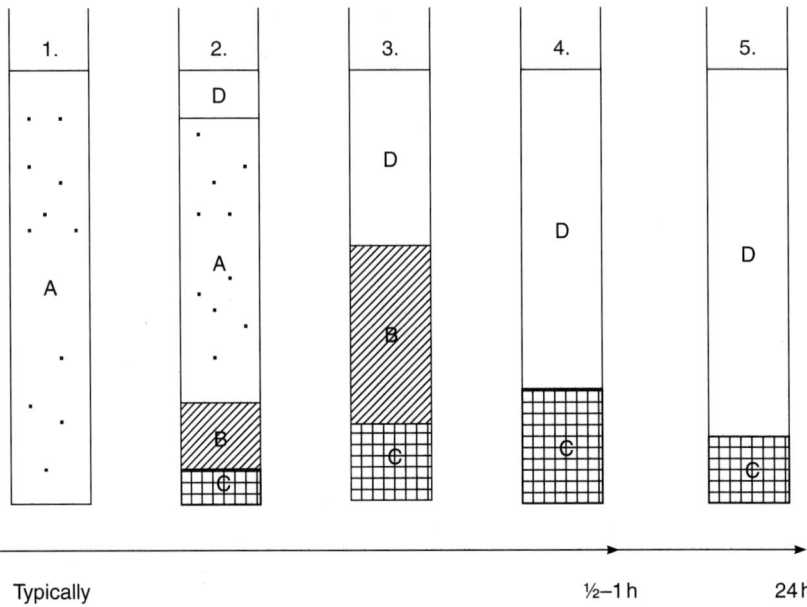

Figure 2.6 *Sedimentation zones during slurry settling (The 'Jar Test'). A, initial uniform concentration; B, zone of increasing concentration; C, sediment or sludge; D, clear liquor*

Typical settling rates

Slow settling ($R < 10$ cm/min, say)
    dense slurries, 3–50% solids of fine particles ($<50\,\mu$m)
Rapid settling ($R > 10$ cm/min, say)
    dilute slurries, <3% solids of coarse crystals ($>50\,\mu$m)
The division arbitrarily corresponds to settling of sand ($\approx 50\,\mu$m) in water (Purchas, 1981).

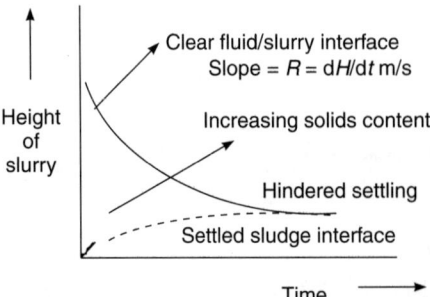

Figure 2.7 *Particle sedimentation rates in slurry settling (schematic)*

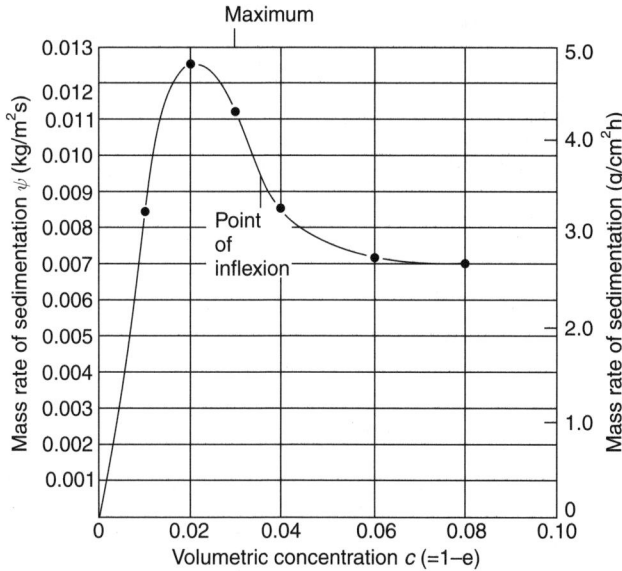

Figure 2.8 *Particle mass flux during slurry sedimentation of precipitated calcium carbonate (Coulson and Richardson, 1991)*

Mass flux

In addition to the particle settling velocity it is also necessary to know the corresponding rate at which mass is transferred during settling and this is best expressed as the mass flux defined by

$$\text{Flux} = \text{Concentration} \times \text{velocity} \quad [\text{kg/m}^3 \times \text{m/s} = \text{kg/m}^2\text{s}] \quad (2.43)$$

Examination of equation 2.43 shows that the mass flux initially increases as concentration increases but then passes through a maximum and finally declines as velocity decreases due to hindered settling (Figure 2.8).

Fluid flow through porous media

In filters etc. the particles become largely static in a 'bed' or 'cake' and in such cases the fluid therefore passes through a fixed array of particles or a porous solid and experiences drag as it does so (Figure 2.9). The particles resist the flow, reduce the velocity and give rise to an enhanced pressure drop compared with that in open channel flow.

Interstitial fluid velocity

$$v = \frac{Q}{A\varepsilon} \quad [\text{m/s}] \quad (2.44)$$

where Q = volumetric flowrate and ε = voidage.

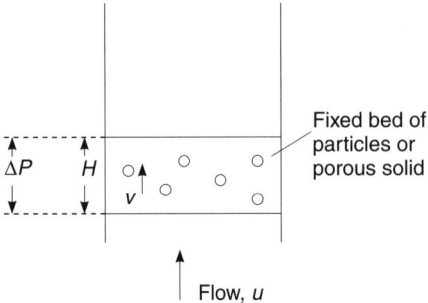

Figure 2.9 *Fluid flow through porous media*

Superficial fluid velocity

$$u = \frac{Q}{A} \text{ [m/s]} \quad (2.45)$$

where A = cross-sectional area.

As usual, two types of flow can occur, laminar and turbulent respectively, and the analyses differ for each case.

Laminar flow

(i) Darcy's law

Darcy's law (Darcy, 1856) relates fluid flowrate to bed pressure drop, depth and permeability

$$u = \frac{1}{A}\frac{dV}{dt} = \frac{K_1 \Delta P}{H} \quad (2.46)$$

where V = Fluid volume collected in time t, K_1 = bed permeability – a measure of the total drag force itself, a function of both the fluid and the particle characteristics.

(ii) Effect of fluid viscosity

Increasing fluid viscosity decreases permeability, thus assuming an inverse linear relationship

$$u = \frac{B \Delta P}{\mu H} \quad (2.47)$$

where B = permeability coefficient (m^2) and is now a function of the particles only i.e.

$$B = K_1 \mu \quad (2.48)$$

where K_1 is a function of the particulate array.

(iii) Effect of bed voidage and particle size

The simplest model ignores tortuosity and assumes the bed equivalent hydrodynamically to a matrix of straight tubes – like a bundle of drinking straws e.g. as in Figure 2.10.

A relationship between bed characteristics and fluid velocity and pressure drop will be derived. Assume

1. Fluid in pore velocity, $v = u/\varepsilon$
2. Hydraulic pore diameter, $\delta \alpha d_1 =$ void volume/total surface area i.e. $d_1 = v_B \varepsilon / v_B S_B = \varepsilon / S_B$
3. Pore length, $H_1 \alpha H$ i.e. the bed depth

Each of these quantities is now considered in turn.

For laminar flow of fluid through a pipe experiencing viscous drag, Carman (1937, 1956) applied the Hagen–Poisseuille Equation

$$v = \frac{d_1^2 \Delta P}{32 \mu H_1} \tag{2.49}$$

Thus for the capillary model, substituting for d_1 and H_1, the interstitial fluid velocity is given by

$$v = \frac{u}{\varepsilon} = \frac{1}{K_2} \left(\frac{\varepsilon}{S_B}\right)^2 \frac{\Delta P}{\mu H} \tag{2.50}$$

Replacing the bed specific surface area by the corresponding particle specific surface area and voidage

$$S_B = S_p (1 - \varepsilon) \tag{2.51}$$

where $S_B =$ bed specific surface area and $S_p =$ particle specific surface area gives

$$u = v\varepsilon = \frac{1}{K_2} \frac{\varepsilon^3}{S_p^2 (1-\varepsilon)^2} \frac{\Delta P}{\mu H} \tag{2.52}$$

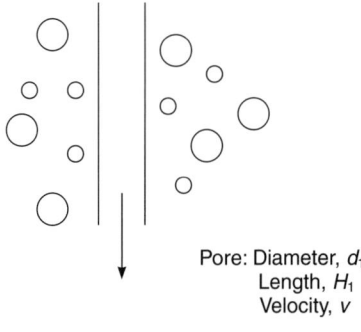

Pore: Diameter, d_1
Length, H_1
Velocity, v

Figure 2.10 *Capillary model for fluid flow in porous media*

which is known as the Carman–Kozeny Equation, whence

$$B = \frac{1}{K_2 S_p^2 (1-\varepsilon)^2} \varepsilon^3 \tag{2.53}$$

where K_2 = Kozeny 'constant' = 5 (but is actually a coefficient in the range 3.5–5.5; see Coulson and Richardson, 1991).

Compressible beds

In practice, virtually all particulate beds and porous media are compressible to a greater or lesser extent be they inorganic solids, organic materials or biological particles. In these cases the control volume occupies not the whole bed (throughout which conditions were assumed to be constant) but simply a differential slice of the matrix, as shown in Figure 2.11, for integration over the whole bed depth.

Since the bed is compressible

$$\text{Voidage} = f(\text{pressure or } \Delta p) \tag{2.54}$$

Thus the Carman–Kozeny equation applicable for the increment dx is given by

$$-\frac{dp}{dx} = \frac{K_2 \mu (1-\varepsilon)^2 S_p^2 u}{\varepsilon^3} \tag{2.55}$$

Integrating over the bed length H gives

$$K_2 u \mu S_p^2 H = \int_{p_1}^{p_2} \frac{\varepsilon^3}{(1-\varepsilon)^2} dp \tag{2.56}$$

In order to evaluate the total fluid flowrate, the integral on the RHS of equation 2.73 must be evaluated over pressure range using a known (usually measured) function of voidage with pressure.

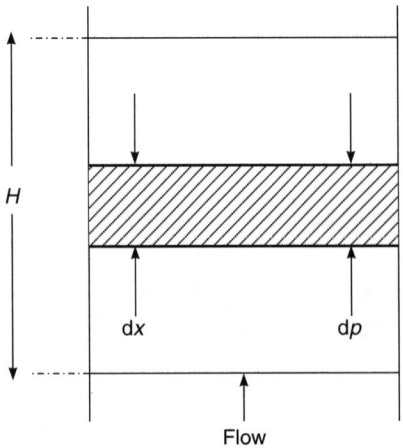

Figure 2.11 *Control volume for compressible porous media*

Empirical methods

Alternatively, the particle bed resistance can be determined experimentally according to

$$r = r' \Delta p^n \tag{2.57}$$

where r is the inverse of permeability (see p. 91) and n is an empirical constant.

Turbulent flow

In this case a kinetic energy (KE) term is added in addition to laminar viscous drag.

Burke and Plummer (1928) proposed that the following semi-empirical equation applies to turbulent flow

$$\frac{\Delta P}{\Delta H} = \frac{K'(1-\varepsilon)\rho u^2}{\varepsilon^3 d_p} \tag{2.58}$$

Whilst Ergun (1952) made a linear sum of the laminar and turbulent terms thus

$$\frac{\Delta P}{\Delta H} = \underbrace{\frac{K_3(1-\varepsilon)^2 \mu u}{\varepsilon^3 d_p^2}}_{\text{Laminar}} + \underbrace{\frac{K_4(1-\varepsilon)\rho u^2}{\varepsilon^3 d_p}}_{\text{Turbulent}} \tag{2.59}$$

where $K_3 = 150$ and $K_4 = 1.75$.

Again, an alternative approach to the prediction of bed pressure drop and fluid flow in porous media is to use friction factors (the analogue of the drag coefficient developed for particle flow above).

Friction factors

Packed bed friction factor

The packed bed friction factor, f_p, is defined by

$$f_p = \frac{\text{shear stress}}{\text{KE}} \tag{2.60}$$

$$= \frac{\frac{F}{A}}{\frac{\rho v^2}{2}} \tag{2.61}$$

From a force balance over the bed

$$\Delta p \varepsilon = f_p \rho v^2 s_p (1-\varepsilon) H \tag{2.62}$$

Therefore

$$f_p = \frac{\Delta p \varepsilon^3}{s_p H (1-\varepsilon)^{\frac{1}{2}} \rho u^2} \tag{2.63}$$

Packed bed Reynolds number

The packed bed Reynolds number, Re_p, is defined by

$$Re_p = \frac{\rho u}{(1-\varepsilon)s_p\mu} \tag{2.64}$$

In deriving expressions for the packed bed friction factor, three separate flow regimes are normally considered (see Figure 2.11) as follows.

Laminar flow ($Re_p < 2$)

By using the Carman–Kozeny Equation (with $K_2 = 5$), the friction factor is given by

$$f_p = \frac{10}{Re_p} \tag{2.65}$$

(*cf* for Poisseuille flow of fluid in pipes: $c_f = 16/Re_{pipe}$)

Intermediate flow ($2 < Re_p < 200$)

The packed bed friction factor is given by

$$f_p = \frac{10}{Re_p} + \frac{0.8}{Re_p^{0.1}} \tag{2.66}$$

after Carman (1956) and Kozeny (1927).

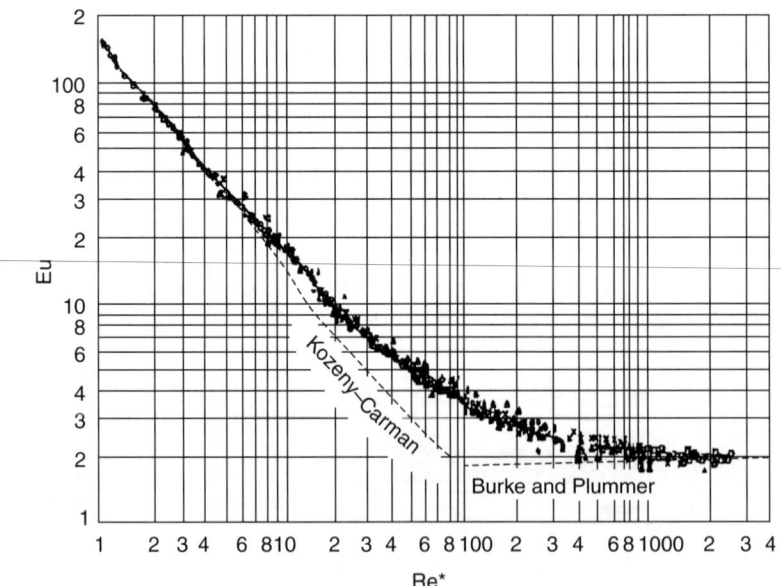

Figure 2.12 *Packed bed pressure loss Euler number versus particle Reynolds number (Ergun, 1952)*

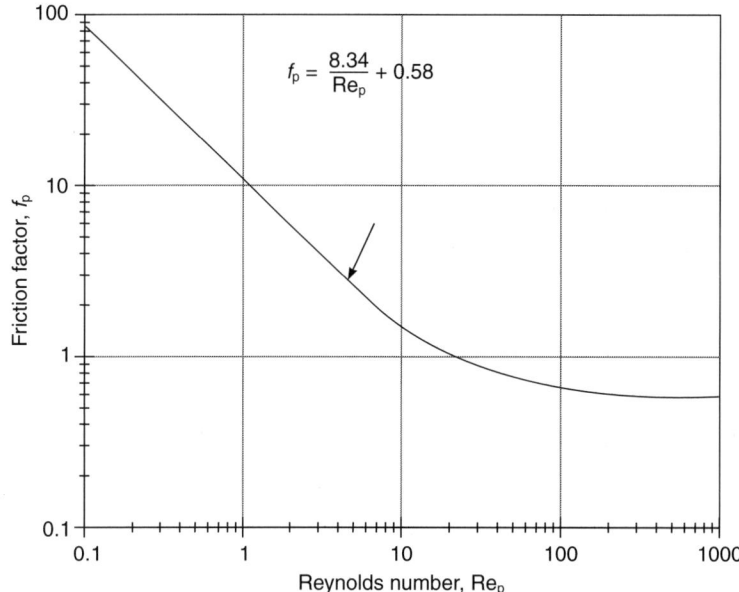

Figure 2.13 *Packed bed pressure loss Euler number versus particle Reynolds number*

General Expression ($1 < \text{Re}_p < 2000+$)

The pressure drop is given by the Ergun equation (Ergun, 1952), expressed by

$$\text{Eu} = \frac{150}{\text{Re}_p} + 1.75 \tag{2.67}$$

where Eu is the Euler number ($\Delta p/\rho f u^2$) as depicted in Figure 2.12 or in terms of friction factor

$$f_p = \frac{8.34}{\text{Re}_p} + 0.58 \tag{2.68}$$

The packed bed friction factor is shown in Figure 2.13 and applies to flow through random packing ($\varepsilon \approx 0.4$).

Suspension of settling particles by agitation

Solid particles in liquids generally tend to settle to the bottom of a vessel under gravity due to their excess density. To maintain a suspension, some form of agitation is normally provided together with wall baffles to prevent vortex formation in the swirling flow (Figure 2.14).

The minimum stirrer speed required to 'just suspend' the particle s, N_{JS}, may be calculated by the semi-empirical equation due to Zweitering (1958)

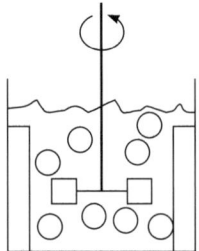

Figure 2.14 *Particle suspension in an agitated vessel*

$$N_{JS} = \frac{S d_p^{0.2} \left[\frac{\Delta \rho}{\rho_L}\right]^{0.45} v^{0.1} x^{0.13}}{D^{0.85}} \quad (2.69)$$

where S is a function of the agitator vessel geometry, and D_P is the size of particles of mass fraction x. Use of the 'just suspended' agitator speed, however, results in a far from homogeneous suspension with axial distribution of the solids volume fraction occurring. In practice therefore, somewhat higher speeds than N_{JS} are adopted. Nienow (1985), however, points out that it is difficult to achieve solid phase suspension homogeneity in large vessels at economic power inputs. Care must also be taken to avoid excessive particle attrition due to collision with vessel walls, impellers etc. by use of too high an agitation rate (see Chapter 5). Further details concerning fluid mixing and particle suspension can be found in Šterbáček and Tausk (1965), Holland and Chapman (1966), Oldshue (1983), Uhl and Gray (1986) and Harnby *et al.* (1992).

Energy dissipation

For turbulent fluid-induced stresses acting on particles, it is necessary to consider the structure and scale of turbulence in relation to particle motion in the flow field. There is, as yet, however, no completely satisfactory theory of turbulent flows, but a great deal has been achieved based on the theory of isotropic turbulence (Kolmogorov, 1941) adapted to solid–liquid systems (Kolař, 1958, 1959; Middleman, 1965; Hughmark, 1969). The energy put into by the impellor is considered to be transferred firstly to large scale eddies and then to larger numbers of smaller, isotropic eddies from which it is dissipated by viscous forces in the form of heat. Energy dissipation, ε, has the effect of enhancing mass transfer coefficient, k_L, according to $k_L \propto \sqrt{\varepsilon}$. The Kolmogorov characteristic time scale is given by

$$\tau_k = \left(\frac{v_L}{\varepsilon}\right)^{0.5} \quad (2.70)$$

and the microscale of turbulence by

$$\lambda_k = \left(\frac{v_L^3}{\varepsilon}\right)^{0.25} \quad (2.71)$$

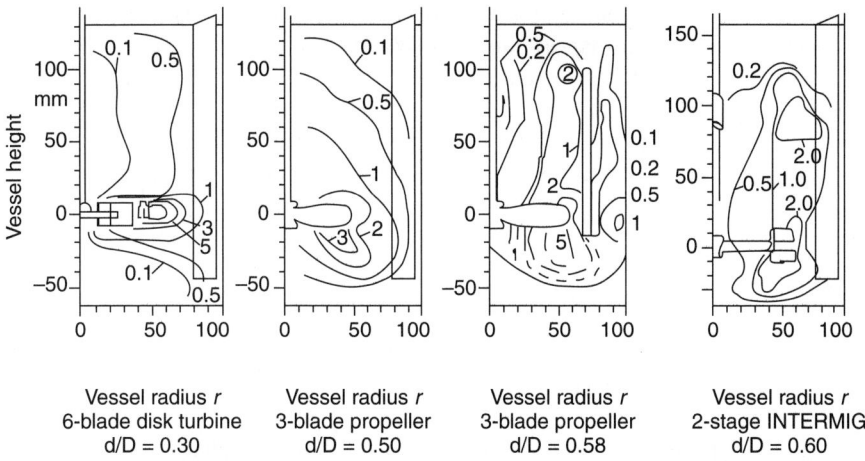

Figure 2.15 *Local specific power input to agitated vessels for several stirrers (after Mersmann and Geisler, 1991)*

Typical quoted values for the Kolmogorov microscale of turbulence for agitated vessels are normally in the range 25–50 μm.

When scaling up mixing operations, it has been observed that the shear rate distribution, fluid velocity and therefore the local energy distribution, changes with increasing vessel size (Oldshue, 1985). With larger scales, the maximum shear rate in the impeller zone increases while the average shear rate in this zone decreases. Therefore, even when scaling up with constant power input per unit volume, the flow field looks different for different scales of operation. For example, the energy very quickly dissipates further away from the Rushton turbine leading to zones with significantly lower local energy dissipation (Geisler *et al.*, 1991; Mersmann *et al.*, 1994). In laser doppler anemometry (LDA) experiments, a zone with high-energy dissipation around the impeller and levels of low energy dissipation in the upper part of the vessel are observed (Figure 2.15).

Rielly and Marquis (2001) present a review of crystallizer fluid mechanics and draw attention to the inconsistency between the dependence of crystallization kinetic rates on local mean and turbulent velocity fields and the averaging assumptions of conventional well-mixed crystallizer models.

Conservation equations

Conservation is a general concept widely used in chemical engineering systems analysis. Normally it relates to accounting for flows of heat, mass or momentum (mainly fluid flow) through control volumes within vessels and pipes. This leads to the formation of conservation equations, which, when coupled with the appropriate rate process (for heat, mass or momentum flux respectively), enables equipment (such as heat exchangers, absorbers and pipes etc.) to be sized and its performance in operation predicted. In analysing crystallization and other particulate systems, however, a further conservation equation is required to account for particle numbers. This is known as the *population balance*. It is another transport equation. When coupled with the corresponding rate processes for particle formation (nucleation, growth, agglomeration etc.), it enables crystallizers to be sized and their performance to be predicted in terms of not only mass yield but also its particle size distribution.

The conservation equations for mass (continuity equation)

$$\frac{D\rho}{Dt} + \rho \nabla \mathbf{v} = 0 \tag{2.72}$$

and momentum (Navier–Stokes equations for incompressible flow with constant viscosity)

$$\rho \frac{D\mathbf{v}}{Dt} = \rho \mathbf{g} - \nabla p + \mu \Delta \mathbf{v} \tag{2.73}$$

are valid for both laminar and turbulent flow. Theoretically, it is possible to solve this set of equations for the latter by direct numerical solution (DNS). As a very high resolution in both time and space is necessary to account for the turbulent fluctuations, however, the computational problem usually becomes very complex. It can only be solved for low Reynolds numbers and simple geometries of the flow domain. Thus, in an alternative approach the variables will be represented by their time average $\bar{\mathbf{v}}$ and their fluctuation component $\mathbf{v}'$ (Reynolds decomposition)

$$\mathbf{v} = \bar{\mathbf{v}} + \mathbf{v}' \tag{2.74}$$

Consequently, six additional unknowns, the Reynolds stresses $-\rho \overline{v'_i v'_j}$, are obtained and the equations for turbulent flow become

$$\frac{D\rho}{Dt} + \rho \nabla \bar{\mathbf{v}} = 0 \tag{2.75}$$

$$\rho \frac{D\bar{\mathbf{v}}}{Dt} = \rho \mathbf{g} - \nabla p + \nabla \tau \tag{2.76}$$

with

$$\tau_{ij} = \mu \frac{d\bar{v}_i}{dx_j} - \rho \overline{v'_i v'_j} \tag{2.77}$$

Using turbulence models, this new system of equations can be closed. The most widely used turbulence model is the k-ε model, which is based on an analogy of viscous and Reynolds stresses. Two additional transport equations for the turbulent kinetic energy k and the turbulent energy dissipation ε describe the influence of turbulence

$$\mu_t = \rho C_\mu \frac{k^2}{\varepsilon} \tag{2.78}$$

where μ_t is the eddy viscosity, and

$$-\rho \overline{v'_i v'_j} = 2\mu_t E_{ij} - \frac{2}{3} \rho k \delta_{ij} \tag{2.79}$$

the so-called extended Boussinesq relationship, with

$$E_{ij} = \frac{1}{2}\left[\frac{\partial \bar{v}_i}{\partial x_j} + \frac{\partial \bar{v}_j}{\partial x_i}\right] \quad (2.80)$$

In these model equations it is assumed that turbulence is isotropic, i.e. it has no favoured direction. The k-ε model frequently offers a good compromise between computational economy and accuracy of the solution. It has been used successfully to model stirred tanks under turbulent conditions (Ranade, 1997). Manninen and Syrjänen (1998) modelled turbulent flow in stirred tanks and tested and compared different turbulence models. They found that the standard k-ε model predicted the experimentally measured flow pattern best.

More advanced models, for example the algebraic stress model (ASM) and the Reynolds stress model (RSM), are not based on the eddy-viscosity concept and can thus account for anisotropic turbulence thereby giving still better predictions of flows. In addition to the transport equations, however, the algebraic equations for the Reynolds stress tensor also have to be solved. These models are therefore computationally far more complex than simple closure models (Kuipers and van Swaaij, 1997).

When using large eddy simulation (LES), the computational mesh is chosen in such a way that it resolves large-scale turbulent eddies. As a direct consequence, only the small-scale turbulence, which is isotropic and can be specified quite easily, has to be modelled with a closure model. Large eddy simulation is therefore a stage between simulations using the closure models mentioned above and the direct numerical solution (DNS), which resolves even small-scale fluctuations of the turbulent motion. Due to the high resolution of the grid, LES is still computationally very demanding.

Cate *et al.* (2001) propose a method for the calculation of crystal–crystal collisions in the turbulent flow field of an industrial crystallizer. It consists of simulating the internal flow of the crystallizer as a whole and of simulating the motion of individual particles suspended in the turbulent flow in a small subdomain (box) of the crystallizer.

Computational fluid dynamics (CFD)

Computational fluid dynamics (CFD) is the numerical analysis of systems involving transport processes and solution by computer simulation. An early application of CFD (FLUENT) to predict flow within cooling crystallizers was made by Brown and Boysan (1987). Elementary equations that describe the conservation of mass, momentum and energy for fluid flow or heat transfer are solved for a number of sub regions of the flow field (Versteeg and Malalasekera, 1995). Various commercial concerns provide ready-to-use CFD codes to perform this task and usually offer a choice of solution methods, model equations (for example turbulence models of turbulent flow) and visualization tools, as reviewed by Zauner (1999) below.

Commercially available CFD codes have three main elements in common. Firstly, the pre-processor enables the user to define the geometry, which is often referred to as the computational domain or flow domain of the problem, for example, of a stirred vessel. The geometry can be created either by using the mesh generation facilities of the code or by converting an external file (e.g. a computer aided design (CAD) file), and consists of a number of blocks, which form the actual geometric layout of the problem. By subdividing the blocks into smaller regions, a grid of computational cells is obtained. The number of grid cells determines the complexity of the problem. A very fine grid resolves the local gradients of the flow variables very accurately, gives a very large number of cells and therefore slows down the simulation. On the other hand, a coarse grid consists of less grid cells, but often causes convergence problems of the numerical solution. There are no hard and fast rules as to how fine the grid has to be made, but it must be ensured that the solution is 'grid-independent', i.e. that a further refinement of the grid has no influence on the results of the simulation. Secondly, the solver performs the actual calculation of the numerical solution.

Three different solution methods can be distinguished

- finite difference (Smith, 1985)
- finite element (Zienkiewicz and Taylor, 1991)
- spectral techniques (Gottlieb and Orszag, 1977).

The finite volume method, a very common method for solving fluid flow problems (Versteeg and Malalasekera, 1995). The balance equations are solved for each grid cell using an iterative solution approach, as the underlying physical phenomena are complex.

Finally, the post-processor helps to visualize the huge amount of data produced by the flow solver. Vector, contour and shaded contour plots of the velocity field, the energy dissipation distribution or other variable fields can be plotted for lines, cross sections or surfaces of the geometry. Some packages even offer particle tracing facilities and animation for transient problems.

Simulation of the fluid dynamics in agitated vessels

Stirred tanks are the most common form of crystallizers. Nevertheless, due to high local gradients of the energy dissipation, the fluid dynamics are not well understood and depend to a large extent on the geometry of the vessel. Different forms of impellers, baffles and draft tubes can produce very different flow fields. As a stirred tank contains a moving impeller, the fluid cells surrounding the impeller are modelled as rotating blocks in CFD (Bakker *et al.*, 1997). A sliding mesh technique has been developed to account for the movement of the rotating impeller grid relative to the surrounding motionless tank cells. Xu and McGrath (1996) compared the sliding mesh simulation results for a stirred tank with experimental laser doppler anemometry (LDA) data and found that the data corresponded very well. An alternative to the sliding mesh technique is the momentum source model, where the impeller region is modelled as a black-box source of momentum. In this case, however, experimental data concerning the forces acting on the impeller are necessary. Furthermore, no local data in the impeller region can be obtained (black box) when using the momentum source model. For a sliding mesh simulation, on the other hand, no experimental data are necessary and the flow field in the vicinity of the impeller is readily obtained.

Guichardon *et al.* (1994) studied the energy dissipation in liquid–solid suspensions and did not observe any effect of the particles on micromixing for solids concentrations up to 5 per cent. Precipitation experiments in research are often carried out at solids concentrations in the range from 0.1 to 5 per cent. Therefore, the stirred tank can then be modelled as a single-phase isothermal system, i.e. only the hydrodynamics of the reactor are simulated. At higher slurry densities, however, the interaction of the solids with the flow must be taken into account.

Mixing models

As the flow of a reacting fluid through a reactor is a very complex process, idealized chemical engineering models are useful in simplifying the interaction of the flow pattern with the chemical reaction. These interactions take place on different scales, ranging from the macroscopic scale (macromixing) to the microscopic scale (micromixing).

In what follows, both macromixing and micromixing models will be introduced and a compartmental mixing model, the segregated feed model (SFM), will be discussed in detail. It will be used in Chapter 8 to model the influence of the hydrodynamics on a meso- and microscale on continuous and semibatch precipitation where using CFD, diffusive and convective mixing parameters in the reactor are determined.

Macromixing and micromixing

The term 'macromixing' refers to the overall mixing performance in a reactor. It is usually described by the residence time distribution (RTD). Originally introduced by Danckwerts (1958), this concept is based on a macroscopic lumped population balance. A fluid element is followed from the time at which it enters the reactor (Lagrangian viewpoint – observer moves with the fluid). The probability that the fluid element will leave the reactor after a residence time τ is expressed as the RTD function. This function characterises the scale of mixedness in a reactor.

An ideal plug flow reactor, for example, has no spread in residence time because the fluid flows like a 'plug' through the reactor (Westerterp *et al.*, 1995). For an ideal continuously stirred reactor, however, the RTD function becomes a decaying exponential function with a wide spread of possible residence times for the fluid elements.

Non-ideal reactors are described by RTD functions between these two extremes and can be approximated by a network of ideal plug flow and continuously stirred reactors. In order to determine the RTD of a non-ideal reactor experimentally, a tracer is introduced into the feed stream. The tracer signal at the output then gives information about the RTD of the reactor. It is thus possible to develop a mathematical model of the system that gives information about flow patterns and mixing.

Baldyga and Bourne (1999) present a comprehensive overview and comparison of macromixing models available in the literature for use in chemical reaction engineering.

Mesomixing and micromixing

The term mesomixing as introduced by Baldyga and Bourne (1992) describes the interaction by mixing between the feed plumes and the bulk, or blending. The reactant entering the reactor is eroded from the plumes and its scale reduced to that of large eddies. In the literature, the terms macromixing and mesomixing are not always considered separately, which has led to confusion when reactions were stated to be macromixing-controlled. In terms of the notation above, these processes are mesomixing-limited, as real macromixing limitation is very rare in precipitation reactors because of the well mixedness on the macroscale and the relatively long residence times of the process. Mesomixing describes the very first moments of a fluid element entering the vessel, whereas macromixing considers the whole lifetime, i.e. the age, of an element in the reactor. In terms of space, mesomixing occurs only in the reaction zone.

Baldyga et al. (1995) proposed the meso time constant defined by

$$t_{meso} = A \frac{\varepsilon_{avg}}{\varepsilon_{loc}} \frac{Q^{\frac{1}{3}}}{N^{\frac{4}{3}} d_s} \qquad (2.81)$$

which can be used to model mesomixing in a stirred tank by including the term $\varepsilon_{avg}/\varepsilon_{loc}$ to take different feed point positions into account. The inverse of the time constant t_{meso} (mesomixing) can be interpreted as a transfer coefficient for mass transfer by convection.

Micromixing is regarded as turbulent mixing on the molecular level. It comprises the viscous-convective deformation of fluid elements, followed by molecular diffusion (Baldyga and Poherecki, 1995). Baldyga et al. (1995) and Baldyga et al. (1997) proposed a characteristic timescale for micromixing based on Kolmogoroff's microscale of eddy lifetime

$$t_{micro} = 17.3 \times \left(\frac{\nu}{\varepsilon_{loc}}\right)^{\frac{1}{2}} \qquad (2.82)$$

The inverse of the time constant t_{micro} (micromixing) can be interpreted as a transfer coefficient for mass transfer by diffusion.

Baldyga and Bourne (1999, Chapter 14) present considerations in relation to choosing an adequate model of mixing for modelling precipitation processes. Micromixing effects are present only at high concentrations with correspondingly low values of the time constants for precipitation compared to time constants for mixing.

Several mesomixing and micromixing models have been proposed to describe the influence of mixing on chemical reactions on the meso- and molecular scale. Most of them fall into one of the three categories discussed below (Villermaux and Falk, 1994).

Phenomenological or mechanistic models

This type of model derives from the RTD concept of macromixing as described above, which is applied on a microscopic level using idealized zones and exchange flows. The mixing parameters do not usually have any physical relevance and are determined experimentally.

The coalescence-redispersion (CRD) model was originally proposed by Curl (1963). It is based on imagining a chemical reactor as a number population of droplets that behave as individual batch reactors. These droplets coalesce (mix) in pairs at random, homogenize their concentration and redisperse. The mixing parameter in this model is the average number of collisions that a droplet undergoes.

Another popular phenomenological mixing model is the interaction by exchange with the mean (IEM) model (Harada *et al.*, 1962; Villermaux and Devillon, 1975). Micromixing takes place by exchange between feed regions (well-mixed batch zones) and a mean environment (bulk) according to a mixing time constant. Garside and Tavare (1985) used the IEM model-to-model extreme cases of micromixing during precipitation.

In the three and four environment (3E and 4E) models (Ritchie and Togby, 1979; Mehta and Tarbell, 1983), the reactor is divided into two segregated entering environments and one or two fully mixed leaving environments. The mixing parameter is the transfer coefficient between the environments.

Pohorecki and Baldyga (1983, 1988) formulated simple multi-environmental or 'compartmental' precipitation models employing mixing parameters based on turbulence theory for batch and continuous operation, respectively, and used for interpretation of experimental data. Garside and Tavare (1985) used the IEM model to predict extreme cases of micromixing during precipitation. In the SFM (see later), the IEM model is reduced to exchange between the feed regions and a mean environment (bulk). Franck *et al.* (1988) developed a two-compartment mixing model with the exchange flow rate (recycle number) between the two compartments as the only adjustable parameter. They applied their model successfully to the precipitation of salicylic acid. Chang *et al.* (1986) compare different phenomenological mixing models and demonstrate their analogies and similarities to the theory of turbulence.

Physical models

As their name suggests, these models are based on the physical principles of diffusion and convection, which govern the mixing process. According to the flow pattern, the reactor is divided into different zones with different flow characteristics.

Baldyga and Bourne (1984a–c, 1989a,b) developed models based on the engulfment-deformation-diffusion (EDD) theory. Entering material is engulfed by bulk material forming vortices, subsequently deformed and stretched to form slabs and finally exchanges mass by molecular diffusion. Bourne (1985) applies a one-dimensional diffusion equation to slabs and shows how vorticity is responsible for mixing by engulfment of fluid. Baldyga and Bourne (1989a,b) show that under some conditions engulfment becomes the rate-determining step in micromixing. They describe the molecular mixing based on the spectral interpretation of mixing in an isotropic turbulent field. The concentration spectrum indicates that molecular diffusion starts between the viscous-convective and the viscous-diffusive subrange and becomes dominant as the scale becomes smaller. Fluid elements in this subrange are laminar deformed by stretching and form slabs.

van Leeuwen (1998) developed a compartmental mixing model for precipitation based on the engulfment theory mentioned above. The author obtains the mixing parameters between the feed and the bulk zone from the flow characteristics in the reactor and subsequently calculates moments and mean sizes of the precipitate.

From the considerations above, it can be concluded that phenomenological models show a lack of predictive quality, as the mixing parameters cannot be determined *a priori*. Furthermore, an analytical solution to the mixing problem in complicated reactor geometries is not yet feasible. Consequently, to date only physical models are suitable for modelling scale-up. The SFM, which originally belonged to the group of phenomenological IEM mixing models, has subsequently been modified and its mixing parameters, the mesomixing and the micromixing time, have become 'meaningful' physical parameters for diffusion and convection in a physical model. They can therefore be related to the flow field in the reactor. The model was first used to investigate micromixing effects of consecutive–competitive semibatch reactions (Villermaux, 1989). It was subsequently applied to predict the effects of mixing on semibatch polymerization (Tosun, 1992) and to model the semibatch precipitation of barium sulphate (Marcant, 1996). The SFM is found to be particularly suitable for modelling mixing effects, as it combines the advantages of both the compartmental IEM model and the physical models.

The population balance

Particle conservation in a vessel is governed by the particle-number continuity equation, essentially a population balance to identify particle numbers in each and every size range and account for any changes due to particle formation, growth and destruction, termed particle 'birth' and 'death' processes reflecting formation and loss of particulate entities, respectively.

The population balance accounts for the number of particles at each size in a continuous distribution and may be thought of as an extension of the more familiar overall mass balance to that of accounting for individual particles.

During the course of crystallization, individual particles may nucleate, grow, agglomerate or break and leave the precipitation reactor in the continuous mode of operation. The mathematical framework used to describe this process is called the population balance. Hulburt and Katz (1964) generalized the concept of population balances for agglomerative processes based on the conservation of population. They derived their mathematical model from statistical mechanics and applied it to processes involving both growth and agglomeration. Randolph and Larson (1988) investigated in detail the modelling of particle size distributions using population balances and developed the mixed-suspension mixed-product-removal (MSMPR) model for batch and continuous crystallizers, anticipated by Bransom and Dunning (1949) and Bransom et al. (1949), respectively.

Accumulation = Input − Output + Net generation

Therefore, in the Lagrangian framework the population balance can be written as

$$\frac{d}{dt}\int_{V_1} n \, dV = \int_{V_1} (B-D) \, dV \tag{2.83}$$

with V_1 as an arbitrarily chosen subvolume. $n(x)$ represents the population density at location x (x, y, z) and is defined as

$$n(x) = \frac{dN(x)}{dL} \tag{2.84}$$

and B and D are the birth and death rate, respectively. The first term can be written as

$$\frac{d}{dt}\int_{V_1} n \, dV = \int_{V_1} \left[\frac{\partial n}{\partial t} + \nabla\left(\frac{dx}{dt} n\right)\right] dV \tag{2.85}$$

with

$$\frac{dx}{dt} = v = v_i + v_e \tag{2.86}$$

and the population balance becomes

$$\int_{V_1} \left[\frac{\partial n}{\partial t} + \nabla(v_i n) + \nabla(v_e n) + D - B\right] dV = 0 \tag{2.87}$$

with the internal velocity v_i and the external velocity v_e. In this context the internal co-ordinate space specifies the characteristics of the particle, while the external co-ordinate space refers to the position of the particle in the physical space. The internal velocity therefore describes the change of particle characteristic e.g. its size, volume or composition, and the external velocity the fluid velocity in the crystallizer.

As the subvolume V_1 was arbitrary, it must be independent of the volume of integration, leading to

$$\frac{\partial n}{\partial t} + \nabla(vn) - B + D = 0 \tag{2.88}$$

This equation is analogous to the continuity equation derived for the conservation of mass in a continuum

$$\frac{\partial \rho}{\partial t} + \nabla(v\rho) = 0 \tag{2.89}$$

The general equation 2.88 can be used when particles are distributed along both internal and external co-ordinate space. For some precipitation and crystallization problems, however, it is sufficient to know only the distribution of particles in the internal phase space. The system is then considered fully backmixed with inputs and outputs Q_k and population densities n_k. The population balance for this simplified case can be derived from the general population balance with the form

$$\int_V \left[\frac{\partial n}{\partial t} + \nabla(v_e n) + \nabla(v_i n) + D - B\right] dV = 0 \qquad (2.90)$$

As n, B and D are only functions of time and internal co-ordinates, these terms can be easily integrated

$$\frac{\partial n}{\partial t} + \nabla(v_i n) + D - B + \frac{1}{V}\int_{V_1} \nabla(v_e n) dV = 0 \qquad (2.91)$$

with

$$\frac{1}{V}\int_{V_1} \nabla(v_e n) dV = n\frac{d(\log V)}{dt} + \sum_k \frac{Q_k n_k}{V} \qquad (2.92)$$

where Q_k and n_k are the flow rate and population density respectively of the flows incoming and outgoing, and using the internal velocity expressed in terms of the crystal growth rate

$$\nabla v_i n = \frac{\partial(Gn)}{\partial L} \qquad (2.93)$$

The *macrodistributed population balance* can then be derived as

$$\frac{\partial n}{\partial t} + \frac{\partial(Gn)}{\partial L} + n\frac{d(\log V)}{dt} = B - D - \sum_k \frac{Q_k n_k}{V} \qquad (2.94)$$

This equation describes the change of population in a well-mixed system and is often used to model fully mixed crystallization and precipitation processes. If the system is imperfectly mixed, however, then the more complicated equation 2.88 can be used provided that the external flow field can be calculated e.g. by use of CFD (see later).

Moment transformation of the population balance

General solution of the population balance is complex and normally requires numerical methods. Using the moment transformation of the population balance, however, it is possible to reduce the dimensionality of the population balance to that of the transport equations. It should also be noted, however, that although the mathematical effort to solve the population balance may therefore decrease considerably by use of a moment transformation, it always leads to a loss of information about the distribution of the variables with the particle size or any other internal co-ordinate. Full crystal size distribution (CSD) information can be recovered by numerical inversion of the leading moments (Pope, 1979; Randolph and Larson, 1988), but often just mean values suffice.

Moments of the distribution are defined by

$$\mu_j = \int_0^\infty nL^j \, dL \quad j = 0, 1, 2, 3\ldots \tag{2.95}$$

with

$$j = 0, 1, 2, \ldots$$

The *micromoment population balance* is derived from the general population balance (Lagrangian framework) and can be written as

$$\frac{\partial m_j}{\partial t} + \nabla \mathbf{v}_e \mu_j = 0^j B^0 + jG\mu_{j-1} + \langle B \rangle - \langle D \rangle \tag{2.96}$$

The micromoment population balance (2.96) averages the internal co-ordinate (i.e. the particle properties) in moments but accounts for their local variations.

The *macromoment population balance* can be obtained from the macrodistributed population balance (2.94). Making the transformation

$$\int_0^\infty L^j \left[\frac{\delta n}{\delta t} + \frac{\delta(nG)}{\delta L} \right] dL = 0 \quad j = 0, 1, 2, 3\ldots \tag{2.97}$$

enables the macrodistributed population balance to be transformed into a set of ordinary differential equations and are thus easier to solve

$$\frac{dm_j}{dt} + m_j \frac{d(\log V)}{dt} = 0^j B^0 + jGm_{j-1} + \langle B \rangle - \langle D \rangle - \sum_k \frac{Q_k n_k}{V} \tag{2.98}$$

It is an ordinary differential equation (time dependence only) where the zeroth moment (μ_0) is the total crystal number (per unit volume of suspension), the first (μ_1) the total crystal length (lined up end to end), the second (μ_2) is related to total surface area, the third (μ_3) total volume (and hence mass) and so on.

The method of moments reduces the computational problem to solution of a set of ordinary differential equations and thus solves for the average properties of the distribution.

While it is possible to recover the original distribution from the leading moments, e.g. by matrix inversion, it is often the case that simply the mean characteristics together with an estimate of the spread of the distribution are sufficient e.g.

$$L_{\mathrm{mm}} = \frac{\mu_4}{\mu_3} \tag{2.99}$$

where L_{mm} is the mass–mean average crystal size. The coefficient of variation, CV, may be similarly derived.

Alternatively, numerical methods may be employed to solve the population balance using discretization, particularly in those cases when crystal breakage and agglomeration processes occur (see below).

These fundamental equations apply to many systems involving 'discrete entities': aerosols, molecules, and particles, even people. A full review of their derivation of these equations is to be found in Randolph and Larson (1988), who have pioneered their application to industrial crystallizers in particular.

Turbulent population balance

The population balance in equation 2.86 employs the local instantaneous values of the velocity and concentration. In turbulent flow, there are fluctuations of the particle velocity as well as fluctuations of species and concentrations (Pope, 1979, 1985, 2000). Baldyga and Orciuch (1997, 2001) provide the appropriate generalization of the moment transformation equation 2.93 for the case of homogeneous and non-homogeneous turbulent particle flow by Reynolds averaging

$$\frac{\partial \langle m_j \rangle}{\partial t} + \frac{\partial \langle u_{pi} m_j \rangle}{\partial x_i} = +0^j \langle R_N \rangle + j \langle G m_{j-1} \rangle + \langle B_f \rangle - \langle D_f \rangle \quad \text{for} \tag{2.100}$$

$$i = 1, 2, 3 \ldots \quad \text{and} \quad j = 1, 2, 3 \ldots$$

Introducing a concept of gradient diffusion for particles and employing a mixture fraction for the non-reacting fluid originating upstream, $f = c_A^0/c_{A0}$, and a probability density function for the statistics of the fluid elements, $\phi(f)$, equation (2.100) becomes

$$\frac{\partial \langle m_j \rangle}{\partial t} + \langle u_{pi} \rangle \frac{\partial \langle m_j \rangle}{\partial x_i} + \langle m_j \rangle \frac{\partial \langle u_{pi} \rangle}{\partial x_i}$$

$$= \frac{\partial}{\partial x_i} \left\{ D_{pT} \frac{\partial \langle m_j \rangle}{\partial x_i} \right\} + 0^j \langle R_N \rangle + j \int_0^1 G(f) m_{j-1}(f) \phi(f) df + \langle B_f \rangle - \langle D_f \rangle \tag{2.101}$$

Equation 2.101 enables calculation of local average quantities such as moments of the particle size distribution. Baldyaga and Orciuch (2001) review expressions for local instantaneous values of particle velocity u_{pi} and diffusivity of particles, D_{pT}, required for its solution and recover the distribution using the method of Pope (1979).

Numerical methods

The population balance is a partial integro-differential equation that is normally solved by numerical methods, except for special simplified cases. Numerical solution of the population balance for the general case is not, therefore, entirely straightforward. Ramkrishna (1985) provides a comprehensive review.

Hounslow et al. (1988), Hounslow (1990a), Hostomský and Jones (1991), Lister et al. (1995), Hill and Ng (1995) and Kumar and Ramkrishna (1996a,b) present numerical discretization schemes for solution of the population balance and compute correction factors in order to preserve total mass and number whilst Wójcik and Jones (1998a) evaluated various methods.

Gelbard and Seinfeld (1978), Nicmanis and Hounslow (1998) and Wulkow et al. (2001) propose alternative finite element methods with improved precision and reduced computational time.

Summary

The behaviour of chemical engineering unit operations is governed by the mathematical equations describing the underlying transport processes. In the analysis of particulate crystallization process systems in particular, equations for both conservation of particle numbers and their motion, and solid–fluid flow (which vary in both particle size and space-time) are required. These take the form of equations for particle sedimentation rate and flow through porous media respectively, the momentum balance, and particle number (population) balance. Use of these equations in the analysis of particulate systems has helped place the design of such unit operations as crystallization and filtration on a rational basis, although many of the 'linking' equations applicable to real polydisperse systems remain empirical at present. Some experimentation is still required for design purposes.

Turbulent fluid flow within agitated vessels is particularly complex leading to gross deviations from the ideal of well mixedness as the size of the vessel increases. Computational fluid dynamics (CFD) shows great promise to help in characterizing and predicting the flow patterns and energy dissipation levels. Progress may be reasonably expected in extending these models to high solids fraction systems of more complex rheology that are commonly encountered industrially. Complete direct numerical solution of coupled population balance and fluid flow models, however, can lead to computationally demanding solutions currently. Alternate moment methods and mixing models, respectively, are available which provide for easier solution. Examples of their application are considered in Chapter 8.

3 Crystallization principles and techniques

This chapter will consider the basic principles governing crystallization and techniques available for achieving it. The performance of any crystallizer is affected by phase equilibria, or solubility, which determines whether particles will form, and influences the mode of process selected. Different types of crystallizer design are available, each having particular performance characteristics. Crystallizer performance can be predicted using the population balance concept of Chapter 2, together with kinetic and solubility data. Precipitation is a particular form of crystallization, one that often happens at high levels of supersaturation giving rise to particular process and product particle characteristics. Each of these aspects is considered in turn below.

Further details can be found in several texts including those on the theory of particulate processes (Randolph and Larson, 1988), crystallization (Van Hook, 1961; Bamforth, 1965; Nývlt, 1970; Jančić and Grootscholten, 1984; Garside et al., 1991; Nývlt, 1992; Tavare, 1995; Mersmann, 2001; Myerson, 2001; Mullin, 2001) and precipitation (Walton, 1967; Söhnel and Garside, 1992).

Phase equilibria

Solubility

The solubility of a substance in a solvent is the maximum concentration that can exist at equilibrium at a given set of conditions and often increases (which sometimes but rarely decreases) with solution temperature (Figure 3.1). In 1878, Gibbs began formulating the thermodynamic conditions for the equilibrium of heterogeneous substances (Gibbs, 1948). It was suggested that the work of formation of a globule of the new phase within an existing phase could be a measure of the stability of the existing phase. Once a critical nucleus of the new phase is formed, the stability of the parent phase is weakened because with the slightest increase in the size of the nucleus, it would tend to increase still further. This concept has been adapted to apply to crystallization, in which nucleation is followed by crystal growth.

The equilibrium phase diagram or solubility–supersolubility plot (Miers and Isaac, 1907), shown in Figure 3.1, provides a useful starting point for considering why crystallization occurs and what type of process might be most suitable for production of a particular substance. It can be divided into three zones (Ostwald, 1897)

Crystallization principles and techniques 59

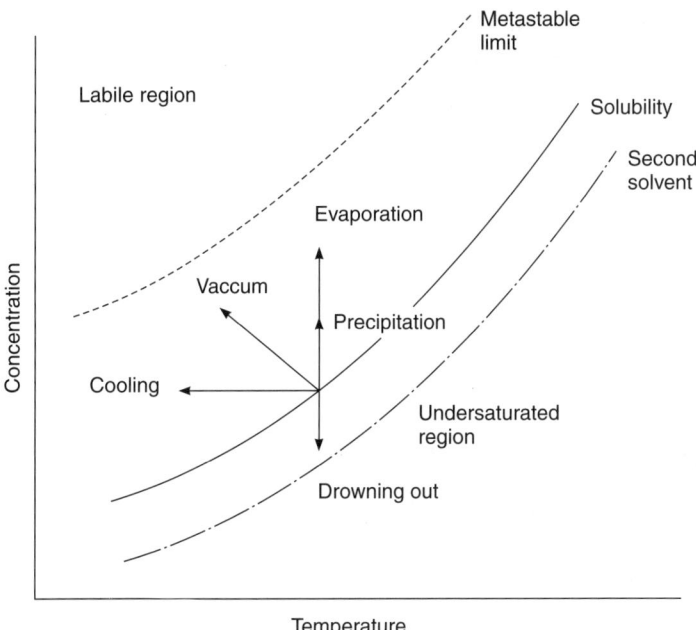

Figure 3.1 *Solubility–supersolubility diagram*

1. *undersaturated* – crystals present will dissolve
2. *metastable* – a supersaturated region in which crystals will grow
3. *labile* – a region in which a solution will nucleate spontaneously.

The term solubility thus denotes the extent to which different substances, in whatever state of aggregation, are miscible in each other. The constituent of the resulting solution present in large excess is known as the solvent, the other constituent being the solute. The 'power' of a solvent is usually expressed as the mass of solute that can be dissolved in a given mass of pure solvent at one specified temperature. The solution's temperature coefficient of solubility is another important factor and determines the crystal yield; if the coefficient is positive then an increase in temperature will increase solute solubility and so solution saturation. An *ideal* solution is one in which interactions between solute and solvent molecules are identical with that between the solute molecules and the solvent molecules themselves. A truly ideal solution, however, is unlikely to exist so the concept is only used as a reference condition.

For a substance to dissolve in a liquid, it must be capable of disrupting the solvent structure and permit the bonding of solvent molecules to the solute or its component ions. The forces binding the ions, atoms or molecules in the lattice oppose the tendency of a crystalline solid to enter solution. The solubility of a solid is thus determined by the resultant of these opposing effects. The solubility of a solute in a given solvent is defined as the concentration of that solute in its saturated solution. A *saturated* solution is one that is in equilibrium with excess solute present. The solution is still referred to as saturated, even after the excess solute is filtered. There are many methods available for measuring solubility and no single one is generally applicable to all system types. One of the major experimental difficulties in any of the methods is the achievement of equilibrium in the solution. Prolonged agitated contact is required between excess solid solute and solution at a constant temperature, usually for several hours or days (according to system temperature and viscosities). Approaching equilibrium from both the under-saturated and the over-saturated states can check the accuracy of the solubility determination

1. *Oversaturated*: In this case, solid in excess of the amount required for saturation is added to the solvent and agitated until apparent equilibrium is reached, i.e. regular sampling is undertaken until solution composition remains constant.
2. *Undersaturated*: The same quantities of solute and solvent are mixed, as for the above case, but the system is then heated for about 20 min above the required temperature (if solubility increases with temperature) so that most, but not all, of the solid is dissolved. The solution is then cooled and agitated at a given temperature for a long period, to allow the excess solid to deposit and an apparent equilibrium to be reached.

Theoretical prediction of the solubility in non-ideal systems (which are the most common) is not yet reliable, however, thus resort is commonly made to empirical expressions e.g. of the form

$$c^* = c_0 + c_1\theta + c_2\theta^2 \tag{3.1}$$

where θ is the prevailing temperature. The solubility of a substance in any given solvent can often be affected by the presence of third component.

Anti-solvents

Other solvents or impurities can have unpredictable results on solubility. For example, addition of a miscible second solvent, sometimes called a co-solvent or diluent, often reduces solubility and is a common means of inducing crystallization (see Figure 3.1). For example, a significant decrease in solubility of potash-alum in aqueous solutions was seen upon the addition of acetone (Mydlarz and Jones, 1989). Again this effect can be correlated by an empirical expression of the form

$$\ln c^* = A + Bx + Cx^2 \tag{3.2}$$

where x is the concentration of the second solvent. The second solvent can be thought of as an impurity and indeed some impurities can affect solubility even at the *ppm* level.

At pressures and temperatures above the critical point, where liquid and vapour phases become indistinguishable, supercritical fluids (SCFs) exhibit very different properties to those of the liquids or gases at ambient. Particle formation in SCFs occurs as a result of a rapid increase in supersaturation, either by means of expansion or by antisolvent mixing processes. Thus Chang and Randolph (1989) demonstrated that small (<1 μm) uniform particles of β-carotene could be precipitated by rapidly depressuring a solution in supercritical ethylene (critical point, 10 °C and 50 bar) by expansion through a nozzle. Currently, carbon dioxide seems to be preferred for supercritical crystallization due to its relatively good availability, cost, safety and toxicity, respectively. Several recent applications are reviewed by Mullin (2001) whilst Shekunov et al. (2001) present a model for particle formation by mixing with supercritical antisolvent to demonstrate and illustrate the industrial potential of this technique with experimental data for the paracetamol/ethanol/CO_2 system.

Solubility is also affected by particle size, small crystals (<1 μm say) exhibiting a greater solubility than large ones. This relationship is quantified in the Gibbs–Thomson, Ostwald–Freundlich equation (see Mullin, 2001)

$$c(r) = c^* \exp\left[\frac{2M\gamma}{nRT\rho r}\right] \tag{3.3}$$

where $c(r)$ is the solubility of a crystal of radius r. This effect, however, is limited to small crystals (<1 μm say) or possibly the corners of larger ones.

Compilations of solubility for various substances are given in various sources including Seidell (1958), Stephen and Stephen (1963), Broul et al. (1981), Wisniak and Herskowitz (1984), Mullin (2001), and the IUPAC series (1980–1991).

Crystallization and precipitation modes

In order to create the solid phase, all industrial crystallizers utilize one or other methods for generating supersaturation e.g. by cooling and/or evaporation (see Figure 3.1). The term 'precipitation' is often applied to crystallizing systems and usually refers to supersaturation being generated by the addition of a third component that induces a chemical reaction to produce the solute or lowers its solubility. A common characteristic of such systems is the rapid formation of the solid phase. Such crystallization modes generally (though not always) create supersaturation at much higher levels than by simple cooling or evaporation. In this context, therefore, the term precipitation is usually meant to imply 'fast crystallization'. Just to complicate matters, some precipitates are amorphous. Crystallization, of course, implies a regular internal array of atoms or ions. Furthermore, the meteorologist regards precipitation as the formation of rain or snow, so perhaps 'dense phase change' would be a more general definition. Whilst noting that it's not a precise definition, it will generally be assumed here that the term precipitation implies fast crystallization, usually brought about as a consequence of chemical reaction or rapid change in solubility by addition of a third component.

The shape of the equilibrium line, or solubility curve, is important in determining the mode of crystallization to be employed in order to crystallize a particular substance. If the curve is steep, i.e. the substance exhibits a strong temperature dependence of solubility (e.g. many salts and organic substances), then a cooling crystallization might be suitable. But if the metastable zone is wide (e.g. sucrose solutions), addition of seed crystal might be necessary. This can be desirable, particularly if a uniformly sized product is required. If on the other hand, the equilibrium line is relatively flat (e.g. for aqueous common salt solutions), then an evaporative process might be necessary. If the yield from either process is low, then perhaps a second solvent can be added to reduce the effectiveness of the first and decrease the residual solution concentration (sometimes called 'drowning-out' or 'watering-out'). If the solute occurs as a consequence of chemical reaction or addition of a common ion, and is relatively insoluble, then precipitation or 'fast crystallization' occurs.

Supersaturation

The fundamental, thermodynamic, driving force for crystallization, or precipitation, is given by the change in chemical potential between standing and equilibrium states. This applies whether the particles formed are organic or inorganic, biochemical or petrochemical. Chemical potential is a quantity that is not easy to measure, however, and the driving force is more conveniently expressed in terms of solution concentration via the following approximation

$$\Delta\mu = \ln\left(\frac{c}{c^*}\right) \cong \frac{c}{c^*} - 1 = \frac{\Delta c}{c^*} = S - 1 = \sigma \qquad (3.4)$$

where $\Delta\mu$ is the change in chemical potential, c is the standing concentration and c^* the equilibrium saturation concentration, S is the supersaturation ratio and σ is the *relative* or *absolute* supersaturation. It is worth noting that, although commonly used, strictly equation 3.4 is valid only for $c \approx c^*$, but many precipitations employ $c \gg c^*$. Supersaturation can be thought of as the concentration of solute in excess of solubility. For practical use, however, supersaturation is generally expressed in terms of concentration

$$\Delta c = c - c^* \qquad (3.5)$$

where c = concentration of solution, c^* = saturation concentration and Δc is sometimes called the 'concentration driving force'.

For ionic systems, however, the definition of the appropriate driving force becomes more complex since the ionic concentrations are not necessarily in stoichiometric ratio and the solubility product generally applies. Thus the supersaturation ratio of sparingly soluble systems can be described by

$$\sigma = \sqrt[r]{\frac{([A]^p[B]^q)}{k_{sp}} - 1)} \qquad (3.6)$$

where $r = p + q$.

In the precipitation of salt systems, the crystallizing species normally exist as free ions in solution and in this case the solubility of each ion has to be taken into account. Thus for a sparingly soluble substance $A^p B^q$, the solubility product is given by: $k = [A][B]$.

In the simplest case $[A] = [B] = c^*$ thus $k = c^{*2}$. For higher concentrations, the concepts of chemical activity are employed (see Mullin, 2001).

It can be seen from these definitions of driving force that sparingly soluble substances can easily exhibit high levels of relative supersaturation. Hence, the crystallization process can be very fast and the precipitation difficult to control, especially on the large scale (see Söhnel and Garside, 1992).

Systems also vary in the extent of the metastable zone width, the point after which spontaneous nucleation is said to occur. Within the metastable zone, however, seed crystals may grow. Metastable zone width is therefore an important factor in assessing the propensity of a system to crystallize and in deciding the appropriate crystallization technique. Kim and Mersmann (2001) provide a review of methods for estimation for metastable zone widths both unseeded and seeded systems.

Types of crystallizer

Many different types of industrial crystallizer exist and the reader is referred to Mullin (2001) for a detailed description. The main features are summarized below.

Unstirred vessels

The simplest type of crystallizer is the non-agitated tank into which hot liquor is poured and the contents allowed to cool naturally in a batch operation. Significant local and transient variations in supersaturation occur and in consequence the product mass can contain large interlocking crystals, agglomerates and fines which are difficult to separate, wash and handle. Solar saltmaking ponds provide the earliest example of this type of crystallizer (see Jones *et al.*, 1981) but, with a few exceptions, most modern crystallizers incorporate some form of agitation.

Agitated vessels

Agitation of slurry to create a suspension in a low tonnage batch crystallizer can be provided in several ways (Figure 3.2). An impeller can be mounted centrally in the vessel and wall baffles added to prevent swirling flow. The agitator can be shrouded with a draft-tube to provide gross circulation from the base up to the top surface. Such agitation improves heat transfer, reduces scale formation, smoothes out supersaturation profiles, suspends crystals and generally gives rise to a more uniform product and reduced batch time. Cooling can be provided via a jacket, hollow internal shroud or coil, or externally via

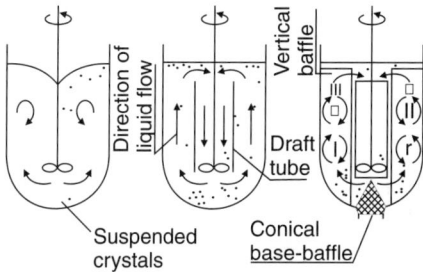

Figure 3.2 *Simple agitated vessels (after Jones and Mullin, 1973)*

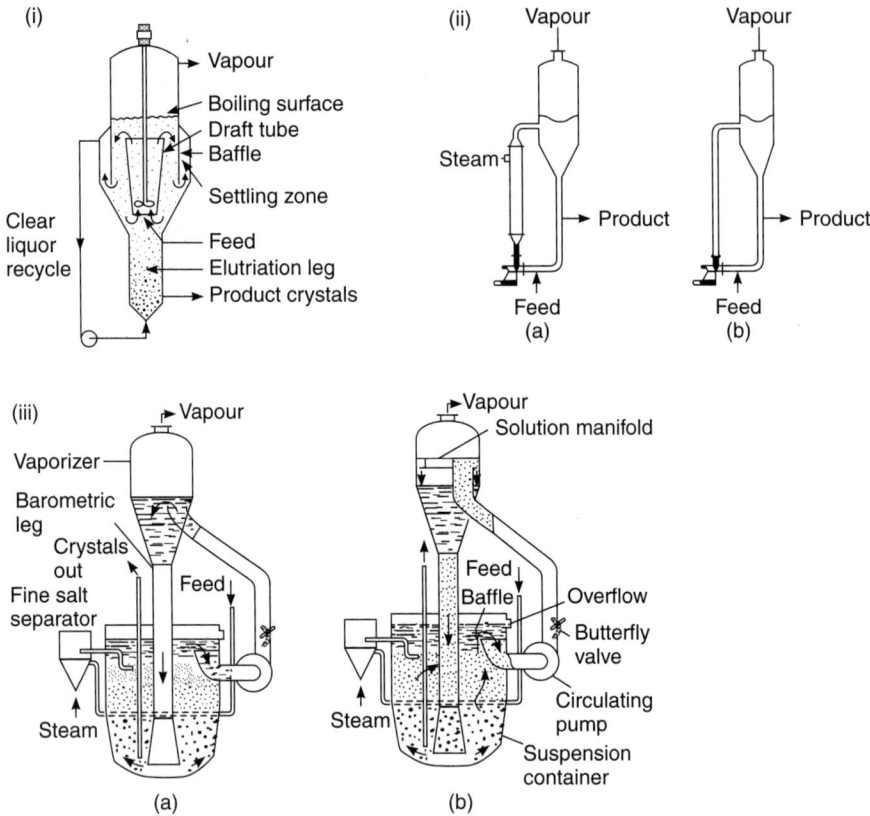

Figure 3.3 *Agitated crystallizers (i) Swenson DTB (Draft tube and baffle); (ii) Forced circulation (a) with and (b) without external heat exchange; (iii) Krystal Oslo (a) circulating liquor (b) circulating slurry*

recirculation of liquor through a heat exchanger. Alternatively, a vacuum may be applied for evaporation and/or cooling or a precipitant added. In order to further improve the product crystal size distribution in a batch crystallizer the creation of supersaturation may be programmed. A 'fines destruction loop' in which tiny crystals are dissolved and the solute liquor returned to the vessel may be added to the vessel in either mode.

For large-scale continuous crystal production, several alternate types of crystallizer vessel are available (Figure 3.3).

Draft-tube and baffled (DTB)

The DTB, crystallizer has a relatively slow-speed propeller agitator located within a draft-tube which draws a fine-crystal suspension up to a boiling zone of wide cross-sectional area, as shown in Figure 3.3(i). The fine-crystal magma then passes through an annular zone in which an additional baffle is located. Liquor flow continues upwards at low velocity while crystals settle out and fall to the base of the vessel. Liquor from the external pumped loop provides an up-flowing stream to an elutriating leg permitting control by size of the product crystal off-take. In consequence, only liquor passes through the pump and crystal sizes larger than from the forced circulation (FC) crystallizer can be produced say in the range 500–1200 μm.

Swenson forced circulation (FC)

The FC crystallizer contains an external circulation loop through a heat exchanger and may also contain an evaporation zone within the main vessel with crystals removed at its base, as shown in Figure 3.3 (ii). Since both crystals and liquor circulate through the pump, secondary nucleation rates and crystal breakage are high. The product crystals are typically in the size range 200–500 μm.

Oslo fluidized-bed

The Oslo fluidized-bed crystallizer has separate sections for the generation of supersaturation and for crystal suspension and growth (Figure 3.3 (iii)). Supersaturation is created externally e.g. by cooling or evaporation and the liquor pumped into a central downcomer in the main vessel. A classified bed of crystals is supported in the vessel by an upward flowing stream of supersaturated liquor in the annular space surrounding the downcomer. Relatively large crystals of mean size say >1000 μm are withdrawn from the base. In an effort to increase efficiency, however, modern practice is frequently to operate Oslo units in the mixed-suspension mode.

The mixed suspension, mixed product removal (MSMPR) crystallizer

The flow of slurry within all the agitated crystallizer vessels illustrated is clearly complex and mixed to a greater or lesser extent at the microscopic level. In order to ease theoretical analysis a new type of vessel therefore had to be invented: This idealized vessel has become known as the continuous MSMPR crystallizer, after Randolph and Lawson (1988). The MSMPR is the crystallization analogue of the CSTR (continuous stirred tank reactor) employed in idealizations of chemical reaction engineering.

The well-mixed continuous crystallizer (Figure 3.4) has received most attention in studies of crystallization and at continuous steady state is the simplest form of crystallizer to analyse for its operation and performance.

Solution enters the vessel (Figure 3.4) and is well-mixed throughout i.e. all conditions – temperature, concentration, velocity, turbulence etc. are uniform (homogeneous). Supersaturation is generated (by evaporation, cooling, etc.) and nuclei form and grow into crystals. Since crystals have varying probabilities of residence time in the vessel, however, the slurry exhibits a crystal size distribution (CSD). Product slurry is continuously withdrawn and has exactly the same composition as the vessel.

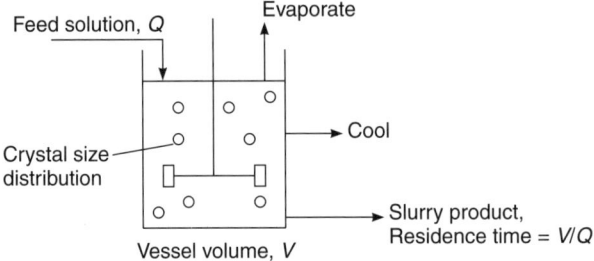

Figure 3.4 *The continuous mixed-suspension, mixed-product-removal (MSMPR) crystallizer*

Crystallizer modelling

Prediction and analysis of crystallizer performance is achieved by constructing models based on conservation equations and rate expressions respectively, as follows. In general form, the conservation equation is given by

Input = Output + Accumulation − Net Generation

Crystal yield

The change in dissolved solute between inlet and outlet of the vessel is matched by the gain in solid crystal mass. Thus, on a unit volume basis

ΔConcentration (inlet − outlet) → Mass Yield

This relationship, of course, only gives the total mass of solids formed. To reveal how that solid matter is distributed across a crystal population, the other conservation equation considered in Chapter 2 viz. the population balance must be invoked. Firstly, however, the crystal yield is considered a little further.

Mass balance

The first, and simplest, step in predicting crystallizer performance is the calculation of crystal yield. This can easily be estimated from knowledge of solution concentration and equilibrium conditions permitting calculation of the overall mass balance

$$Y = M_h(c_i - c_o) \tag{3.7}$$

where c_i and c_o are this inlet and outlet concentrations respectively. The magnitude of c_o is generally unknown, of course, but if it is assumed that $c_o = c^*$ then this calculation gives the maximum mass that can be deposited. Whilst knowledge of crystal yield is important it is of only limited utility in predicting the form of a crystalline product, in particular its mean size and size distribution. Prediction of the crystal size distribution then requires knowledge of crystallization kinetics and residence time, which determine total crystal numbers and their size. These quantities are embodied within the mathematical framework known as the population balance. When coupled with a mass

Crystallization principles and techniques 67

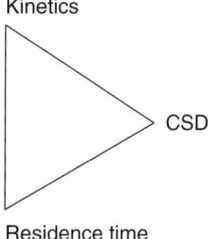

Figure 3.5 *The crystallization triangle*

balance on the liquid and solid phases, the crystal size distribution can be predicted in a manner akin to prediction of a reactor product spectrum in the field of chemical reaction engineering.

Prediction and analysis of CSD

The CSD from the continuous MSMPR may thus be predicted by a combination of crystallization kinetics and crystallizer residence time (see Figure 3.5). This fact has been widely used 'in reverse' as a means to determine crystallization kinetics – by analysis of the CSD from a well-mixed vessel of known mean residence time. Whether used for performance prediction or kinetics determination, these three quantities, (CSD, kinetics and residence time), are linked by the population balance.

Given expressions for the crystallization kinetics and solubility of the system, the population balance (equation 2.4) can, in principle, be solved to predict the performance of both batch and of continuous crystallizers, at either steady- or unsteady-state

$$\frac{\partial n}{\partial t} + \frac{\partial (nG)}{\partial L} + \frac{n - n_o}{\tau} = B_a - D_a + D_d - D_d + \int_0^\infty \delta(L - L_u) dL \qquad (3.8)$$

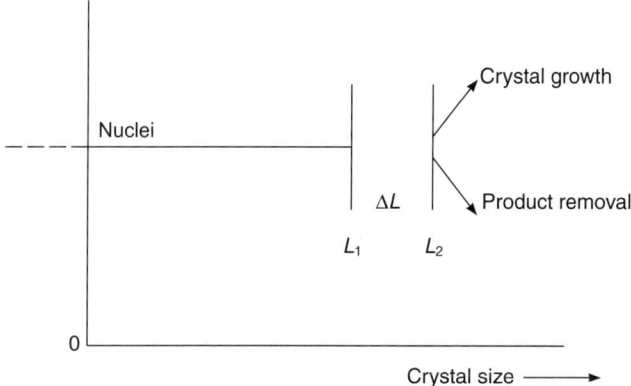

Figure 3.6 *Schematic particle flows in the ideal MSMPR crystallizer at steady state*

As is evident, however, the general population balance equations are complex and thus numerical methods are required for their general solution. Nevertheless, some useful analytic solutions are available for particular cases.

Continuous MSMPR operation

The MSMPR crystallizer provides a particularly elegant way to illustrate the derivation of the population balance under certain assumptions. Within the MSMPR crystallizer crystals leave any given size range either by growth or by outflow in the product stream (Figure 3.6).

Population balance

Population density is a fundamental statistic of a size distribution and is defined by

$$n = \lim_{\Delta L \to 0} \frac{\Delta N}{\Delta L} \quad \text{i.e. number per unit size} \tag{3.9}$$

where ΔN = number of crystals (per unit volume of slurry) in size range ΔL.

Crystal growth rate

$$G = \frac{dL}{dt} \cong \frac{\Delta L}{\Delta t} \quad \left[\frac{\text{size increment}}{\text{time increment}}\right] \tag{3.10}$$

and
$$\Delta N = nG\Delta t \tag{3.11}$$

Crystal number balance

Over size interval ΔL in time Δt

Number entering due to = Number leaving − Number removed
growth from smaller   due to growth into in outflow
size range   next size range

i.e.
$$\Delta N_1 \cdot V = \Delta N_2 \cdot V + \Delta \bar{N} \cdot \underbrace{Q\Delta t}_{\substack{\text{Output}\\\text{Volume}}} \tag{3.12}$$

so, substituting for ΔN gives

$$n_1 G_1 \Delta t \cdot V = n_2 G_2 \Delta t \cdot V + \bar{n} Q \Delta L \Delta t \tag{3.13}$$

and as $\Delta L \to 0$, and assuming $G_1 = G_2 = G$ (constant) gives the population balance as follows.

$$\frac{dn}{dL} + \frac{n}{G\tau} = 0 \tag{3.14}$$

where $\tau = V/Q$, the mean residence time.

Equation (3.14) is thus consistent with the general population balance (equation 3.8) when $B = D = 0$ i.e. nucleation occurs at 'zero' size and both the 'birth' and 'death' terms due to agglomeration and breakage are neglected, and the feed is crystal-free.

MSMPR crystal size distribution (CSD)

Integrating of the entire possible size range $(0, \infty)$ gives the crystal size distribution thus

$$n(L) = n^0 \exp\left(\frac{-L}{G\tau}\right) \tag{3.15}$$

Equation 3.18 reflects the reality that as the population of nuclei grow there is increasing likelihood that they will have been washed out with the product, therefore fewer larger crystals than nuclei are present in the product.

A log-linear plot of the idealized continuous MSMPR population density versus crystal size is shown in Figure 3.7.

Now the special utility of the MSMPR population balance model equation at steady state can be clearly seen. Firstly, at known residence time, τ, the *Growth rate*, G, may be obtained from the slope $(= -1/G\tau)$ of the plot in Figure 3.7.

Secondly, since the *Nucleation rate*, $B = \dfrac{dN}{dt} = \dfrac{dN}{dL}\dfrac{dL}{dt}$, is given by

$$B = n^0 G \tag{3.16}$$

The nucleation rate (B) can therefore subsequently be obtained from the intercept (n^0) having evaluated G from the slope. Thus, analysis of crystal size distribution yields both growth and nucleation rates *simultaneously*, a great advantage experimentally.

The CSD of the product from a continuous crystallizer is determined by a direct relationship between nucleation and crystal growth rates and magma residence time distributions. Since *a priori* prediction of crystallization kinetics is not yet possible, however, experimentally determined and statistically correlated nucleation and growth rates are needed for the design and analysis of industrial crystallizers. Analysis of the CSD from continuous MSMPR (mixed-suspension, mixed-product-removal) crystallizers has proved to be a popular way of inferring such crystallization kinetics. This approach has been widely

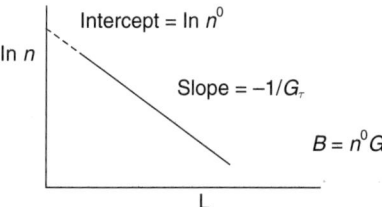

Figure 3.7 *Crystal population distribution from the MSMPR crystallizer*

Crystallization Process Systems

used to determine crystal growth and nucleation kinetics and for continuous crystallizer design (see Randolph and Larson, 1988).

Example

A pilot-scale continuous MSMPR crystallizer of 10 litre capacity is used to crystallize potash alum from aqueous solution, supersaturation. This is being achieved using a 15-min residence time, a 100-ml slurry sample was taken and the crystals contained in this sample subjected to a size analysis. The results of this analysis are given below

BS sieve size (μm)	Weight on sieve (g)
850	–
710	0.23
500	1.63
355	2.09
250	2.89
180	1.80
125	1.07
90	0.37
63	0.15
45	0.06
through 45	0.02
	10.13 g

Taking the crystal density as 1770 kg/m^3 and the volume shape factor, f_v, as 0.47 calculate
(i) The crystal growth rate, G (m/s)
(ii) The nuclei population density, n^0 (particles/μm m^3 slurry)
(iii) The nucleation rate, B (particles/s m^3 slurry)

Solution

First convert the mass distribution data to number density distributions: 0.23 g retained between 850 and 710 μm; $\bar{L} = 780$ μm, $\bar{L}^3 = 4.74 \times 10^8$ μm^3, $\Delta L = 140$ μm.

$$\text{Mass of one crystal} = f_v \rho \bar{L}^3 = 0.47 \times 1.77 \text{ g/cm}^3 \times 4.74 \times 10^8 \times 10^{-12} \text{ cm}^3/\text{μm}^3$$
$$= 3.94 \times 10^{-4} \text{ g}$$

Number of crystals retained

$$= \frac{0.23}{3.94 \times 10^{-4}} = 5.84 \times 10^2 \text{ per 100 ml slurry}$$

$$n = \frac{5.84 \times 10^2}{140} = 4.17 \text{ \#/μm 100 ml slurry}$$

$$= 4.17 \times 10^4 \text{ \#/μm m}^3 \text{ slurry}$$

BS sieve size (μm)	Weight on sieve (g)	$\bar{L}$ (μm)	$\bar{L}^3$ (μm^3)	ΔL (μm)	n (#/μm m^3 slurry)
850	—				
710	0.23	780	4.74×10^8	140	4.17×10^4
500	1.63	605	2.21×10^8	210	4.20×10^5
355	2.09	428	7.84×10^7	145	2.21×10^6
250	2.89	303	2.78×10^7	105	1.19×10^7
180	1.80	215	9.94×10^6	70	3.10×10^7
125	1.07	153	3.58×10^6	55	6.53×10^7
90	0.37	108	1.26×10^6	35	1.01×10^8
63	0.15	77	4.57×10^5	27	1.46×10^8
45	0.06	54	1.58×10^5	18	2.54×10^8
through 45	0.02	(22)	(1.06×10^4)	(45)	(5.04×10^8)
	10.13 g				

Slurry density or 'magma density', $M_T = 10.13$ g/100 ml slurry $= 103.1$ kg/m^3. A plot of $\ln n$ versus $\bar{L}$ from the table is given below.

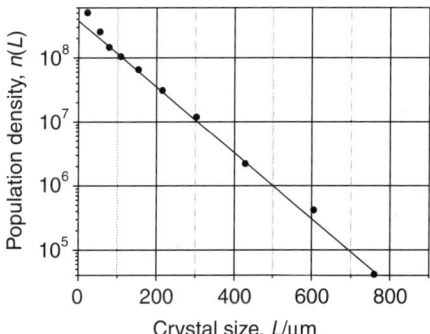

The slope of population density plot enables the growth rate to be inferred, thus

$$= -\frac{1}{G\tau} = -0.0115\,\mu\text{m}^{-1}$$

$$G = \frac{1}{0.0115\,\mu\text{m}} \times \frac{1}{15\,\text{min}^{-1} \times 60\,\text{min/s}} = 0.097\,\mu\text{m/s} = 9.7 \times 10^{-8}\,\text{m/s}$$

While from intercept of population density plot, $n^0 = 3.6 \times 10^8$ #/μm m^3, thus

$$B = n^0 G = 3.6 \times 10^8\ \#/\mu\text{m m}^3 \times 9.7 \times 10^{-8}\,\text{m/s} \times 10^6 = 3.5 \times 10^7\ \#/\text{s m}^3\ \text{slurry}$$

Crystal number, area and mass

Mass distributions

The general form of the population density function from the ideal MSMPR crystallizer (equation 3.15) has rather fortunate statistical properties such that mass distributions can easily be determined analytically together with values of mean crystal sizes and their coefficient of variation.

Putting $x = L/G\tau$, dimensionless crystal size in $n = n^0 \exp(-L/G\tau)$ and integrating gives

Differential mass distribution

$$m(x) = \frac{1}{6} x^3 \exp(-x) \tag{3.17}$$

The mass distribution from the idealized MSMPR crystallizer is thus a Gamma function, as shown in Figure 3.8b.

Cumulative mass distribution

Integrating over all possible crystal sizes $(0, \infty)$

$$M(x) = 1 - \left(1 + x + \frac{1}{2}x^2 + \frac{1}{6}x^3\right)(\exp - x) \tag{3.18}$$

This produces an inflected curve as shown in Figure 3.8a.

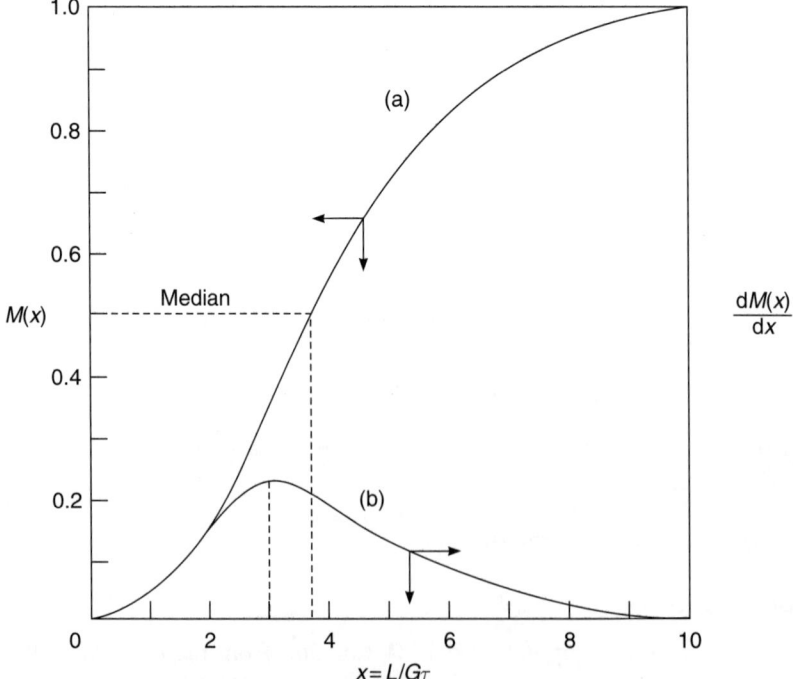

Figure 3.8 *MSMPR mass distributions (a) cumulative; (b) differential*

Mass modal size

The cumulative mass of crystal per unit volume of suspension up to size L is given by

$$M(L) = f_v \rho_c \int_0^D n L^3 \, dL \qquad (3.19)$$

The total mass in suspension (per unit volume) i.e. up to size ∞ is given by

$$M_T = f_v \rho_c \int_0^D n L^3 \, dL \qquad (3.20)$$

Now, for MSMPR crystallizer

$$n(L) = n^0 \exp\left(\frac{-L}{G\tau}\right) \qquad (3.21)$$

thus

$$M_T = f_v \rho_c \int_0^\infty \exp\left(\frac{-L}{G\tau}\right) L^3 \, dL) \qquad (3.22)$$

making the substitution $p = L/G\tau$ and noting that $\int_0^\infty e^{-p} p^j \, dp = j! = \Gamma(j+1)$, the gamma function. Then

$$M_T = 6 f_v \rho_c n^0 (G\tau)^4 \qquad (3.23)$$

Thus, the cumulative mass fraction is given by

$$M(L) = \frac{f_v \rho_c \int_0^D n^0 \exp\left(\frac{-L}{G\tau}\right) L^3 \, dL}{6 f_v \rho_c n^0 (G\tau)^4} \qquad (3.24)$$

Therefore the differential mass fraction is given by

$$x(L) = \frac{dm(L)}{dL} = \frac{f_v \rho_c n^0 \exp\left(\frac{-L}{G\tau}\right) L^3 \, dL}{6 n^0 f_v \rho_c (G\tau)^4 \, dL} \qquad (3.25)$$

$$= \frac{\exp\left(\frac{-L}{G\tau}\right) L^3}{6 (G\tau)^4} \qquad (3.26)$$

The mode occurs at the maximum in the differential mass fraction distribution curve. By differentiation and putting $L = L_M$ at the mode.

$$\frac{d}{dL}(x(L)) = 3L_m^2 \exp\left(\frac{-L_M}{G\tau}\right) - \left(\frac{-L_M^3}{G\tau}\right) \exp\left(\frac{-L_M}{G\tau}\right) = 0 \qquad (3.27)$$

$$L_M = 3G\tau \qquad (3.28)$$

Note that the modal size also happens to equal the Sauter mean size, L_{sm}, in this case.

Often, it is not necessary to know the total CSD since some mean particle characteristic and possibly its variance may suffice. In general, such characteristics of the CSD can be obtained from the moment equation, below, by substitution of the population density from the size distribution equation. For the special case of the continuous MSMPR at steady state, the CSD is given analytically by equation 3.15. When this expression for the CSD is integrated within the moment equations, corresponding analytic expressions for the total crystal number, length, area volume etc. and mean properties are obtained, as follows

$$\text{Particle characteristics} = \text{Moment equation} + \text{Size distribution}$$
$$\text{for } \mu_j \qquad \text{equation for } n(L)$$

i.e.

$$\mu_j = \int_0^\infty n(L) L^j \mathrm{d}L \quad j = 0, 1, 2, 3\ldots \tag{3.29}$$

where

$$n(L) = n^0 \exp\left(\frac{-L}{G\tau}\right) \tag{3.30}$$

The more useful ones are the total number, surface and volume, from which may the total mass also be obtained. (Note that $\int_0^\infty e^{-p} p^j \mathrm{d}p = p!$).

Total crystal number (j=0)

$$N_T = n^0 G \tau = B\tau \tag{3.31}$$

since $B = n^0 G$. Thus the total crystal number is the product of the nucleation rate and mean residence time.

Total surface area (j=2)

$$A_T = 2 f_s n^0 (G\tau)^3 \tag{3.32}$$

Note here that the surface shape factor, f_s, has been introduced to determine the total crystal surface area.

Total crystal mass (j=3)

$$M_T = 6 f_v \rho_c n^0 (G\tau)^4 \tag{3.33}$$

where to obtain crystal volume and mass respectively, the volume shape factor, f_v, and density, ρ_c, are included.

Mean sizes of the CSD

From the moments various mean sizes can be calculated. Two particularly useful ones are as follows

Sauter (surface:volume) mean

$$L_{sm} = \frac{\mu_3}{\mu_2} \tag{3.34}$$

$$\Rightarrow \frac{\int_0^\infty L[L^2 n(L)] \mathrm{d}L}{\int_0^\infty L^2 n(L) \mathrm{d}L} \tag{3.35}$$

by inspection

$$L_{sm} = 3G\tau \qquad (3.36)$$

i.e. The surface:volume mean size is *three* times the product of crystal growth rate and mean residence time. It is particularly useful for calculating the total crystal surface area within the crystallizer.

Mass mean size

It can also be shown analytically that the mass mean size is given by

$$L_{wm} = 4G\tau \qquad (3.37)$$

Mass median size

The mass median size (above and below which half the distribution lies) is given by

$$L_{50} = 3.67G\tau \qquad (3.38)$$

numerically.

Coefficient of variation

For the ideal MSMPR crystallizer at steady-state with $G \neq G(L)$

$$CV = 50\% \qquad (3.39)$$

always, as shown by Randolph and Larson (1988).

Anomalous crystal growth

It has often been observed that the plot of $\ln(L)$ versus L results in curvature rendering the method of determining the growth rate from the slope strictly inappropriate, but ways to accommodate such deviations have also been proposed. Thus, if $G = G(L)$ integration of equation 3.15 leads to the following expression for determining crystal growth rates (Sikdar, 1977)

$$G(L) = \frac{N(L)}{n(L)\tau} \qquad (3.40)$$

where $N(L)$ is the cumulative oversize population (number) distribution.

Similarly, the corresponding nucleation rate is obtained from the relation

$$B = \frac{N_T}{\tau} \qquad (3.41)$$

Such cases of curvature can arise due to so-called anomalous growth. A variety of mechanistic causes for this behaviour have been proposed which fall into two broad classes viz. growth rate dispersion and size-dependent crystal growth. Both classes have been analysed in detail but unfortunately lead to indistinguishable crystal size distributions. The simplest mathematically, however, is size-dependent growth such that an 'effective' crystal growth rate may be determined for practical purposes.

In comparative studies, Mydlarz and Jones (1991, 1994) point out that use of different size-dependent crystal growth rate models and methods for the analysis of curved log-linear MSMPR CSD data can lead to gross variations in the magnitudes of crystallization kinetics so determined. The common occurrence of such anomalous MSMPR CSD data and their analysis by differing methods may in large part explain variations in reported crystallization kinetics. The use of exponential size-dependent equations provides improved data fitting and CSD prediction compared with all other models tested, but given their purely empirical basis they should nevertheless still be used with caution.

In addition, other causes of non-log linearity include crystal agglomeration, disruption and attrition (see Chapter 4). These also need to be carefully eliminated during model identification or otherwise accommodated. Neglect of crystal agglomeration, for example, can lead to significantly overestimating crystal growth rates whilst underestimating those of nucleation (Hostomský and Jones, 1991). The best way to see if agglomeration exists is simply to view the crystals under the microscope at an appropriately high magnification.

Precipitation

The behaviour of a precipitation reaction depends greatly on the particular system under consideration and the conditions employed (see Söhnel and Garside, 1992 for a detailed treatment). A qualitative illustration of the main processes generally involved in precipitate particle formation is shown in Figure 3.9.

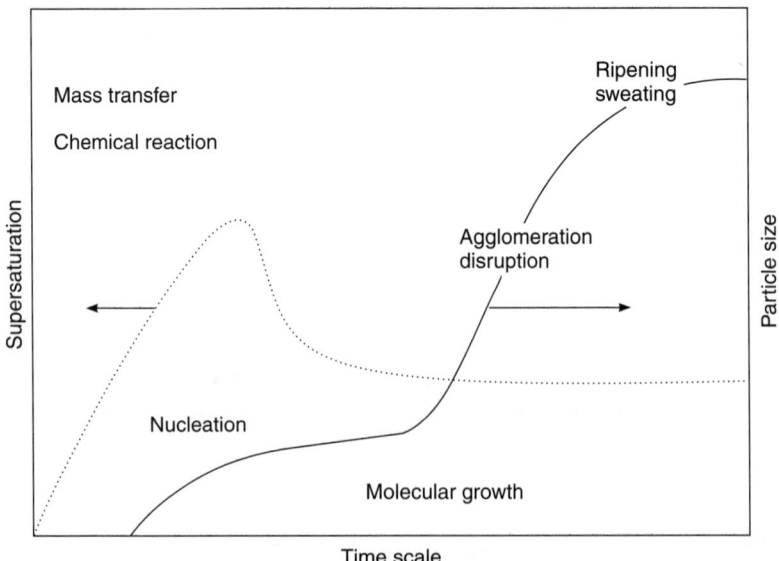

Figure 3.9 *Schematic precipitation process transients (Wachi and Jones, 1995)*

In precipitation systems, supersaturation is generated following mass transfer occurring when the reactants are mixed. Such supersaturation levels can be very high, particularly if the reagents are relatively soluble whilst the product is insoluble. This puts increased emphasis on the requirement for good mixing since the prospects for spatial variations in supersaturation are correspondingly high – true MSMPR conditions may not then be attainable at realistic power inputs. Nucleation and crystal growth then proceed as normal, but with the high level of supersaturation go high particle numbers, often giving rise to crystal agglomeration, which can become the dominant particle growth mechanism and is considered in detail in Chapter 6. Finally, the size-dependence of solubility can give rise to Ostwald ripening in which the larger particles grow at the expense of smaller ones (of higher solubility) dissolving.

Phase transformation

During precipitate ageing, a gradual transformation of an initially precipitated metastable phase into a final crystalline form often occurs. The metastable phase may be an amorphous precipitate, a polymorph of the final material, a hydrated species or some system-contaminated substance (Mullin, 2001). In 1896, Ostwald promulgated his 'rule of stages' which states that an unstable

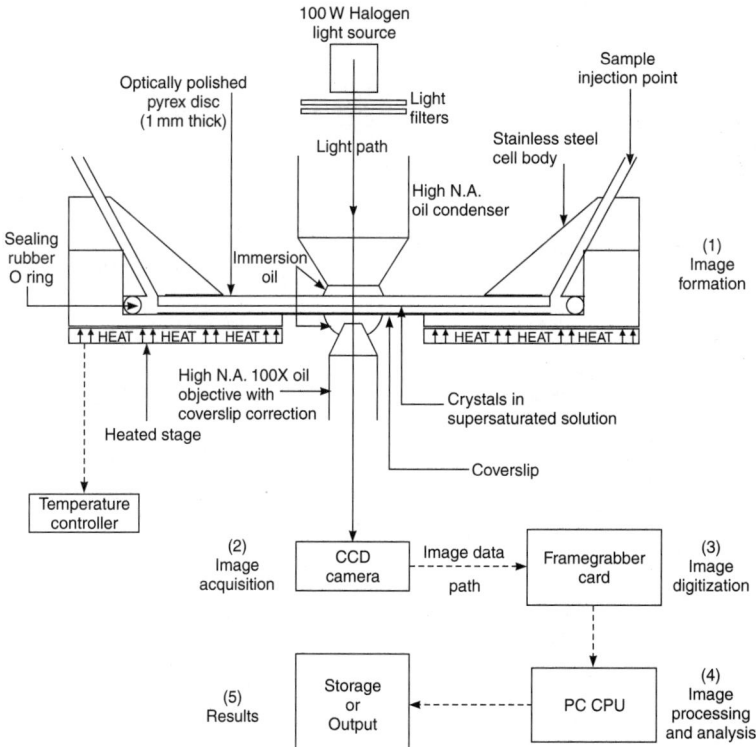

Figure 3.10 *Schematic diagram of microscope and precipitation cell (Jones et al., 1999)*

78 Crystallization Process Systems

system does not necessarily transform directly into the most thermodynamically stable state, but into another transient state whose formation is accompanied by the smallest loss of free energy and is so formed the fastest. Jones *et al.* (1999), have observed such transformation from an initial amorphous state to crystalline form during the precipitation of zeolite directly under the microscope.

Microscopic analysis

A schematic diagram of the microscope cell is shown in Figure 3.10. An Olympus IX40 inverted transmitted light microscope, with a temperature

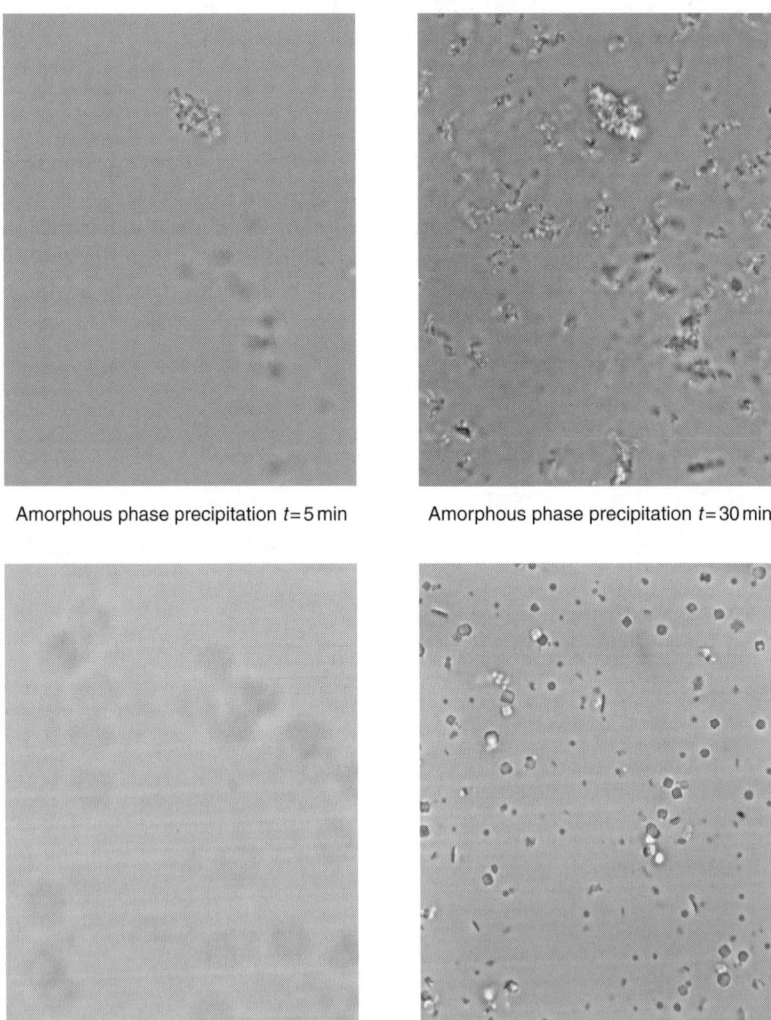

Figure 3.11 *Time lapse sequence of zeolite amorphous to crystalline phase transformation (60 °C and 3.2 w/w % solids) (Jones et al., 1999)*

controlled heated stage (accurate to 0.1 °C) is employed in combination with a planfluorite oil objective with 1.3 numerical aperture (NA) and 100 times magnification. A flat sealed stainless-steel cell is used to produce a shallow heated supersaturated solution of variable depth between 0.5–1.0 mm. A Sony 2/3-inch CCD camera is used to capture the formed image, which is then digitized to a spatial resolution of 768 by 576. A combination of Matrox Meteor^{TM} Framegrabber and Millennium^{TM} graphics adapter on a PCI bus, Intel Pentium^{TM} Pro 200-based PC provides the live display with the image analysed using Kontron^{TM} KS300 software (Ejaz, 1997).

Figure 3.11 presents a time-lapse sequence of photomicrographs of zeolite phase transformation taken over a number of hours at a temperature of 60 °C and NaOH concentration of 3.4 M. These results show that the large flocs of initially precipitated amorphous particles dissolve prior to smaller individual zeolite particles crystallizing elsewhere from the solution. Subsequent SEM analysis showed that a phase transformation had occurred as mainly cubic zeolite. A crystals were present together with small amounts of amorphous material. Image analysis is being used increasingly as a particle characterization technique, and may have application in crystallizer control (see Chapter 9).

Summary

Crystallization occurs in thermodynamically unstable solutions in which the driving force for particle formation is the concentration of solute (or strictly, its chemical potential) in excess of its equilibrium or saturation value. Techniques for achieving this include cooling, evaporation and precipitation, for which various crystallizer designs are available. The two primary particle formation processes during crystallization are nucleation – giving rise to large numbers of tiny crystals – and their subsequent crystal growth. The product crystal size distribution is governed by a combination of these kinetic processes and the crystallizer residence time distribution, the analysis of which is facilitated by the population balance.

The population balance analysis of the idealized MSMPR crystallizer is a particularly elegant method for analysing crystal size distributions at steady state in order to determine crystal growth and nucleation kinetics. Unfortunately, the latter cannot currently be predicted *a priori* and must be measured, as considered in Chapter 5. Anomalies can occur in the data and their subsequent analysis, however, if the assumptions of the MSMPR crystallizer are not strictly met.

A secondary particle formation process, which can increase crystal size dramatically, is crystal agglomeration. This process is particularly prevalent in systems exhibiting high levels of supersaturation, such as from precipitation reactions, and is considered along with its opposite viz. particle disruption in Chapter 6. Such high levels of supersaturation can markedly accentuate the effects of spatial variations due to imperfect mixing within a crystallizer. This aspect is considered further in Chapter 8.

4 Solid–liquid separation processes

The process step following crystallization in suspension is often that of solid–liquid separation. The type of operation ranges from simple gravity settling of suspensions containing low concentrations of solids through to slurry thickeners and finally the filtration of dense slurries. In each case, the effect of gravity can be substantially enhanced by centrifugation. These operations thus depend for their performance on the principles of solid–liquid flow and liquid percolation through particulate arrays described in Chapter 2. In this chapter gravity settlers and thickeners, filters and centrifuges are considered, respectively, followed by driers.

For derivations that are more complex and a more detailed description of solid–liquid separation applications, the reader is referred to specialist texts e.g. Coulson and Richardson (1991), Purchas (1981), Purchas and Wakeman (1986), Matteson and Orr (1987), Wakeman (1990a,b), Cheremisinoff (1998), Sinnott (1999) and Svarovsky (2000). Several types of drier and associated drying theory are considered in detail in a number of texts, including Nonhebel and Moss (1971), Keey (1972, 1978, 1991), Masters (1985) and Coulson and Richardson (1991).

Settling tanks

Settling tanks are used to separate low solids concentration suspensions (<1% v/v). They are the simplest solid–liquid separation devices usually comprising a vessel containing a single baffle which directs the suspension to the base from where it rises to the outlet, as illustrated in Figure 4.1. So long as the

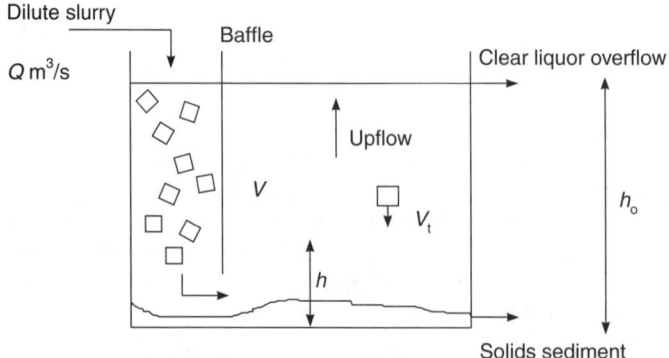

Figure 4.1 *Continuous gravity settling tank*

settling velocity is greater than the superficial liquor up-flow velocity then particles will settle out and the over-flow liquid will be clear. Small particles of low settling velocity or dense suspensions exhibiting hindered settling will overflow.

Terminal settling velocity (v_t). The terminal settling velocity is that achieved under free-fall conditions (see Chapter 2).

Up-flow velocity of liquid (u_0). The upflow velocity is given by

$$u_0 = \frac{Q}{A} \tag{4.1}$$

where Q and A are volumetric flowrate and cross-sectional area of tank respectively.

Operating principle. Particles of terminal velocity $v_t > u_0$ will tend to settle therefore design for $v_t < u_0$ of the smallest particle present in the feed stream. In other words, the settling time should be less than the mean residence time of the up-flowing fluid.

The mean residence time of fluid τ is defined as

$$\tau = \frac{V}{Q} \tag{4.2}$$

Whilst the settling time of particle T is given by

$$T = \frac{h}{v_t} \tag{4.3}$$

Design basis. Clearly, the settling time of the particles should be less than the mean residence time of the liquid. Thus

$$\tau > T \tag{4.4}$$

or

$$\frac{V}{Q} = \frac{Ah_0}{Au_0} = \frac{h_0}{u_0} > \frac{h}{v_t} \tag{4.5}$$

i.e.

$$\frac{v_t}{u_0} > \frac{h}{h_0} \tag{4.6}$$

Therefore, design to keep

$$\frac{v_t}{u_0} > \frac{h}{h_0} \tag{4.7}$$

Thus at each arbitrary point h, keep $v_t > u_0$ and the particles will settle out.

Since v_t is fixed by the fluid and particle characteristics, u_0 needs to be kept low

⇒ Large cross-sectional tank area

82 *Crystallization Process Systems*

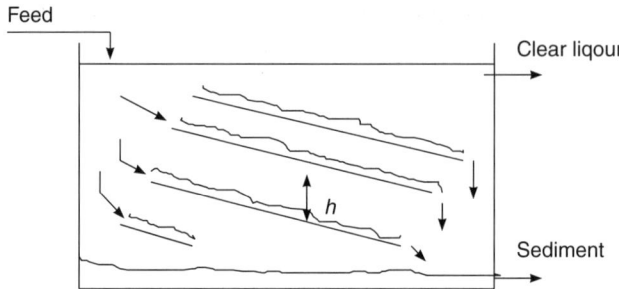

Figure 4.2 *Lamella settler*

and to keep $\dfrac{h}{h_0}$ small

⇒ Short separation distance i.e. a shallow vessel.

Thus the cross-sectional area of the settler is most important, while the separation distance should be kept small. This has led to the introduction of designs having numerous baffles stacked within the settler as in the Lamella settler.

Lamella settler. The lamella settler provides an improved design on the simple gravity-settling tank by use of 'louvres', or slats, slightly inclined to the horizontal (Figure 4.2). The effective cross-sectional area of the settler is increased while decreasing the separation distance between the bulk suspension and the settled sludge. The louvres collect the concentrated sediment in the manner of dust settling on Venetian blinds. The result is reduced residence time and smaller volume of the unit.

Thickeners

Thickeners are normally used for higher density feed slurries (say 10–20% v/v). They generally comprise cylindrical vessels with a conical base (Figure 4.3). In

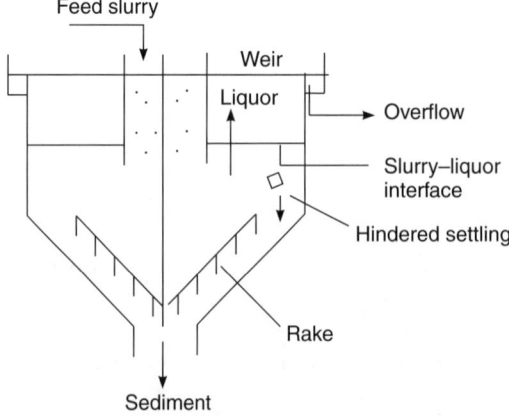

Figure 4.3 *Continuous gravity thickener*

batch operation the solid–liquid interface falls at a rate R due to particle settling (see Figure 2.6). In a continuous unit, however, there is a product off-take at the base and overflow and the rate of fall is enhanced due to slurry removal. At steady state this rate of fall is balanced by the rate of addition of fresh feed. A steady-state interface develops and under ideal conditions the solid phase passes down through the vessel in plug flow. In order to minimize mixing within the vessel, the feed suspension is introduced continuously via a central down-pipe to a region below the slurry–liquor interface of similar concentration. Because the slurry becomes more concentrated, however, settling is generally slower and reduces as the sediment proceeds to the outlet. There is therefore a corresponding vertical distribution of mass flux. At the base the sediment is helped towards the exit by a slow moving rake. At the top the clear liquor overflow passes over a weir to the outlet.

Design basis. Particles in suspension will settle when the upward flow velocity of the fluid, u_u, is less than the settling rate of the particles, R. i.e.

$$u_u < R \tag{4.8}$$

Let Q_f, Q_o, Q_s = Volumetric flow-rates of feed, overflow and sediment (m³/s), W = Mass flowrate of solids (kg/s), F = Mass ratio of liquid to solid in feed (kg/kg) and S = Mass ratio of liquid to solid in sediment (kg/kg).

Taking the basis as a unit mass of solid, the unit can be analysed by simple mass balance on the suspension as follows:

Mass balance. An overall mass balance is stated as

$$\text{Overflow} = \text{Input} - \text{Underflow}$$

i.e.

$$Q_o \rho = (F - S)W \tag{4.9}$$

Therefore, volumetric flowrate to overflow

$$Q_o = \frac{(F - S)W}{\rho} \tag{4.10}$$

The upward liquor velocity, u_u, is given by

$$u_u = \frac{(F - S)W}{A\rho} \tag{4.11}$$

Applying the design criterion

$$R \geq u_u \tag{4.12}$$

then

$$R \geq \frac{(F - S)W}{A\rho} \tag{4.13}$$

Coe and Clevenger method. Since the slurry settling rate R varies with concentration of solids in the thickener batch settling tests are used to determine R over the concentration range of operation ($F \rightarrow S$) and the corresponding values of A are calculated for each point in the vessel (Coe and Clevenger, 1916). The value of A is then selected to cope with the condition of minimum mass flux of settling solids i.e. maximum A. Thus, the minimum mass flux over range of $F - S$ and R applying in the vessel is given by

$$\frac{W}{A} = \frac{R\rho}{F - S} \qquad (4.14)$$

Thus the cross-sectional area of the settler required at the minimum mass flux is

$$A = \frac{W(F - S)}{R\rho} \qquad (4.15)$$

Depth of thickening zone. Feed slurry entering the unit at concentration F has to reside in the vessel for time t_R to reach concentration S. The sediment therefore has to remain in the thickening zone for at least a period t_R defined by t_R = Volume of thickening zone/Volumetric flowrate of slurry.

Thus

$$t_R = \frac{V}{Q_s} \qquad (4.16)$$

i.e.

$$t_R = \frac{AH_T}{\left(\dfrac{W}{\rho_s} + \dfrac{WS}{\rho}\right)} \qquad (4.17)$$

where H_T = depth of thickening zone, ρ_s = density of solid particles and ρ = density of liquor in sediment.

Rearranging equation (4.17)

$$H_T = \frac{Wt_R}{A\rho_s} + \frac{WSt_R}{A\rho} \qquad (4.18)$$

i.e.
The depth of the settling zone is given by

$$H_T = \frac{Wt_R}{A\rho_s}\left(1 + \frac{S\rho_s}{\rho}\right) \qquad (4.19)$$

The settling rate and settling time can be estimated using e.g. the Richardson and Zaki equation (2.42). For slurries of irregular particles, however, the assumptions in the correlation are exceeded and the settling rate then becomes more difficult to calculate. Consequently, the 'Jar Test' (see Chapter 2) is frequently used to determine R and t_R in practice.

Example

Slurry of crystalline fines in water contains 15% w/w solids. It is desired to thicken this to a sediment containing 25% w/w solids at a rate of 100 ton/day (dry basis).

What would be the dimensions of an appropriate thickener, the fraction of water taken out of the slurry and the size of the largest particle in the overflow?

Laboratory test data

Y   Slurry concentration (kg H_2O/kg solids)
R   Settling rate (mm/s)

Test	1	2	3	4	5
Y	5.5	5.0	4.5	4.0	3.5
R	0.3	0.20	0.14	0.11	0.085

$t_R = 30$ min
$\rho = 10^3$ kg/m^3
$\rho_s = 2.2 \times 10^3$ kg/m^3
$\mu = 1 \times 10^{-3}$ kg/ms

Solution

Basis: 1 kg crystals

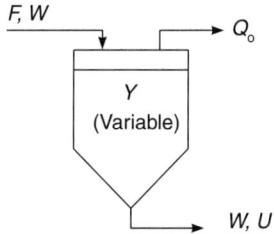

Mass feed-rate of solids

$$W = \frac{100 \times 10^3}{24 \times 3600} = 1.16 \text{ kg/s}$$

Concentration of feed = 15% solids [w/w]

$$F = \frac{85}{15} = 5.67 \text{ kg } H_2O/\text{kg solids}$$

Concentration of slurry = 25% w/w

$$S = \frac{75}{25} = 3.0 \text{ kg/kg}$$

Underflow

$$U = 3.0 \text{ kg } H_2O/\text{kg solids}$$

Therefore
  Overflow

$$= 5.67 - 3.0 = 2.67 \text{ kg/kg}$$

Water removal

$$= \frac{2.67}{5.67} \times 100 = 47\%$$

Range of concentration within the settling zone, $Y = 3.0$ (slurry) -5.67 (feed). First, calculate the variation of mass flux through the unit between inlet and outlet

Concentration Y (kg/kg)	Water to overflow $(Y - U)$ (kg/kg)	Settling rate R (m/s $\times 10^{-4}$)	Mass flux $R\rho/(Y - U)$ (kg/m²s)
5.5	2.5	3.0	0.12
5.0	2.0	2.0	0.10
4.5	1.5	1.4	0.09*
4.0	1.0	1.1	0.11
3.5	0.5	0.85	0.22

From the table (or graphically): Minimum mass flux $= 0.09 \, \text{kg/m}^2\text{s}$
Therefore

$$\text{Area} = \frac{(Y - U)W}{R\rho} = \frac{1.16}{0.09} = 12.8 \, \text{m}^2 \text{ at the minimum mass flux}$$

$$D = \sqrt{\frac{4 \times 12.8}{\pi}} = 4.05$$

Therefore
Diameter $= 4 \, \text{m}$
Height of thickening zone

$$H_T = \frac{W t_R}{A\rho_s}\left(1 + \frac{S\rho_s}{\rho}\right)$$

i.e.

$$H_T = \frac{1.16 \times 1800}{12.4 \times 2.2 \times 10^3}\left(1 + \frac{3 \times 2.2 \times 10^3}{10^3}\right) = 0.58$$

i.e.

$$H_T = 1 \, \text{m}$$

Add say 1 m (arbitrary) for clarifying zone, total height $= 2 \, \text{m}$. Thus a suitable thickener would be $2 \, \text{m} \times 4 \, \text{m}$ dia above a discharge cone.

Particle carry-over. To determine whether particles will be carried over, the superficial liquid up-flowrate must be determined and compared with the settling velocity of the slowest settling particles i.e. the fines.

Overflow rate $= (F - S)W$ [kg clarified slurry/s]

Therefore

$$Q_o = \frac{(F-S)W}{\rho} \quad [\text{m}^3/\text{s}]$$

Assume 'liquor' is clear

$$Q_o = \frac{(5.67 - 3.0)1.16}{10^3} = 3.10 \times 10^{-3} \, \text{m}^3/\text{s}$$

Up-flow velocity in clarifying zone must balance settling velocity of smallest particle. Thus

$$u_o = \frac{Q_o}{A} = \frac{3.10 \times 10^{-3}}{12.4} = 2.50 \times 10^{-4} \, \text{m/s} \equiv u_t \, \text{(particle terminal velocity)}$$

Assuming that Stokes' law holds for the settling of fine particles in 'clear' liquor

$$d_{st} = \sqrt{\frac{18\mu u_t}{g(\rho_s - \rho)}} = \sqrt{\frac{18 \times 10^{-3} \times 2.50 \times 10^{-4}}{9.81 \times (2.2 - 1.0) \times 10^3}} = 1.96 \times 10^{-5} \, \text{m}$$

Thus size of the largest particle in overflow = 20 μm.

In practice, a size analysis of the dried material would enable a calculation of the mass of material lost in the overflow to be made (i.e. mass below 20 μm) and a check should finally be made to see whether this would affect the mass balance significantly.

Application of the Coe and Clevenger equations tends to underestimate the required thickener area as reported by Scott (1968b). In an alternative approach to using batch settling tests for continuous thickener design, Kynch (1952) presented a graphical method based on the concentrations and fluxes in the settling zones which Talmage and Fitch (1955) demonstrated to overestimate the required thickener area. The various methods are compared and contrasted by Osborne (1990). In order to extend the range of parameter values from batch settling tests, Font and Laveda (1996) presented an extension of a method using a semi-batch test whereby periodically the supernatant clear liquor is withdrawn and a volume of suspension is added.

Filters and filtration

Filtration is the concentration of solids (or clarification of liquor) from slurry by fluid flow through a permeable medium. This normally takes the form of a membrane, filter 'leaf' or packed bed, which restricts the particles, more than the fluid (Figure 4.4).

Filters generally achieve a lower final moisture content than obtained by gravity sedimentation and are often fed from thickeners, as indicated in the schematic particulate process shown in Figure 9.2. In this chapter the principles of slurry filtration will be described and certain simplified filter design equations derived. For more complex derivations the reader is referred to specialist texts e.g. Coulson and Richardson (1991), Wakeman (1990a) and Purchas (1981).

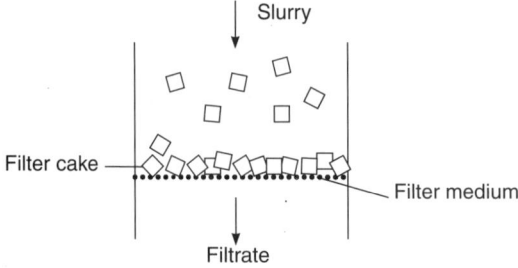

Figure 4.4 *Principle of filtration*

Methods of filtration

Four different methods of achieving filtration are depicted in Figure 4.5. Gravity filtration is the simplest method. Filtration occurs unaided apart from the effect of gravity causing clarified fluid flow through the medium. A filter cake builds up and slowly reduces the liquor flow rate as the resistance to flow increases correspondingly.

Filtration rates can be enhanced by applying a vacuum, as in a laboratory Buchner funnel, and this principle can be adapted, with a different mechanical design, to the industrial scale. Much higher filtration rates can be achieved, however, by the application of pressure. Simply pumping the slurry into an enclosed vessel containing the filter medium can do this. It is the principle employed by the filter press. Finally, filtration rates can be substantially enhanced by the application of a centrifugal force – effectively another type of pressure filter forcing liquor through the filter medium at high flow rates.

Types of filter

Filter press

The plate and frame press is a type of pressure filter (Figure 4.6). Filter plates are held within a frame and slurry is pumped under pressure into the spaces between successive plates. As filtration proceeds filter cake builds up and the pressure drop increases, finally reaching the limit of the pump. Eventually, the plate spaces become filled with cake and the plates are released in the frame and the cake discharged. A wash cycle may also be introduced to remove liquor held up in the filter cake. The operation is thus essentially batch-wise and can be labour intensive.

Rotary vacuum filter

The rotary vacuum filter (RVF, Rovac) comprises a revolving perforated drum that supports a filter medium and is suspended into a vessel containing slurry (Figure 4.7). Liquor is drawn through the drum by the application of an

Solid–liquid separation processes 89

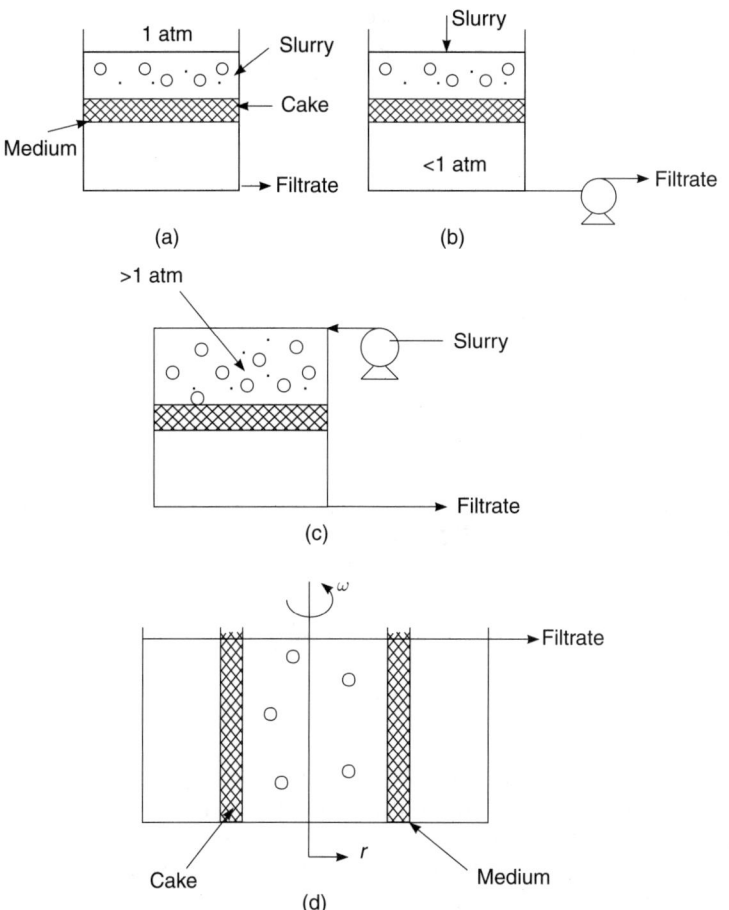

Figure 4.5 *Some methods of filtration (schematic):* (*a*) *Gravity (atmospheric pressure);* (*b*) *Vacuum;* (*c*) *Pressure;* (*d*) *Centrifugal (pressure)* $\Delta p \, \alpha \, \omega^2$

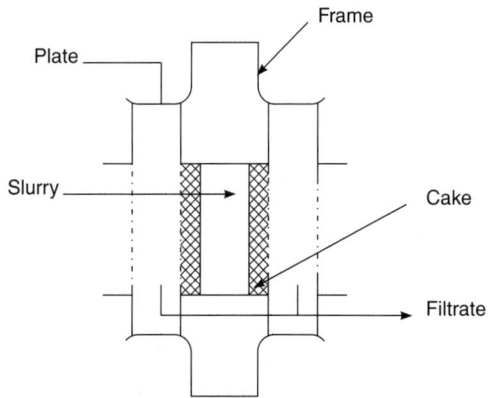

Figure 4.6 *Plate and frame press (schematic)*

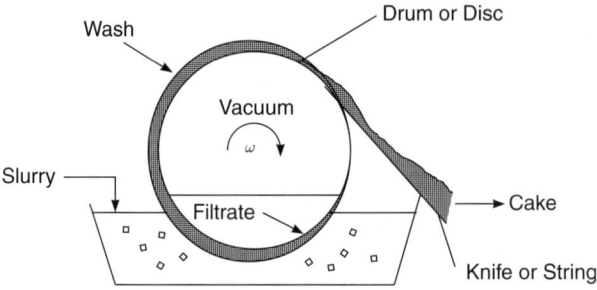

Figure 4.7 *Rotary vacuum filter (schematic)*

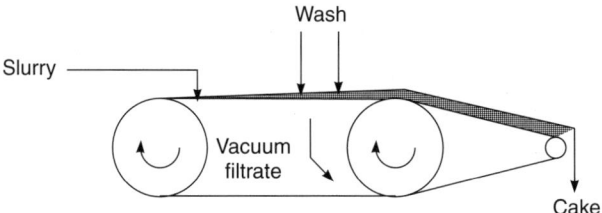

Figure 4.8 *Schematic belt filter*

internal vacuum. Filter cake builds up on the outer surface of the filter medium surface and is removed each revolution of the drum by a string or knife in close proximity. Interestingly, although operating continuously, each revolution of the drum is itself batch-wise in operation. Again, a wash cycle can be introduced.

Belt filter

The belt filter also works on the vacuum principle, but this time two drums rotate (Figure 4.8). The cake is thus held in a horizontal position during filtration thereby potentially increasing its stability.

Theory of filtration

For filter design and performance prediction it is necessary to predict the rate of filtration (velocity or volumetric flowrate) as a function of pressure drop, and the properties of the fluid and particulate bed. This can be achieved using the modified Darcy equation developed in Chapter 3.

Modified Darcy equation

In this simple model, the superficial liquor velocity through a filter cake (Figure 4.9) increases linearly with applied pressure drop and inversely with respect to

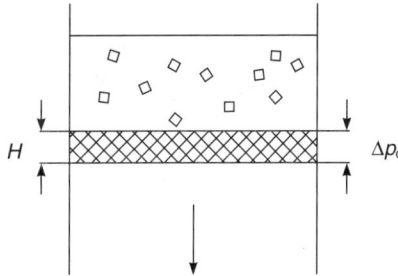

Figure 4.9 *Schematic filter cake and slurry*

liquor viscosity, bed depth and specific resistance (which is the inverse of permeability). Thus

$$u = \frac{1}{A}\frac{dV}{dt} = \frac{\Delta p_c}{r\mu H} \tag{4.20}$$

where V is the volume of filtrate collected.

Filtration equation. The basis for the analysis of filters is normally chosen as the 'unit volume of clear filtrate collected'.

Specific cake volume. Where $\bar{v}$ = volume of particulate cake deposited when unit volume of clear filtrate is collected.

Then, the total volume of cake is given by

$$V\bar{v} = HA \tag{4.21}$$

where V = volume of clear filtrate collected, H = depth of particulate cake and A = cross-sectional area of particulate cake.

Substituting for H in equation 4.20 gives

$$\frac{dV}{dt} = \frac{A^2 \Delta p_c}{r\mu \bar{v} V} \tag{4.22}$$

The specific cake resistance $r(m^{-2})$ depends on particulate bed characteristics ε and d_{sm}. According to the Carman–Kozeny equation for packed beds (Chapter 2)

$$u = \frac{1}{A}\frac{dV}{dt} = \frac{\varepsilon^3 \Delta p_c}{5(1-\varepsilon)^2 S_p^2 \mu H} \tag{4.23}$$

Thus solving for r between equations 4.22 and 4.23 gives

$$r = \frac{5(1-\varepsilon)^2 S_p^2}{\varepsilon^3} \tag{4.24}$$

where 5 is the magnitude of the Kozeny 'constant' and S_p is the specific surface area of the particles, $= F/d_{sm}$. Equation 8.4 thus enables the filtration rate to be calculated as a function of fluid and particulate bed properties, as required.

Example

Relate the filter cake deposited per unit volume of clear filtrate collected $\bar{v}$ to feed slurry concentration, w.

Solution

Mass balance
Let

w = mass fraction of solids in slurry feed [mass solids/(mass solids + liquid)]
$\bar{v}$ = volume of particulate cake deposited per unit volume of clear filtrate

$$\text{Mass of solids in cake} = (1-\varepsilon)AH\rho_s \tag{4.25}$$

$$\text{Mass of liquid in cake} = \varepsilon AH\rho_s \tag{4.26}$$

$$\text{Mass of liquid in filtrate} = V\rho = \frac{AH}{\bar{v}}\rho \tag{4.27}$$

Now
[Mass of solids in slurry feed/(Mass of particulate solids + clear liquid in feed)] = [Mass of particulate solids in cake/(Mass of particulate solids in cake + Mass of clear liquor in cake + Mass of liquor in clear filtrate)]
i.e.

$$w = \frac{(1-\varepsilon)AH\rho_s}{(1-\varepsilon)AH\rho_s + \varepsilon AH\rho + \frac{AH\rho}{\bar{v}}} \tag{4.28}$$

whence

$$\bar{v} = \frac{w\rho}{(1-\varepsilon)(1-w)\rho_s - w\varepsilon\rho} \tag{4.29}$$

thereby relating $\bar{v}$ to the feed concentration, w, and cake voidage, ε.
It can also be shown that

$$\bar{v} = \frac{\left[\dfrac{1}{\rho_s} + \dfrac{m}{\rho}\right]}{\left[\dfrac{1}{c} - \dfrac{m}{\rho}\right]} \tag{4.30}$$

where m = kg liquid/kg solid in slurry and c = kg solid/kg liquid in slurry.

Modes of operation

There are two modes of filter operation representing extreme cases of filtration viz. (1) constant rate; and (2) constant pressure.

(a) Constant rate filtration

In this mode the feed pump is opened up slowly thereby overcoming the increased resistance to flow as solids build-up in the filter cake i.e.

$$\frac{dV}{dt} = \text{constant} \tag{4.31}$$

$$\Rightarrow \quad \frac{V}{t} = \text{constant} \tag{4.32}$$

From the filtration equation 4.20

$$\frac{V}{t} = \frac{A^2 \Delta p_c}{r\mu \bar{v} V} = \text{constant} \tag{4.33}$$

Therefore

$$\frac{\Delta p_c}{V} = \text{constant} \tag{4.34}$$

or

$$\frac{t}{V} = \frac{r\mu \bar{v} V}{A^2 \Delta p_c} \tag{4.35}$$

Thus, Δp_c is directly proportional to V. (Figure 4.10)

(b) Constant pressure filtration

Following the usual initial build-up period, the pump is fully on. Thus the pressure drop in the filter is held constant at the maximum head delivered by the pump.

$$\Delta p_c = \text{constant} = \Delta p_{max} \tag{4.36}$$

The flow path through the filter cake continues to increase, however, thereby increasing the total resistance to flow and the filtrate flowrate therefore slowly declines.

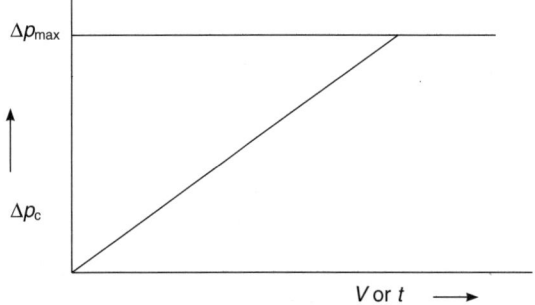

Figure 4.10 *Constant rate filtration*

Integrating the filtration equation now gives a linear relation between V^2 and t, or, conventionally, between t/V and V (Figure 4.11).

$$V^2 = \frac{2A^2 \Delta p_c t}{r\mu \bar{v}} \tag{4.37}$$

i.e.

$$V^2 = k_1 t \tag{4.38}$$

or

$$\frac{t}{V} = \frac{r\mu \bar{v} V}{2A^2 \Delta p_c} \tag{4.39}$$

Thus

$$\frac{t}{V} = k_2 V \tag{4.40}$$

Following an initial build-up, constant pressure is the more usual case.

The slope of the t/V versus t plot indicates the filter cake resistance. Note, however, that in practice the t/V versus t plots do not pass through the origin, due to the resistance of the filter cloth and any associated particles entrapped within it (Figure 4.8).

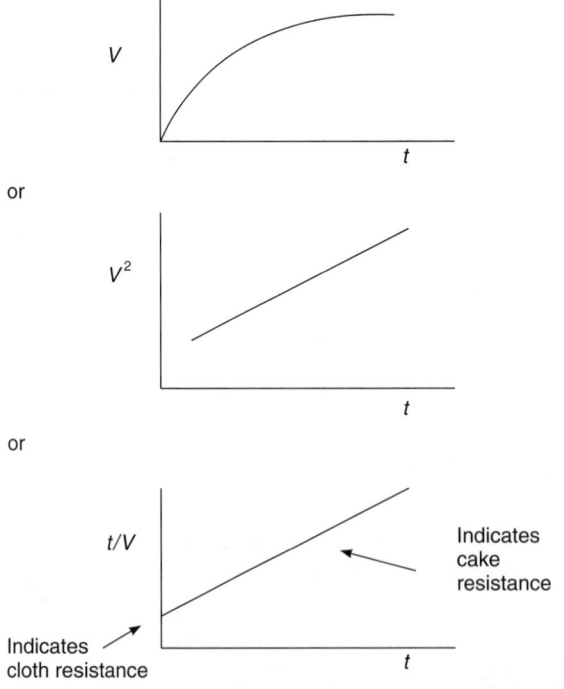

Figure 4.11 *Constant pressure filtration*

Effect of the filter medium

The preceding analysis can be extended such that the effect of the filter medium can be taken into account (Figure 4.12).

Pressure drop

$$\Delta p = \Delta p \text{ cloth} + \text{cake in cloth} + \Delta p \text{ cake} \qquad (4.41)$$

After a rapid initial increase, pressure drop gradually builds up due to cloth 'blinding', thereby creating a 'filter cloth plus cake held-in-cloth' resistance.

Resistance (inverse of permeability)

$$R = r(H + L) \qquad (4.42)$$

where L = equivalent cake thickness of cloth.
Thus the filtration equation 4.19 becomes

$$\frac{1}{A}\frac{dV}{dt} = \frac{\Delta p}{r\mu(H+L)} \qquad (4.43)$$

substituting for H in equation 4.43 gives

$$\frac{dV}{dt} = \frac{A^2 \Delta p}{r\mu\left(\dfrac{V\bar{v}}{A} + L\right)} \qquad (4.44)$$

i.e.

$$\frac{dV}{dt} = \frac{A^2 \Delta p}{r\mu\bar{v}\left(V + \dfrac{LA}{\bar{v}}\right)} \qquad (4.45)$$

Thus equation 4.45 is an ordinary differential equation (ODE) which can easily be solved for filter area, A (in the design problem) or filtrate collected, V (for performance mode calculation).

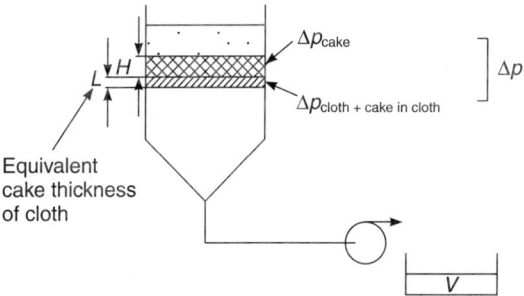

Figure 4.12 *Schematic filter cake and cloth*

(a) Constant rate

Integrating equation 4.26

$$\text{from } t = 0, V = 0, \Delta p = \Delta p_0$$
$$\text{to } t = t_1, V = V_1, \Delta p = \Delta p_1$$

gives

$$\frac{dV}{dt} = \text{constant} = \frac{V_1}{t_1} = \frac{A^2 \Delta p_1}{r\mu\bar{v}\left(V_1 + \frac{LA}{\bar{v}}\right)} \tag{4.46}$$

i.e.

$$V_1^2 + \frac{LA}{\bar{v}} V_1 = \frac{A^2 \Delta p}{r\mu\bar{v}} t_1 \tag{4.47}$$

(b) Constant pressure

Integrating equation 4.26

$$\text{from } t = t_1, V = V_1$$
$$\text{to } t = t, V = V$$
$$\text{with } \Delta p = \Delta p_1 = \Delta p_2$$

gives

$$\frac{1}{2}(V^2 - V_1^2) + \frac{LA}{\bar{v}}(V - V_1) = \frac{A^2 \Delta p}{r\mu\bar{v}}(t - t_1) \tag{4.48}$$

i.e.

$$\frac{t - t_1}{V - V_1} = \frac{r\mu\bar{v}}{2A^2\Delta p}(V - V_1) + \frac{r\mu\bar{v}V_1}{A^2\Delta p} + \frac{r\mu L}{A\Delta p} \tag{4.49}$$

| term due to cake build-up | initial condition at t_1 | term due to cloth resistance |

Simplified filtration equation

A simplified form of the constant pressure filtration equation can be written if it is assumed that $\Delta p = $ constant from $t = 0$ i.e. $V_1 = 0$.

$$\frac{t}{V} = k_1 V + k_2$$

$$k_1 = \frac{r\mu\bar{v}}{2A^2\Delta p} \tag{4.50}$$

$$k_2 = \frac{r\mu L}{A\Delta p}$$

enabling k_1 and k_2, hence $r\mu\bar{v}$ and $r\mu L$ to be determined experimentally, as follows.

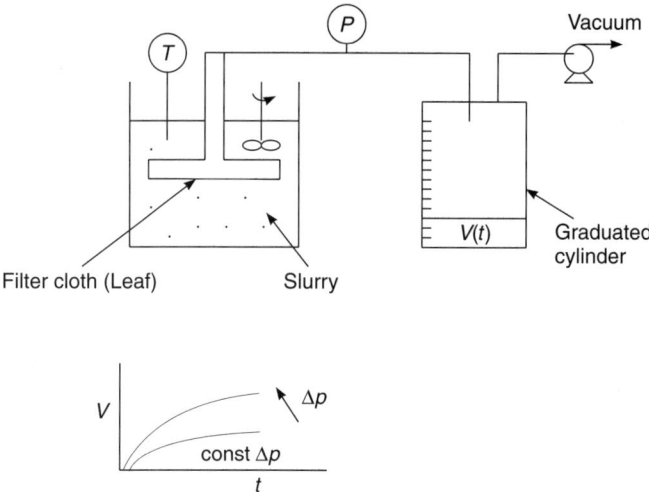

Figure 4.13 Filter 'leaf test'

Leaf test

In principle, filter bed permeabilities can be calculated using the Carman–Kozeny equation 2.53. For slurries containing irregular particles, however, cake filtrabilities together with filter medium resistance are determined using the 'Leaf Test' (Figure 4.13). In this technique, a sample of suspended slurry is drawn through a sample test filter leaf at a fixed pressure drop and the transient volumetric flowrate of clear filtrate collected determined.

A simple laboratory apparatus for the measurement of crystal bed permeability is shown in Figure 4.14, the standard procedure for which is described in detail in Purchas (1981).

Since for constant pressure filtration, the t/V versus V data can be linearized, as shown in Figure 4.15, the resistances of cake and cloth plus cake held up in cloth can be determined. The former value is usually fairly reproducible while the latter is often variable, being particularly sensitive to start up conditions when cloth blinding occurs. Such tests can be rerun at different pressures and the extent of cake compressibility determined. Similarly, a wash cycle can be introduced.

Example

A sample of fine catalyst crystals for use in a reactor was found to have the following particle characteristics

Sauter mean size	120 μm
Particle voidage	0.5
Sphericity	0.8

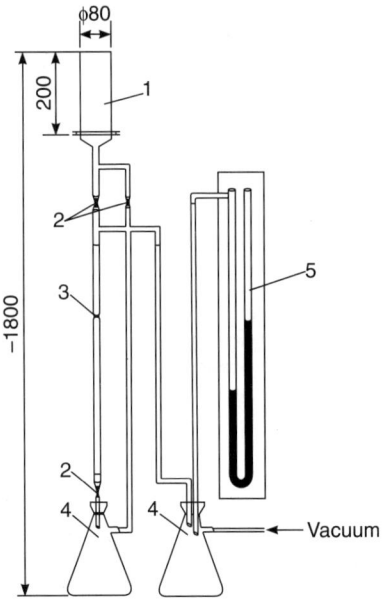

Figure 4.14 *Permeability test cell: 1 – slurry vessel, 2 – valves, 3 – pipette, 4 – vessels, 5 – U-tube manometer (Mydlarz and Jones, 1989)*

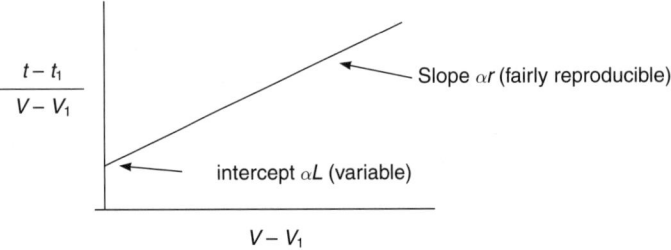

Figure 4.15 *'Leaf Test' filtration analysis*

Assuming a Kozeny constant of 5, what would be the permeability of the reactor packed with the sand?

Solution

$$D_{sm} = 120\,\mu m \equiv 1.20 \times 10^{-2}\,cm$$
$$\varepsilon = 0.5$$
$$\psi = 0.8$$
$$K_2 = 5$$

Specific surface area of particles

$$S_p = \frac{6}{\psi D_{sm}} = \frac{6}{0.8 \times D_{sm}}$$

Permeability coefficient

$$B = \frac{\varepsilon^3}{K_2(1-\varepsilon)^2 S_p^2} = \frac{0.5^3 \times 0.8^2 \times D_{sm}^2}{5.0 \times 0.5^2 \times 6^2} = 1.78 \times 10^{-3} D_{sm}^2$$

Since the particle size = 120 μm, then

$$B = 1.78 \times 10^{-3}(1.2 \times 10^{-2})^2 = 25.6 \times 10^{-8}\, cm^2 \equiv 25.6 \times 10^{-12}\, m^2$$
$$B = 26\, \text{Darcy}$$

Note: Permeability is strongly dependent on both particle size and voidage of bed.

Example

It is proposed to try out a new filter cloth material in a filter press in use to separate a dispersed dyestuff from aqueous slurry. The press has 10 plates with leaves of $0.5\,m^2$ each.

A leaf-test was carried out on the new medium with slurry giving the following results:

Time (min)	Filtrate collected (ml)
0	0
5	250
10	400

Test filter leaf area = $0.05\,m^2$
Vacuum applied = 0.70 bar

The pressure in the press rises up to a final value of 4 bar after 3 min and continues for a further 15 min at constant pressure.
Predict for the average filtration rate using the new filter cloth material.

Solution

Leaf test – Assume constant Δp
Integrated filtration equation of the form

$$\frac{t}{V} = k_1 V + k_2$$

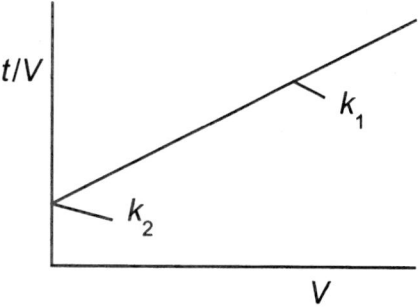

where

$$k_1 = \frac{\mu r \bar{v}}{2A^2 \Delta p} \text{(cake)}$$

$$k_2 = \frac{\mu r L}{A \Delta p} \text{(cloth)}$$

Convert data to SI units

t		V		t/V
min	s	m²l	m³	s/m³
0	0	0	0	0
5	300	250	250×10^{-6}	1.2×10^6
10	600	400	400×10^{-6}	1.5×10^6

$A = 0.05 \, \text{m}^2$ and $\Delta p = 70 \, \text{kN/m}^2$

Solving equation 4.50, or plot t/V versus V (not shown).

$$\Rightarrow \text{intercept} = 0.7 \times 10^6, \quad \text{slope} = 2 \times 10^9$$

$$r\mu L = A\Delta p \times 0.7 \times 10^6; \quad r\mu\bar{v} = 2A^2\Delta p \times 2 \times 10^9$$

Therefore

$$r\mu \, L(\text{cloth}) = 2.45 \times 10^6 \text{ and } r\mu\bar{v} \, (\text{cake}) = 7 \times 10^8$$

Filter press

Assume $\mu r v$ and $\mu r L$ as for leaf test

$$10 \text{ plates} \equiv 20 \text{ leaves}$$

$$\text{Total area} = 20 \times 0.5$$

i.e.

$$A = 10 \, \text{m}^2; \quad A^2 = 10^2 \, \text{m}^4$$

(a) First period
Assume constant flowrate; 0–4 bar.
Integrated filtration equation (constant rate)

$$t_1 = \frac{(\mu r \bar{v})V_1^2}{A^2 \Delta p} + \frac{(\mu r L)V_1}{A \Delta p}$$

Therefore

$$3 \times 60 = \frac{7 \times 10^8}{10^2 \times 4 \times 10^2} V_1^2 + \frac{2.45 \times 10^6}{10 \times 4 \times 10^2} V_1$$

$$1.75 \times 10^4 V_1^2 + 6.125 \times 10^2 V_1 - 180 = 0$$

Noting that this equation is in the form

$$ax^2 + bx + c = 0; \quad x = \frac{-b \pm \sqrt{b^2 - 4ac}}{2a}$$

$$\Rightarrow x = 0.0854 \, \text{m}^3$$

i.e.

$$V_1 = 85.4 \, \text{L (in 3 min)}$$

(b) Second period
Constant pressure drop $\Delta p = 4$ bar for 15 min.
Applying the integrated filtration equation. (constant Δp)

$$t_2 - t_1 = \frac{\mu r \bar{v}(V_2^2 - V_1^2)}{2A^2 \Delta p} + \frac{\mu r L (V_2 - V_1)}{A \Delta p}$$

$$18 - 3 \times 60 = \frac{7.80 \times (V_2^2 - (8.54 \times 10^2)^2)}{2 \times 10^2 \times 4 \times 10^2} + \frac{2.45 \times 10^6 (V_2 - 8.54 \times 10^{-2})}{10 \times 4 \times 10^2}$$

which is similarly in the form of a quadratic

$$\Rightarrow x = 0.308 \, \text{m}^3$$

i.e.

$$V_2 = 308 \, \text{L (in 18 min total)}$$

Thus the average filtration rate is

$$\frac{V_2}{t_2} = \frac{308}{18}$$

$$= 17 \, \text{L/minute}$$

Optimum cycle time during batch filtration

As a batch filtration proceeds, so the rate of filtrate collection decreases due to the increased cake thickness. While operation with thin filter cake results in a higher instantaneous filtration rate, however, it also requires more frequent dismantling of the filter and discharge of the filter cake. There is thus an optimum balance between *filtration time* and *down time*.

Assume

The constant pressure period dominates the cycle. Thus

$$\frac{t}{V} = \frac{r \mu \bar{v}}{2A^2 \Delta p} V + \frac{r \mu L}{A \Delta p} \tag{4.51}$$

$$\Rightarrow k_1 V + k_2 \tag{4.52}$$

and the time of filtration is given by

$$t = k_1 V^2 + k_2 V \tag{4.53}$$

Let the down time be t' and assume that it is substantially independent of cake thickness. The total time, T, of a cycle in which a volume V is collected is then

$$T = t + t' \tag{4.54}$$

The average rate of filtration is given by

$$Q = \frac{d\bar{V}}{dt} = \frac{V}{T} = \frac{V}{t+t'} \tag{4.55}$$

Substituting for t

$$Q = \frac{V}{k_1 V^2 + k_2 V + t'} \tag{4.56}$$

Clearly, the flowrate $\frac{d\bar{V}}{dt}$ is a maximum when $\frac{dQ}{dt} = 0$
Thus

$$k_1 V^2 + k_2 V + t' - V(2k_1 V + k_2) = 0 \tag{4.57}$$

i.e.

$$t' = k_1 V^2 \tag{4.58}$$

or

$$V = \sqrt{\frac{t'}{k_1}} \tag{4.59}$$

Low r ($\propto k_1$) implies a wide frame (high capacity).
If the resistance of the filter medium is neglected

$$t = k_1 V^2 \tag{4.60}$$

$$t = t' \tag{4.61}$$

Thus for the optimal operation of a batch filter, the filtration time and the down time are equal. In practice $t > t'$ due to cloth resistance, which is neglected in this simple analysis.

Maximizing filter throughput

Consider the case of constant pressure with negligible cloth resistance. The filtration time is given by

$$t = \frac{r\mu\bar{v}}{2A^2 \Delta p} V^2 \tag{4.62}$$

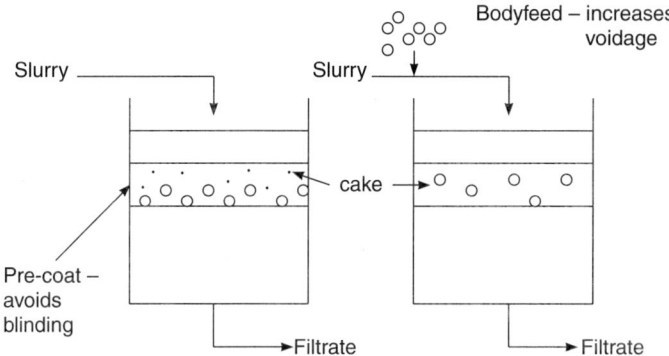

Figure 4.16 *Use of pre-coat and bodyfeed to improve slurry filtrability*

Therefore in order to reduce filtration time any of the following measures can be adopted

1. reduce μ – e.g. higher temperature operation.
2. increase Δp – but beware compressible cakes (e.g. biologicals)
3. maximize A – by suitable mechanical design
4. minimize r and L – avoid cloth blinding and increase filter cake permeability, use 'precoat' and 'bodyfeed' e.g. Keisulguhr or diatomaceous earth.

Pre-coat and bodyfeed

Pre-coat and bodyfeed are often used to improve slurry filterability (Figure 4.16). Given that the bed permeability is related to the characteristics of the filter cake by the equation

$$B = \varepsilon^3 / K_2 (1 - \varepsilon)^2 S_p^2 \tag{4.63}$$

where $S_p^2 \propto 1/d_{sm}^2$ and $K_2 =$ Kozeny constant

Thus to maximize flow regime require large particles in porous bed e.g. decrease S_p by increasing d_{sm}; e.g. increasing ε from 0.5 to 0.6 increases B by ~ 3.

Continuous rotary vacuum filtration

Operating principle

The operation of the continuous rotary vacuum filter is illustrated in Figure 4.17. The filter drum is immersed in a bath of slurry from where the liquor is drawn through the filter medium by the pressure drop caused by the application of a vacuum within the drum. During filtration cake builds up on the outside surface of the drum between points A and B as if by constant pressure batch operation. It then travels to point C where it is removed and the cycle repeated.

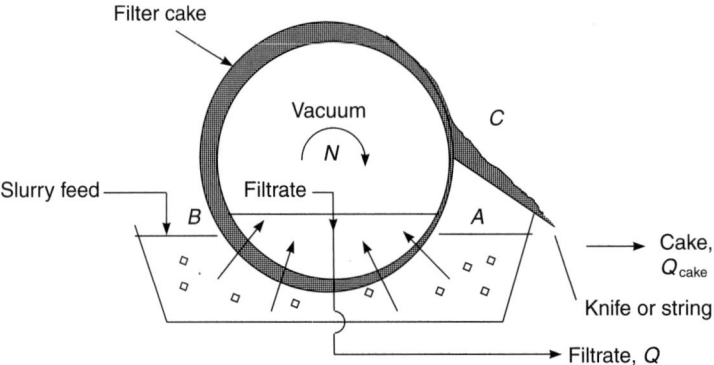

Figure 4.17 *Continuous rotary vacuum filtration*

Submergence fraction

Filtration only proceeds on that part of the drum that is submerged at any time

$$a = \frac{A_c}{A_d} = \frac{\text{area of cake submerged}}{\text{area of drum}} \quad (4.64)$$

Recalling the filtration equation (ignoring R_{cloth}) is given by

$$Q = \frac{dV}{dt} = \frac{A\Delta p}{r\mu l} = \frac{A^2 \Delta p}{r\mu \bar{v} V} \quad (4.65)$$

where l is the particulate cake thickness, and

$$\bar{v}V = lA \quad (4.65)$$

i.e. the cake volume.

Operating equations

Constant pressure drop is maintained during operation. Taking as the basis one revolution of the drum, the time per revolution and submergence fraction, respectively, can be defined as

$$\text{Time per revolution } t_r = \frac{1}{N} \quad (4.66)$$

$$\text{Submergence time } t_s = \frac{a}{N} \quad (4.67)$$

The integrated filtration equation is

$$V^2 = \frac{2A_d^2 \Delta p}{r\mu \bar{v}} t_s \quad (4.68)$$

since each element of drum is submerged for time t_s only.

i.e.

$$V^2 = k\Delta p A_d^2 t_s \tag{4.69}$$

where $k = \dfrac{2}{r\mu\bar{v}}$

or

$$V = \sqrt{\dfrac{k\Delta p A_d^2 a}{N}} \tag{4.70}$$

an expression in terms of drum speed and submergence.

Effect of drum speed

At constant pressure drop, Δp, the volume of filtrate collected depends on the drum speed according to the relation

$$V \propto \dfrac{1}{\sqrt{N}} \tag{4.71}$$

Thus the volume of filtrate collected per revolution is *inversely* proportional to the square root of drum speed. But at higher speeds there are more revolutions made in unit time. Thus

(i) Average flowrate of filtrate (Q)

$$Q = \dfrac{V}{t_r} = VN \tag{4.72}$$

i.e.

$$Q_{\Delta p} \propto \dfrac{N}{\sqrt{N}} \propto \sqrt{N} \tag{4.73}$$

Therefore, the clear filtrate flowrate varies *directly* with the square root of the drum speed.

(ii) Average cake thickness (l)

Recalling that by definition $\bar{v}$ = volume of cake deposited/volume of clear filtrate collected, then the rate of deposition of cake, Q_{cake}, is given by

$$Q_{\text{cake}} = Q\bar{v} \tag{4.74}$$

Since filter cake is deposited in one revolution before removal, the filtration time per revolution is effectively t_r. Hence

$$l = \dfrac{Q_{\text{cake}}}{A_{\text{drum}}} t_r = \dfrac{Q\bar{v}}{\pi DlN} \tag{4.75}$$

Recalling that $Q \propto \sqrt{N}$ then

$$l \propto \frac{1}{\sqrt{N}} \qquad (4.76)$$

Thus, cake depth l varies *inversely* with the square root of drum speed N. In practice a minimum cake thickness is reached, typically ~5 mm.

Centrifuges

Centrifugation is the application of rotational velocity to enhance the effect of gravity and intensify the separation of phases. It can be applied in two distinct ways

1. sedimentation – primarily used where the clear liquor is required as a product
2. filtration – primarily used where the filter 'cake' is the product.

Centrifuges are machines that spin at high speed to induce angular fluid velocities within. They are widely used in industry for the separation of particles from liquids, and frequently form part of an industrial particulate processing system, often as an alternative to filtration where appropriate. Hydrocyclones are a special type of sedimenting centrifuge in which the fluid is induced to swirl in a static vessel by virtue of its geometry rather than by spinning the vessel itself. They are often used as ancillaries to crystallizer vessels in order to help control the fines in the CSD. In this chapter we will describe and analyse slurry centrifugation and derive some simplified design equations. For more complex derivations the reader is referred to specialist texts e.g. Backhurst and Harker (1973), Coulson and Richardson (1991), Purchas (1981) and Zeitsch (2000).

Classification of centrifuge types

There are numerous types of centrifuge and several ways of classifying them.

Construction

1. solid bowl (sedimenting)
2. perforated bowl (filtering).

Centrifuge operation

1. high speed
2. low speed
3. batch
4. continuous etc.

Particle size

Another way of selecting a centrifuge is to consider the particle size in the feed. In practice properties of the slurry, cake, throughput and costs are also considered.

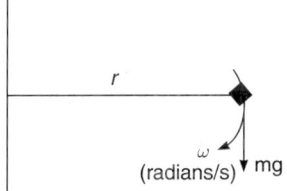

Figure 4.18 *Principle of centrifugation*

Mechanical construction

All centrifuges require

1. inlet for introduction of feed
2. rotating separating chambers
3. solid and liquid discharge systems for product off-take:
   (a) solids by mechanical action
   (b) liquor by weir overflow.

Theory of centrifugation

Separating effect

The separating effect, G, or 'power' of centrifuge is the ratio of acceleration obtained in the machine: terrestrial gravity, as illustrated in Figure 4.18.
  Noting that

$$\text{Centrifugal Force} = m\omega^2 r \tag{4.77}$$

and

$$\text{Gravitational Force} = mg \tag{4.78}$$

Then the separating power of a centrifuge is given by

$$G = \frac{m\omega^2 r}{mg} = \frac{\omega^2 r}{g} \tag{4.79}$$

Solid bowl centrifuge

The principle of the solid bowl centrifuge is illustrated in Figure 4.19. It is essentially a device for enhancing the rate of sedimentation. The application of high G, however, creates stress in the machine and is equivalent to the application of pressure (cf. pressure filtration). There is therefore a balance to be struck between separating effect and stress.

Pressure

$$\text{Mass of fluid in annular cylinder} = 2\pi r dr H \rho \tag{4.80}$$

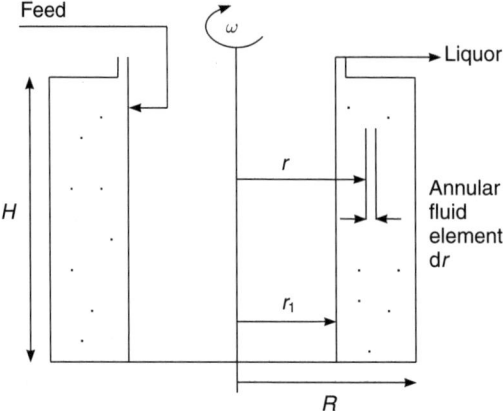

Figure 4.19 *Solid bowl centrifuge*

Differential centrifugal force

$$dF = (2\pi r \, dr \, H \rho)\omega^2 r \tag{4.81}$$

Differential pressure

$$dP = \frac{dF}{A} = \frac{2\pi r \, dr \, H \rho \omega^2 r}{2\pi r H} = \rho \omega^2 r \, dr \tag{4.82}$$

Total pressure

$$P = \int_{r_1}^{R} \rho \omega^2 r \, dr = \frac{1}{2} \rho \omega^2 (R^2 - r_1^2) \tag{4.83}$$

The maximum value of the pressure occurs when the bowl is full ($r_1 \to 0$) thus

$$P_{max} = \frac{1}{2} \rho \omega^2 R^2 \tag{4.84}$$

Stress

The stress in the bowl wall is $\propto R$ i.e.

$$\text{Maximum Stress} \propto R^2 \omega^2 \tag{4.85}$$

The design stress is limited in practice by the material of construction. For a given material of construction, the separating effect at the maximum stress can be derived

$$\frac{\text{Separating Effect}}{\text{Maximum Stress}} \propto \frac{\omega^2 R}{\omega^2 R^2} \tag{4.86}$$

i.e.

$$\propto \frac{1}{R}$$

Hence for good separation, R must be kept small as possible (thin cylinder) in order that high speeds may be used, but this inevitably induces a cost penalty (area:volume).

Example

A centrifuge on a crystallization process plant is to be scaled down for tests. If the plant centrifuge is 1.0 m diameter and rotates at 25 Hz at what speed should a 100 mm laboratory centrifuge run if it is to duplicate plant conditions?

Solution

Separation power

$$G = \frac{\text{Centrifuge Force}}{\text{Gravitational Force}} = \frac{\omega^2 r}{g}$$

where $\omega = 2\pi N$.
For plant

$$G_1 = \frac{\left(2\pi \frac{1500}{60}\right)^2 \times 0.5}{9.81} = 1258$$

For laboratory

$$G_2 = \frac{(2\pi x)^2 \times 0.05}{9.81} = 0.2012 x^2$$

Thus for the same separation effect in plant and laboratory $G_1 = G_2$
Therefore

$$x = \sqrt{\frac{1258}{0.2012}} = 79 \text{ Hz}$$

The required speed of laboratory centrifuge $= 79 \times 60 = 4740$ rpm.

Thin layer solid bowl centrifuge

Consider a thin layer solid bowl centrifuge as shown in Figure 4.20. In this device, particles are flung to the wall of the vessel by centrifugal force while liquor either remains stationary in batch operation or overflows a weir in continuous operation. Separation of solid from liquid will be a function of several quantities including particle and fluid densities, particle size, flowrate of slurry, and machine size and design (speed, diameter, separation distance, etc.). A relationship between them can be derived using the transport equations that were derived in Chapter 3, as follows.

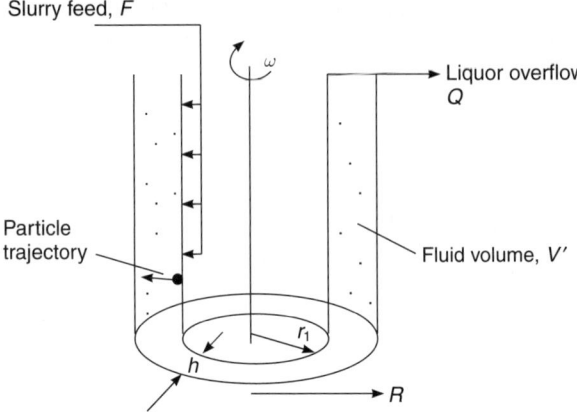

Figure 4.20 *Thin layer sedimenting centrifuge*

Transport equations

Assume that Stokes' Law gives the particle equation of motion (for $Re_p < 0.2$). Thus for flow across the thin cylinder $R - r_1$

$$v = \frac{dr}{dt} = \frac{D^2(\rho_s - \rho)}{18\mu} \times \omega^2 r \tag{4.87}$$

where D = particle diameter, m.
Defining the 'mean residence time', τ, of the fluid

$$\tau = \frac{V'}{Q} \tag{4.88}$$

where V' = volume of slurry in centrifuge, m³ and Q = liquor flowrate, m³/s.
The particle settling rate within the slurry v is

$$v = \frac{dr}{dt} \ [m/s] \tag{4.89}$$

The particle settling distance h is

$$h = R - r_1 \ [m] \tag{4.90}$$

Therefore, if $h \ll R$ (thin bowl), then $v \approx$ constant, and the particle settling time t_s is

$$t_s = \frac{h}{v} \ [s] \tag{4.91}$$

Now, particles will separate out if the settling time is less than the mean residence time of the fluid i.e.

$$t_s \leq \tau \leq \frac{V'}{Q} \tag{4.92}$$

where τ = mean residence time.

That is, the particles will be separated if

$$\frac{h}{v} \leq \frac{V'}{Q} \tag{4.93}$$

Therefore for the worst case i.e. to prevent particle 'washout'

$$Q_{max} = \frac{vV'}{h} \tag{4.94}$$

The effective area of the centrifuge, known as the Sigma value, Σ, is obtained by integration of equation 4.18, as follows.

Sigma value, Σ

Substituting for v in equation 4.94 gives

$$Q_{max} = \left[\frac{D^2(\rho_s - \rho)g}{18\mu}\right]\left[\frac{V'\omega^2 R}{gh}\right] \tag{4.95}$$

i.e.

$$Q_{max} = v_t \times \Sigma \tag{4.96}$$

where two distinct quantities have now been separated out viz. that concerning the slurry and another concerning the machine

v_t = function of particle and fluid only

Σ = function of machine only

Σ is effectively an area term [m^2] equivalent to the cross-sectional area of a sedimentation tank required to perform a similar separation duty. It is normally many orders of magnitude greater than the floor area occupied by the centrifuge itself and is a classic example of 'intensification'. Similar expressions for Σ can be derived for other centrifuge configurations with results depending on their geometry e.g. $\Sigma_{disc} > \Sigma_{bowl}$.

Effect of particle size distribution

For particle size distributions a 'cut size' for a 50% probability of removal is defined, whence

$$Q_{max} = \frac{2v_t \Sigma}{0.5} \tag{4.97}$$

Scale-up

To determine the required machine size

$$\Sigma_{plant} = \frac{Q_{plant}}{Q_{lab}} \times \Sigma_{lab} \tag{4.98}$$

Note, since different types of machine have different sigma values, always take care in checking the precise definition of Σ.

Centrifugal filtration

Consider a continuous screen bowl centrifuge as illustrated in Figure 4.21. In this device liquor passes through a filter medium and cake builds up inside as before.

Transport equations

The velocity of liquor through the cake is given by

$$u = \frac{1}{A}\frac{dV}{dt} = \frac{Q}{A} \tag{4.99}$$

Assuming Darcy's Law applies

$$u = \frac{\left(\frac{dp}{dR}\right)}{r\mu} \tag{4.100}$$

During build-up of the cake, the flow area changes, thus integrating equation 4.100 from R_2 to R_i gives

$$\Delta p_f = \frac{r\mu}{2\pi b} \ln \frac{R_2}{R_i} \frac{dV}{dt} \tag{4.101}$$

The pressure drop due to centrifugal action is given by

$$\Delta p = \frac{1}{2}\rho\omega^2(R_2^2 - R_1^2) \tag{4.102}$$

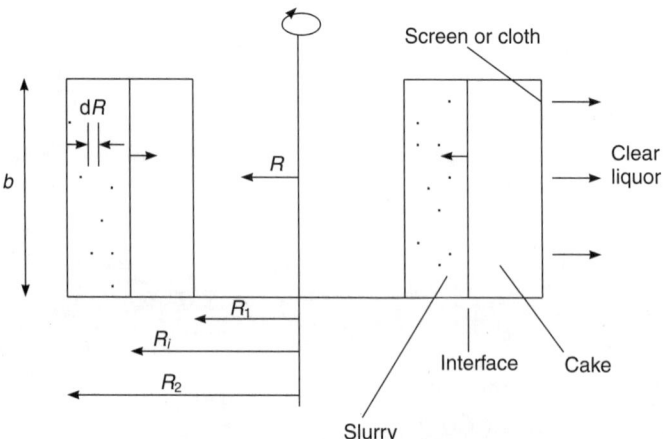

Figure 4.21 *Perforated basket filtering centrifuge*

Therefore if the level of slurry is maintained constant with addition of fresh feed, then the pressure drop due to the drag in the liquid flowing through the cake can be equated to the pressure drop due to centrifugal action viz. from equations 4.84 and 4.85

$$\frac{r\mu}{2\pi b} \times \ln\frac{R_2}{R_i}\frac{dV}{dt} = \frac{1}{2}\rho\omega^2(R_2^2 - R_1^2) \qquad (4.103)$$

Solving for the flow rate through the filter gives the Storrow equation

$$\frac{dV}{dt} = \frac{\pi\rho\omega^2 b(R_2^2 - R_i^2)}{r\mu\ln\left(\dfrac{R_2}{R_i}\right)} \qquad (4.104)$$

Thereby relating the filtration rate to the geometry and speed of the centrifuge and the properties of the cake (see Grace, 1953).

More complex forms can be derived e.g. allowing for a change in radial position of the slurry (as in batch operation), or allowing for the resistance of the filter cloth.

Filtration time

During operation, the filter cake builds up and is periodically discharged. The time taken to build up cake can be estimated by integrating equation 4.87 but first it is necessary to derive a relationship between cake thickness and filtrate volume, viz. by definition

$$\bar{v}\,dV = -2\pi R_i b\,dR \qquad (4.105)$$

i.e.

$$\frac{dV}{dt} = \frac{-2\pi R_i b}{\bar{v}}\frac{dR}{dt} \qquad (4.106)$$

Thus equating

$$\frac{\pi\rho\omega^2 b(R_2^2 - R_i^2)}{r\mu\ln\left(\dfrac{R_2}{R_i}\right)} = \frac{-2R_i b}{\bar{v}}\frac{dR}{dt} \qquad (4.107)$$

i.e.

$$\frac{1}{2}\frac{\bar{v}}{r\mu}\rho\omega^2\,dt = \frac{-2\pi R_i b}{\bar{v}}\frac{dR}{dt} \qquad (4.108)$$

Integrating

R_i from $R_2 \to R_1$

t from $0 \to t$

gives

$$\frac{1}{2}\frac{\bar{v}}{r\mu}\rho\omega^2 t = \frac{1}{(R_2^2 - R_1^2)}\left[\frac{1}{4}(R_2^2 - R_i^2) + \frac{1}{2}R_i^2\ln\left(\frac{R_i}{R_2}\right)\right] \qquad (4.109)$$

i.e. the filtration time is given by

$$t = \frac{r\mu}{\bar{v}\rho\omega^2}\left[\frac{1(R_2^2 - R_i^2)}{2(R_2^2 - R_1^2)} + \frac{R_i^2}{(R_2^2 - R_1^2)}\ln\left(\frac{R_i}{R_2}\right)\right] \qquad (4.110)$$

Note that the volume of cake is given by

$$V\bar{v} = \pi(R_2^2 - R_i^2)b \qquad (4.111)$$

and the volume of liquor by

$$V = \frac{\pi}{\bar{v}}(R_2^2 - R_i^2)b \qquad (4.112)$$

Hydrocyclones

Hydrocyclones are used for product classification whereby the exit stream from the crystallizer is split in to two, each with crystals either predominantly above or below a cut size. The cut size is determined by the design parameters and the properties of the stream itself. These devices are extensively used because of the simplicity of their design, effectiveness and robustness. The basic principle is similar to that of a centrifuge. Slurry is caused to spin via the tangential entry to a cylindrical cone (Figure 4.22) thereby creating a high pressure at the wall and a low-pressure zone at the axis. Solid particles, if they are heavier than the liquid, are flung outwards whilst liquid migrates to the centre. A vortex finder

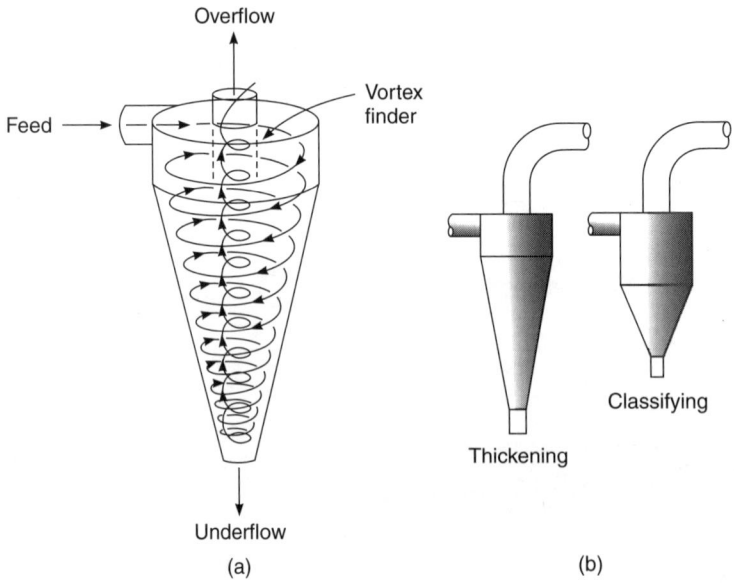

Figure 4.22 *Schematic diagram of hydrocyclones: (a) Flow pattern; (b) Alternate aspect ratios adapted for service (after Wallas, 1988)*

withdraws 'clear' liquid out of the top of the unit whilst concentrated slurry is discharged below.

More elongated devices are used for slurry thickening whilst those of more squat aspect ratio devices are suitable for product classification by size, as in fines-destruction circuits. The pressure drop in a hydrocyclone varies with the feed rate raised to a power between 2 and 3.3. The cut size is a weak function of pressure drop, varying with $\Delta p^{-0.25}$ for dilute feeds. Large pressure drops are not therefore economical. Large hydrocyclones (say 1 m diameter) operate with a pressure drop of about 1 atm. whilst small ones (10 mm diameter) operate at about 4–5 atm.

It should be noted, however, that the 'cut size' is an idealization. In practice perfect separation does not occur; some smaller particles than expected will be present in the underflow whilst some larger than predicted occur in the overflow leading to a 'blurring' of the separation in a 'grade efficiency curve' (Figure 4.23).

Theoretical representation of the behaviour of a hydrocyclone requires adequate analysis of three distinct physical phenomenon taking place in these devices, viz. the understanding of fluid flow, its interactions with the dispersed solid phase and the quantification of shear induced attrition of crystals. Simplified analytical solutions to conservation of mass and momentum equations derived from the Navier–Stokes equation can be used to quantify fluid flow in the hydrocyclone. For dilute slurries, once bulk flow has been quantified in terms of spatial components of velocity, crystal motion can then be traced by balancing forces on the crystals themselves to map out their trajectories. The trajectories for different sizes can then be used to develop a separation efficiency curve, which quantifies performance of the vessel (Bloor and Ingham, 1987). In principle, population balances can be included for crystal attrition in the above description for developing a thorough mathematical model.

Because of the complexity of the flow within hydrocyclones, however, various, largely empirical, methods for prediction of the cut size have been proposed for use in practice, as reviewed by Svarovsky (2000).

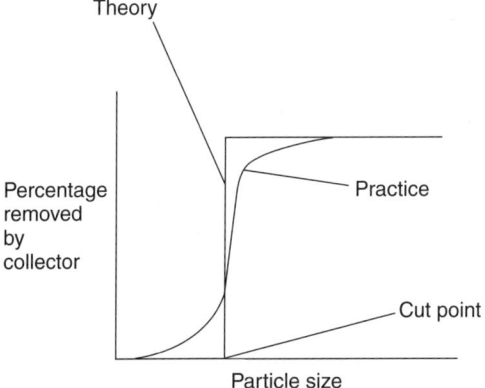

Figure 4.23 *Hydrocylone grade efficiency curve*

Crystal washing

Liquors used in industrial crystallization processes often contain impurities or other unwanted components that must be separated. Impurities may also be retained adsorbed on the crystal surface or occluded within. Indeed, crystallization is often used as a purification process, a pure crystal being the requirement. Crystal cakes and pastes arising from solid–liquid separation operations using filters and centrifuges range in content from 5 to 50 per cent, depending on the crystal size and shape characteristics, and the pressure employed, fine, irregular and agglomerated crystals having the higher values. Direct drying of such materials by evaporation will simply leave behind the impurities in the dried crystal mass. Further details concerning washing mechanics and processes are dealt with by Wakeman (1990b).

In these cases, the product is reslurried with pure liquor or fresh solvent if the solubility is not too high, and refiltered. In order to meet the required product impurity level, several such washings may take place in series. See Coulson and Richardson (1991) and Mullin (2001) for design guides and examples calculations. It is noted that impurities retained within liquid occlusions are particularly difficult to remove without first crushing the crystals.

Crystal drying

Drying usually follows dewatering (thickening, centrifugation, filtration etc.). The purposes of drying can be several, including.

- to meet customer requirements
- to facilitate handling of solids
- to facilitate product formulation (mixing, granulation etc.)
- to reduce costs of transport
- to provide particular properties (free flowing, rigidity etc.)
- to preserve product quality during storage
- to avoid corrosion of containers etc.

Solids drying often forms the final stage in an industrial crystallization process and precedes solids handling and finishing operations (conveying, milling, packing etc.). In order to minimize energy costs, drying normally concerns the removal of (relatively) small amounts of moisture from solids by evaporation.

Often, it is essential that crystals are not damaged unduly by the drying operation. And in the case of food and pharmaceutical grade materials, care should be taken to avoid contamination and powder leakage via dust formation.

Equilibrium moisture content

Crystalline materials coming forward from the filter vary in their moisture content, X, usually expressed as a percentage of the mass of the dry substance.

$$X, \text{ moisture content} = \frac{\text{mass of moisture}}{\text{mass of dry solid}}$$

If the wet material is exposed to air at a give temperature, it will either lose or absorb solvent until an equilibrium condition is established. The equilibrium vapour pressure, P_A^*, exerted by moisture in a wet solid depends on:

1. the temperature of the crystal mass
2. the moisture content of the solid
3. the nature of the solid.

Any moisture may be present in two forms

1. *Free moisture*: This is solvent in excess of the equilibrium moisture content and is relatively easily removed.
2. *Bound moisture*: This is moisture retained within the solid such that it exerts a vapour pressure less than that of free solvent (Figure 4.24). Such solvent may be adsorbed on the surface, retained in capillaries or within cells or occlusions of liquor. The latter can be difficult to remove without resorting to high temperatures, which may damage the crystals.

Bound moisture may be located in various positions within a crystal

- may be trapped in capillaries (concave interface → lower vapour pressure)
- may be trapped in cells
- may be chemically or physically bound
- may contain dissolved solids.

Trapped liquor may cause crystal fracture due to vapour pressure build-up on heating, or crystallize on drying giving rise to caking (see Chapter 9).

Drying tests

The design of driers is dependent on experimental data. Simple tests provide useful information. The normal procedure is to dry a sample (continuously

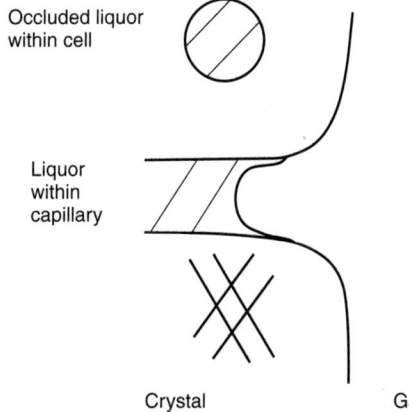

Figure 4.24 *Bound moisture within a crystal*

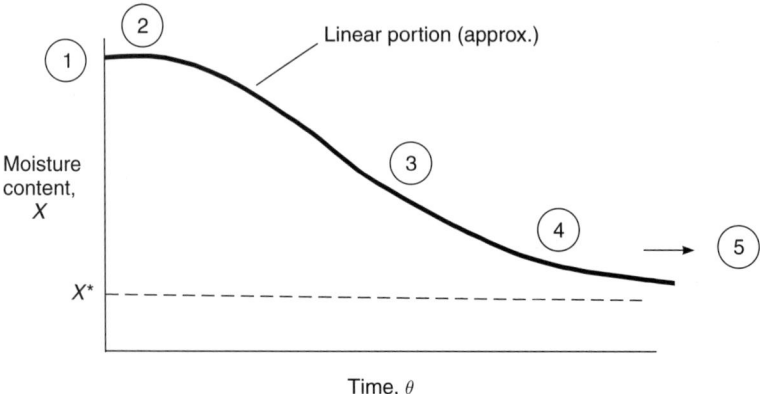

Figure 4.25 Transient moisture content during drying

weighed) that is exposed to constant drying conditions (air temperature, velocity, humidity etc.). As the drying progresses, differing drying rates are observed (Figure 4.25).

Drying rates

The rate of drying is defined by

$$N = \frac{L_S}{S}\frac{\Delta X}{\Delta t} = \frac{L_S}{S}\frac{dX}{dt} \qquad (4.113)$$

where L_S = mass of dry solid and S = drying surface area.

The rate of drying is high for a period initially (Figure 4.26) and then declines.

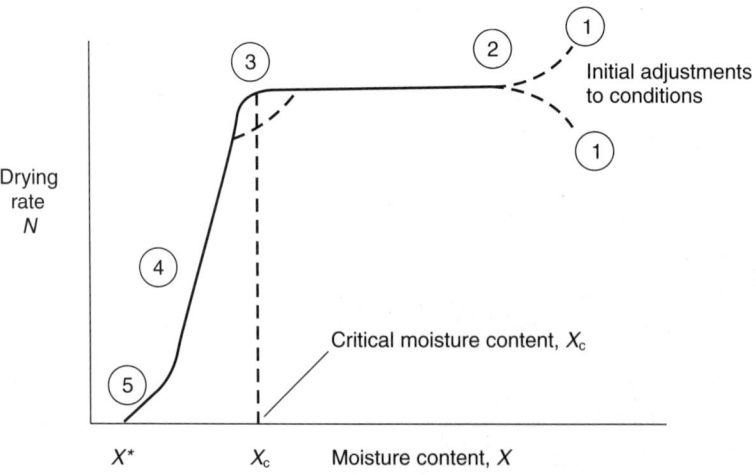

Figure 4.26 Transient drying rates

Drying periods

Referring to Figure 4.26, three main drying periods can be identified.

②⟶③ *Constant rate period*
— rate of drying dependent on rate of moisture movement within body of solid.

③⟶④ *First falling rate period*
— rate of moisture movement within the solid mass not sufficiently rapid to keep surface wet.

④⟶⑤ *Second falling rate period*
— rate of drying controlled by movement of moisture within solid mass. Often lumped with first FRP (3–4).

Total drying time

The total drying time is the sum of the durations of the constant rate and falling rate periods and is often simplified into two linear portions (Figure 4.27). The period taken for X_1 to reach X_C is known as the constant rate time, θ_{CR}, whilst that to go from X_C to X^* is the falling rate time, θ_{FR}. The sum of these two quantities gives the total drying time θ_T.

Drier design equations

In the constant drying rate period, the rate of drying is determined by the humidity driving force

$$N_{CR} = k_y(Y^* - Y) \tag{4.114}$$

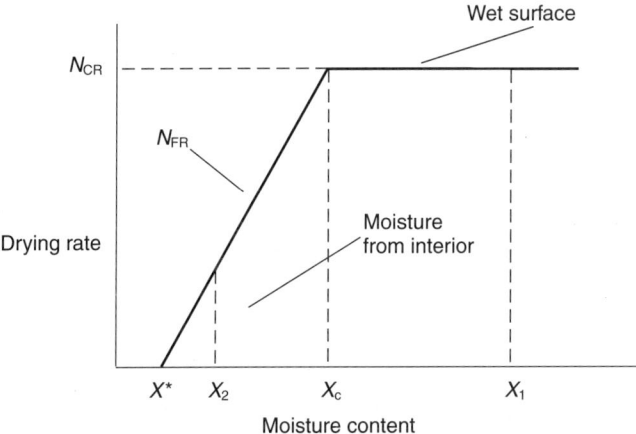

Figure 4.27 *Transient drying rates during drying*: X_1, *initial moisture content of wet solids*; X_2, *final moisture content*; X_C, *critical moisture content of wet solids and* X^*, *equilibrium moisture content of solids*

where Y is the humidity of the drying gas stream, Y^* is the saturation humidity of the gas stream at temperature T_i (temperature of wet surface) and k_y is the gas phase mass transfer coefficient.

Wet-bulb protection

In many cases the solid assumes the wet-bulb temperature of the gas stream (wet-bulb protection), particularly if the solid is in the form of granules or is spray dried. This is particularly useful for heat sensitive materials. So, at constant drying rate

$$N_{CR} = k_y(Y_W - Y) \tag{4.115}$$

where k_y is the gas phase mass transfer coefficient (which varies with velocity$^{0.8}$) and Y_W is the wet-bulb humidity of the wet solids.

Drying time

The total drying time can be obtained by integration of the drying rate equations. Thus for a batch drier:

(i) Constant rate period

$$N = \frac{L_s}{S}\frac{dX}{dt} \tag{4.116}$$

by integration

$$\tau_{CR} = \frac{L_s}{Sk_y}\frac{(X_1 - X_c)}{(Y_w - Y)} \tag{4.117}$$

(ii) Falling rate period (linear)

$$\tau_{CR} = \frac{L_s}{Sk_y}\frac{(X_1 - X^*)}{(Y_w - Y)}\ln\frac{(X_c - X^*)}{(X_2 - X^*)} \tag{4.118}$$

Total drying time

$$\tau_T = \tau_{CR} + \tau_{FR} \tag{4.119}$$

Once the drying time is known, equations for the required drier area can be formulated for various configurations.

Types of drier

The heat required to evaporate moisture may be obtained by convection from the gas stream, by conduction e.g. on heated trays, in pneumatic risers, fluidized beds, spray towers, or by radiation. A typical fluidized bed drier is depicted in Figure 4.28. The hydrodynamic principles of fluidization were summarized in Chapter 2. Within a fluid bed drier, heated air, or other hot

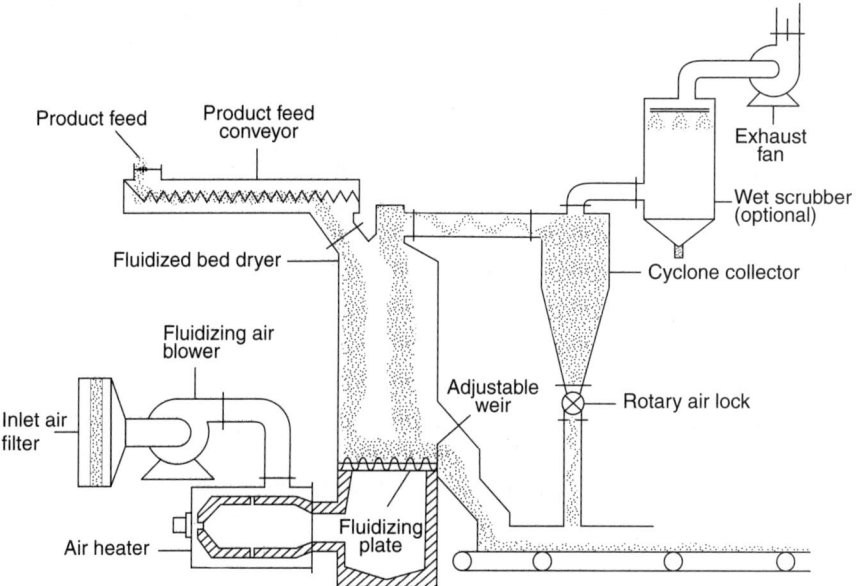

Figure 4.28 *Fluidized bed drier system (after Coulson and Richardson, 1991)*

gas, is passed into the base of a vessel into which wet crystals are fed from the top. The particles are suspended in the gas stream and quickly mix with the already dried material within the vessel. Product material flows out over a weir. Any fine crystals entrained within the gas stream are collected by a cyclone and added to the product stream. Clark (1967) proposed a simple method for calculation of fluidized bed cross-sectional area, with an example calculation presented by Coulson and Richardson (1991). Often, however, it is essential that crystals are not damaged unduly by the drying operation. Clearly, passage of crystals in an air stream through a drier may induce unwanted crystal fragmentation and attrition (see Chapter 4). For fragile crystals, simple drum, shelf or tray driers may be the equipment of choice. It is common practice to test wet crystal samples on a variety of suppliers' equipment before deciding on a particular solution.

In the case of food and pharmaceutical grade materials, care should be taken to avoid contamination between batches and powder leakage via dust formation.

Summary

In this chapter simple design equations have been derived for solid–liquid gravity separation using the transport equations of Chapter 2. The design of simple settling tanks used to clarify relatively dilute suspensions is based on the particle terminal settling velocity, whereas for thickeners relationships appropriate to hindered settling are employed. While analytic expressions are available for settling rates of ideal dilute suspensions of spherical particles, in practice settling rates of real slurries are determined experimentally for use in design calculations.

Simple design equations have also been derived for solid–liquid separation. For idealized systems of uniform particles, filter cake resistance can be calculated from the equations for flow through packed beds. Again in practice, however, resistance of both cake and cloth are determined experimentally, using real slurry. For solid bowl type centrifuges and hydrocyclones used to clarify relatively dilute suspensions, the design is based on the terminal particle settling velocity while for perforated basket/screen type centrifuges filtration relationships appropriate to flow through packed beds are appropriate. Whilst analytic expressions are available for ideal dilute suspensions and beds of spherical particles respectively, in practice settling rates, filtrabilities and separation efficiency curves of real slurries are again often best determined experimentally.

Similarly, a variety of drier types are available to remove residual moisture content for which tests are normally required prior to final equipment selection. In each of these operations, care should be taken to ascertain the extent of particle size change due to attrition and fragmentation, or, in some cases, granulation within the downstream units.

5 Crystal formation and breakage

The primary particle formation processes occurring during crystallization and precipitation are firstly nucleation, which determines the initial formation of crystals, and secondly crystal growth, which determines their subsequent size. Important secondary processes can also occur, however. Particle disruption can occur by which existing particles are broken down into a larger number of smaller fragments. The latter process is also related to secondary nucleation, an additional mode of nucleation particularly prevalent in industrial crystallizers. A further growth process is that known as agglomeration and is considered in Chapter 6. Study of the kinetics of these particle formation processes is important since, when coupled with residence time distribution, they determine product particle characteristics from industrial crystallizers, akin to chemical kinetics determining the product spectrum from reactors.

Nucleation and growth mechanisms are described in detail in texts on precipitation (Söhnel and Garside, 1992), and crystallization (Mullin, 2001; Mersmann, 2001 and Myerson, 2001). Granulation is considered in Sherrington and Oliver (1981), Stanley-Wood (1990) and Pietsch (1991) whilst comminution is dealt with in Prior *et al.* (1990).

Crystallization and precipitation kinetics

Initially, crystallization is a two-step process viz. nucleation and crystal growth requiring a change of free energy (Gibbs, 1928), as shown schematically in Figure 5.1.

Both nucleation and growth of crystals depend on the degree of supersaturation, but usually to different 'orders', nucleation exhibiting the stronger dependence and higher free energy requirement. Crystals can subsequently be disrupted or agglomerate, however, thereby further changing both their particle size and numbers.

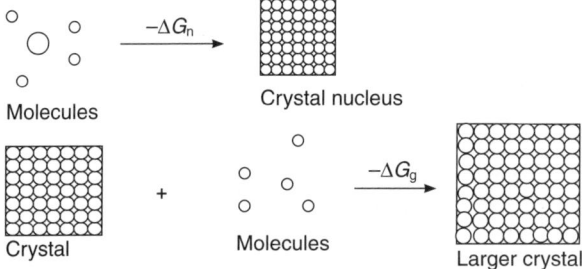

Figure 5.1 *Schematic representation of crystallization*

Nucleation

Nuclei are the first formed embryos, possibly of only a few nanometers in size, which subsequently grow to produce tangible crystals. Nucleation is thus the first formation of the solid phase. It occurs due to clustering and aggregation of molecules or ions in a supersaturated melt, solution or vapour to a size at which such entities become viable i.e. they will grow rather than redissolve. Two modes of nucleation are distinguished (viz. primary and secondary) together with several mechanisms, as illustrated in Figure 5.2.

Some examples, which illustrate these different nucleation modes, and the conditions under which they apply, are given by the ubiquitous ice-water system:

1. carefully purified water, distilled and filtered: cool to below $-30\,°C$ before ice forms due to *primary homogeneous nucleation*
2. tap water: ice forms at about $-6\,°C$ as a consequence of *primary heterogeneous nucleation*
3. continuous crystallization of ice in retained bed crystallizer will operate at -2 or $-3\,°C$ giving rise to *secondary nucleation*.

Nucleation can be divided into two distinct types:

Primary nucleation (nucleation without crystalline matter)

- homogeneous nucleation (spontaneous nucleation from clear solution)
- heterogeneous nucleation (induced by foreign particles)

Secondary nucleation (induced by presence of existing crystals)

- contact (with other crystals or the crystallizer parts)
- shear (due to fluid flow)
- fracture (due to particle impact)
- attrition (due to particle impact or fluid flow)
- needle (due to particle disruption).

Each particle formation process is now considered in turn.

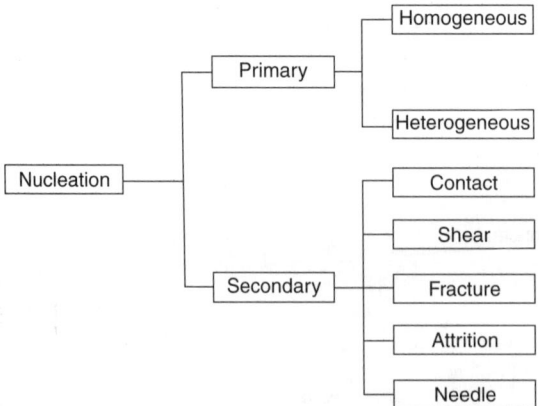

Figure 5.2 *Modes and mechanisms of nucleation*

Primary nucleation

Primary nucleation is the 'classical' form of nucleation. It occurs mainly at high levels of supersaturation and is thus most prevalent during unseeded crystallization or precipitation. This mode of nucleation may be subdivided into either *homogeneous* viz. spontaneously from clear solution, or *heterogeneous* viz. in the presence of 'dust' particles in suspension, or solid surfaces.

Homogeneous nucleation

The process of homogeneous nucleation is determined by the formation of stable nuclei in a supersaturated solution.

Gibbs considered the change of free energy during homogeneous nucleation, which leads to the classical nucleation theory and to the Gibbs–Thompson relationship (Mullin, 2001).

$$B^0_{\text{hom}} = A_{\text{hom}} \exp\left[-\frac{16\pi\gamma^3 v^2}{3k^3 T^3 (\ln S)^2}\right] \tag{5.1}$$

where S is the supersaturation ratio c/c^*, c is the solution concentration and c^* is the equilibrium saturation concentration.

Heterogeneous nucleation

Heterogeneous nucleation is induced by foreign nuclei or surfaces present in the solution and becomes significant at lower supersaturation levels. However, it is often difficult to distinguish between homogeneous and heterogeneous nucleation. In analogy to homogeneous nucleation, heterogeneous nucleation can be described by means of a relationship similar to equation 5.1 (Söhnel and Garside, 1992)

$$B^0_{\text{het}} = A_{\text{het}} \exp\left[-\frac{16\pi\gamma^3 v^2 f(\varphi)}{3k^3 T^3 (\ln S)^2}\right] \tag{5.2}$$

with the factor $f(\varphi)$ accounting for the decreased energy barrier to nucleation due to a foreign solid phase.

Unfortunately, both primary nucleation parameters cannot be predicted *a priori* as yet and in practice the nucleation rate must be measured and correlated empirically for each system.

Crystal growth

Crystal growth is a diffusion and integration process, modified by the effect of the solid surfaces on which it occurs (Figure 5.3). Solute molecules/ions reach the growing faces of a crystal by diffusion through the liquid phase. At the surface, they must become organized into the space lattice through an

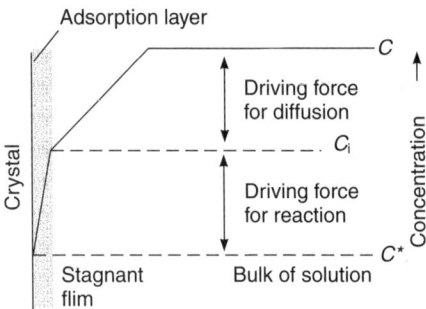

Figure 5.3 *Growing crystal-solution interface*

adsorbed layer (Volmer, 1939). Neither the diffusion step nor the interfacial step, however, will proceed unless the solution is supersaturated. The rate of crystal growth can be expressed as the rate of displacement of a given crystal surface in the direction perpendicular to the face. Different crystallographic faces of a crystal, however, usually have different linear growth rates. Variations occur in the shape of crystals when individual faces grow at different rates, the overall crystal habit being determined by the slowest growing face (Mullin, 2001).

Growth of crystals from solution thus involves two major processes:

1. Mass transport from the solution to the crystal surface by diffusion, convection, or a combination of both mechanisms.
2. Incorporation of material into the crystal lattice through surface integration also described as the surface reaction process.

The second step can be subdivided into a number of stages:

(i) Adsorption of the growth unit first on to the crystal surface.
(ii) Release of part of its solvation shell, after which the growth unit diffuses into the adsorption layer until it is either incorporated into the lattice or leaves the adsorption layer and returns to the solution.
(iii) If the growth unit reaches a point where it can be built into the lattice, it loses the remaining of its solvation shell before final lattice incorporation.

Since the kinetic processes occur consecutively, the solution concentration adjusts itself so that the rates of the two steps are equal at steady state. In most cases, more than one mechanism influences a crystal's growth rate. If the different mechanisms take place in parallel, then the mechanism resulting in the faster growth controls the overall rate. If the processes take place in series, as in the case of bulk diffusion followed by surface reaction, then the slower mechanism will control the overall rate.

Crystal growth rate may be expressed either as a rate of linear increase of characteristic dimension (i.e. velocity) or as a mass deposition rate (i.e. mass flux). Expressed as a velocity, the overall linear crystal growth rate, G ($=dL/dt$ where L is the characteristic dimension that is increasing). The rate of change of a crystal mass M_c is related to the rate of change in crystal characteristic dimension L by the equation

$$\frac{dM_c}{dt} = \frac{d(\rho_c f_v L^3)}{dt} = 3\rho_c f_v L^2 \frac{dL}{dt} \tag{5.3}$$

where ρ_c is the crystal density and f_v is the volume shape factor, assumed constant. Since the area shape factor, f_a is defined by the equation

$$f_a = \frac{A_c}{L^2} \tag{5.4}$$

and G is defined as dL/dt, then

$$\frac{dM_c}{dt} = 3\rho_c \left(\frac{k_v}{k_a}\right) A_c G \tag{5.5}$$

where the growth rate, G, is the rate of change of the characteristic dimension

$$G = \frac{dL}{dt} \tag{5.6}$$

Systems are said to follow McCabe's ΔL law (McCabe, 1929a,b) if they exhibit this behaviour, if not they are said to exhibit anomalous growth.

Crystal growth may occur by a variety of mechanisms for which several theoretical models have been proposed as reviewed in detail by O'Hara and Reid (1973), Söhnel and Garside (1992) and Mullin (2001).

Diffusion-reaction model

Berthoud (1912) proposed a model in which growth is divided into two stages, i.e. diffusion through the bulk up to the interface

$$\frac{dM_c}{dt} = k_d A(c - c_i) \tag{5.7}$$

and subsequent 'reaction' incorporating the growth units into the crystal

$$\frac{dM_c}{dt} = k_r A(c_i - c^*)^r \tag{5.8}$$

Elimination of the unknown interfacial concentration c_i leads to

$$\frac{dM_c}{dt} = k_g A(c - c^*)^g \tag{5.9}$$

whence

$$\frac{1}{k_g} = \frac{1}{k_d} + \frac{1}{k_r} \tag{5.10}$$

giving the familiar relationship between two resistances in series.

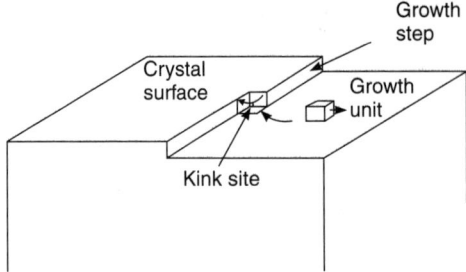

Figure 5.4 *Attachment of a growth unit into a kink site*

Surface integration models

Several theories of crystal growth by surface integration have been proposed.

Continuous growth model

In this model, integration of growth units on a rough surface takes place where their energy demand for orientation is lowest, such as at a kink site (Kossel, 1934), as depicted in Figure 5.4. Continuous crystal growth at kink sites will cease, however, as soon as the deformities heal over, so another mechanism is required to reconcile the observed higher crystal growth rates.

Screw dislocation, or BCF, model

In order to overcome the limitation of the continuous growth model, Frank (1949) recognized the significance of the screw dislocation, which presents a continuous spiral during growth (Figure 5.5). The BCF model (Burton *et al.*, 1951) is based on the assumption that growth occurs along such screw dislocations forming spiral staircases. Following a kinematic analysis of the motion of the growth steps, the crystal growth rate is given by the expression

$$G \propto \sigma^2 \tan h\left(\frac{k_{BCF}}{\sigma}\right) \qquad (5.11)$$

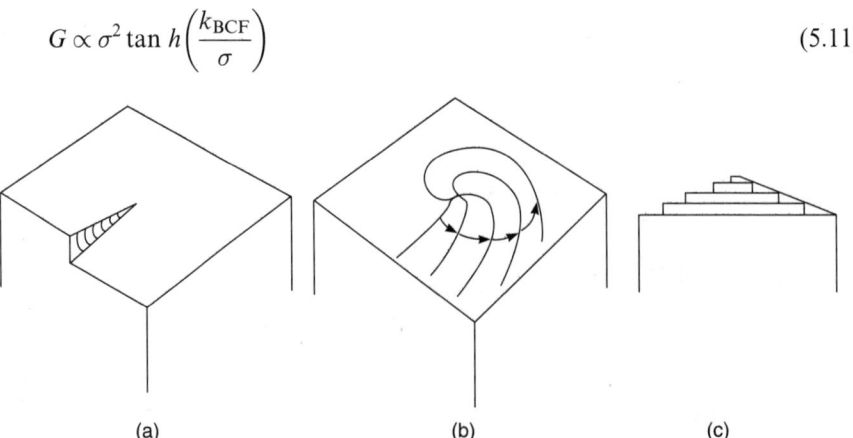

Figure 5.5 *Development of a crystal growth spiral staring from a screw dislocation*

The form of equation 5.13 varies with the level of supersaturation. At low supersaturations it leads to a parabolic (second order) relationship whilst a linear one is predicted at higher supersaturation.

Birth and spread (B&S) model

O'Hara and Reid (1973) proposed that the activation energy for growth on a smooth surface can be overcome if it is assumed that the formation and growth of surface nuclei occur (Figure 5.6).

$$G \propto \sigma^g \exp\left(-\frac{k_{B+S}}{\sigma}\right) \tag{5.12}$$

This mechanism is therefore more likely to occur at high levels of solution supersaturation.

The difficulty with each of the theoretical approaches to date, however, is that they cannot yet predict crystal growth rate coefficients and exponents for a particular substance *a priori*. Thus as with nucleation kinetics, crystal growth rate data from industrial crystallizers are usually correlated empirically with environmental conditions, such as concentration and temperature using a power law model of the form

$$\frac{dL}{dt} = G = k_g \sigma^g \tag{5.13}$$

From the theoretical considerations above, for diffusion controlled growth $g = 1$, for crystal growth originating from screw dislocations $g = 1 - 2$ and for polynuclear growth $g > 2$.

Temperature strongly affects crystal growth rate as it can significantly affect the relative rates of the diffusion and surface integration steps. The effect can be sufficient to produce diffusion-controlled growth at high temperatures, compared to integration step controlled at low temperature. The rate of precipitation often increases at high temperature, while the crystal size, shape and type can all change with temperature.

Temperature of the system has a pronounced effect on the growth rate. The relation between growth kinetics and temperature is often given by Arrhenius expression

$$k_G = k_G^0 \exp\left[-\frac{\Delta E_G}{RT}\right] \tag{5.14}$$

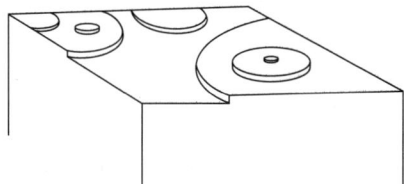

Figure 5.6 *Crystal growth arising from a surface nucleation ('birth and spread') mechanism*

where k_G is a growth rate coefficient of the type required in equation 5.14, k_G^0 is a constant and E_G is the activation energy for growth.

Impurities usually cause a reduction in the growth rates of crystalline materials due to blocking of kink sites, thereby leading to smaller crystals than required. This is a common problem, which is often related to contamination of the feed solution. In some other cases, however, impurities can enhance growth rates, thought to be due to a reduction in interfacial tension and hence increase in surface nucleation rates.

Similarly a change of solvent, or addition of a co-solvent can lead to a change in crystal growth rates. One mechanism by which this may occur is as a consequence of the effects of solvent on mass transfer of the solute via changes in solution viscosity, density and diffusivity. Another mechanism is via changes to the structure of the interface between crystal and solvent, as with some impurities. A solute–solvent system with a higher solubility is likely to produce a rougher interface and concomitantly larger crystal growth rates. Lahav and Leiserowitz (2001) provide an extensive account of the effect of solvent on crystal growth and morphology.

McCabe's (1929a,b) ΔL law states that crystals of the same substance growing under the same conditions should grow at the same rate. Experimental evidence has shown that this law is frequently violated. The growth rate of a crystal face, for example, and the instantaneous velocity of steps spreading across the surface of a crystal have been shown to fluctuate with time, even though external conditions, e.g. temperature, supersaturation and hydrodynamics, remain constant.

Growth rate fluctuations appear to increase with an increase in temperature and supersaturation leading to crystals of the same substance, in the same solution at identical supersaturation, exhibiting different growth rates; this is thought to be a manifestation of the phenomenon of either *size-dependent crystal growth* or alternatively, *growth rate dispersion*.

Possible reasons for anomalous growth include size dependence of solubility, changes in the activity of the dominant group of dislocations during crystal growth; changes in the position of dislocations relative to the faces during growth and also due to mechanical stress on the crystal.

Correlation of the apparent effect of crystal size on growth rate is often attempted by use of empirical expressions of the form

$$G = G^0(1 + \gamma L)^b \quad b < 1 \tag{5.15}$$

where the parameters G^0, γ and b are determined from experimental data.

Hydrodynamic determinants of crystal growth rates

Liquid velocity around a particle affects its mass transfer boundary layer (Figure 5.3) and hence the mass flux. In forced convection, the dependence of the mass transfer coefficient on hydrodynamics is given by the Frössling equation

$$\mathrm{Sh} = 2 + \phi \mathrm{Re}_p^a \mathrm{Sc}^b \tag{5.16}$$

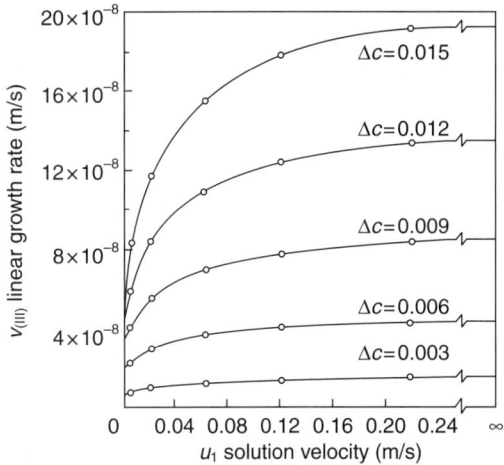

Figure 5.7 *Effect of solution velocity on crystal growth rates (after Mullin and Garside, 1967)*

where the Sherwood number $Sh = kL/D$, the particle Reynolds number $Re_p = \rho u D/\mu$, the Schmidt number $Sc = \mu/\rho D$, k is the mass transfer coefficient, L is the particle size, D the solution diffusivity, ρ the solution density, μ the solution viscosity and u the solution relative velocity.

This dependence of the mass flux can be reflected in observed velocity dependence of crystal growth rates (Figure 5.7). Thus, if a crystals growth rate is determined by volume diffusion, it is affected by solution velocity and is thus subject to change during scale up if local solution velocities vary. This effect is most noticeable at high supersaturation during which the surface integration step controls the overall crystal growth rate.

Induction periods

The time elapsed from the creation of the initial supersaturation to the detection of the first crystals formed in the system is known as the induction period. The level of supersaturation attained is then akin to the 'metastable limit'. Neither quantity (viz. the induction time and metastable limit) is therefore a fundamental quantity. Both are useful measures, however, of the propensity of a solution to nucleate. Measurement of the induction time as a function of supersaturation can be used to help determine crystallization kinetics and mechanism. Thus, the induction time may be expressed by (Walton, 1967)

$$t_i = t_n + t_g \tag{5.17}$$

where t_i is the time taken for a critical nucleus to grow to a detectable size. The time to grow to a crystal of detectable size, t_g, can be estimated, at least in principle, from a kinetic expression describing crystal growth by a particular mechanism. The nucleation time, t_n, is harder to estimate but it can be assumed that supersaturation is attained instantaneously and does not change throughout the induction period. This is only valid, however, if the duration of the non-steady state nucleation is negligible, i.e. when the steady-state nucleation rate is reached very quickly. Indeed, Söhnel and Mullin (1988) have shown that non-steady state nucleation is not an important factor during the formation of crystal electrolytes from aqueous solutions, at least at moderate supersaturation and viscosity, irrespective of whether there is heterogeneous or homogeneous nucleation occurring.

The average time for the formation of a critical nucleus following the statistical concept of nucleation may be expressed as

$$\bar{t} = J^{-1} \qquad (5.18)$$

where J is the rate of nucleation and $\bar{t}$ is the average induction time.

Note, however, that when the non-steady state nucleation period cannot be ignored then equation 5.18 is no longer applicable (Janse and de Jong, 1978), and it is then necessary to include the non-steady state nucleation period as part of the overall induction time.

From equation 5.18 three separate cases may exist:

1. $t_n \gg t_g$
2. $t_g \gg t_n$
3. $t_n \cong t_g$

These cases will now be briefly considered in turn.

Case 1: $t_n \gg t_g$

The induction time is determined by the time needed for the formation of a critical nucleus in the system. Then assuming a steady-state nucleation rate, J, (Walton, 1967)

$$J = A \exp\left[\frac{\beta V_m^2 \gamma_{SL}^3 f(v) N_A}{v^2 (RT)^3 \ln^2 S}\right] \qquad (5.19)$$

where β is a geometrical shape factor; V_m is the molar volume of the solid; γ_{SL}, the solid–liquid interfacial tension; $f(v)$, the correction factor of heterogeneous nucleation; N_A, the Avagadro's number; v, the number of ions in a molecular unit; R, the universal gas constant; T, the absolute temperature and S, the supersaturation ratio.

When $f(v) = 1$, nucleation is said to be homogeneous, and when $f(v) < 1$, heterogeneous nucleation occurs. If a steady-state nucleation rate occurs in the system, substitution of equation 5.18 into equation 5.19 and subsequent rearrangement gives

$$\log t = \frac{B}{T^3} \log^2 S - A \qquad (5.20)$$

where

$$B = \frac{\beta \gamma_{SL} V_m^2 N_A f(v)}{(23R)^3 v^2} \qquad (5.21)$$

and
$$A = vkT \ln S \tag{5.22}$$
where k is the Boltzmann constant.

Case 2: $t_g \gg t_n$

Here the induction period is determined by the time needed for a critical nucleus to grow to a detectable size. An expression for the induction period may be found by integrating the respective kinetic expressions for crystal growth between the characteristic dimensions of a critical nucleus, r_x, and a detectable crystal, r_y. And since $r_y \gg r_x$, the following equations may be defined.

Screw-dislocation growth

$$t_i = \left(\frac{r_y}{k_g}\right)(c - c^*)^{-g} \tag{5.23}$$

where $c - c^*$ is the concentration driving force.

Polynuclear growth

$$t_i = \frac{r_y}{Cf(S)} \exp\left[\frac{\beta' V_m^{\frac{4}{3}} \gamma_{SL}^3 N_A^{\frac{2}{3}}}{3(RT)^2 v \ln S}\right] \tag{5.24}$$

where

$$\beta' = \frac{p^2}{4\beta_S} \tag{5.25}$$

$$C = DV_m^{\frac{1}{3}} c^{*\frac{2}{3}} N_A \tag{5.26}$$

and

$$f(S) = S^{\frac{7}{6}}(S-1)^{\frac{2}{3}}(\ln S)^{\frac{1}{6}} \tag{5.27}$$

where β' is a geometrical factor; p, the perimeter shape factor; β_S, the surface shape factor; and D, the diffusion coefficient in solution.

Mononuclear growth

$$t_i = \left(\frac{d^3}{6D_S r_x}\right) \exp\left[\frac{\beta' V_m^{\frac{4}{3}} \gamma_{SL}^2 N_A^{\frac{2}{3}}}{(RT)^2 v \ln S}\right] \tag{5.28}$$

where d is the interplanar distance in solid phase; and D_S, the diffusion coefficient on a surface.

Case 3: $t_n \cong t_g$

In this case the induction period is determined by both nucleation and crystal growth (Nielsen, 1969; Söhnel and Mullin, 1978).

Nucleation followed by diffusional growth

$$t_i = \frac{V_m^{\frac{2}{3}}}{2Dx^{\frac{1}{3}}N_A^{\frac{2}{3}}} \exp\left[\frac{2\beta V_m^2 \gamma_{SL}^3 N_A}{5v^2(RT)^3 \ln^2 S}\right] \tag{5.29}$$

where x is the number of building units arriving on a unit surface of a nucleus per unit time (Nielsen, 1969).

Nucleation followed by polynuclear growth (Nielsen, 1969; Söhnel and Mullin, 1978)

$$t_i = \left(\frac{3}{2\pi}\right)^{\frac{1}{4}} \left(\frac{V_m^{\frac{5}{3}}}{N_A^{\frac{8}{3}} D^4 c^*}\right)^{\frac{1}{4}} \left(\frac{S}{(S-1)^2}\right)^{\frac{1}{4}} \exp\left[\frac{\beta V_m^2 \gamma_{SL}^3 N_A}{4v^2(RT)^3 \ln S} + \frac{\beta' V_m^{\frac{4}{3}} \gamma_{SL}^2 N_A^{\frac{2}{3}}}{4v(RT)^2 \ln S}\right] \tag{5.30}$$

For a limited supersaturation range, linearity is generally found on a log-log plot of t_i versus $f(S)$, but when a wide range is investigated a break in the induction period curve usually occurs (Nielsen, 1969; Nielsen and Söhnel, 1971; Söhnel and Mullin, 1988). This change of slope is normally attributed to the fact that at low supersaturation nucleation is mainly heterogeneous ($f(v) < 1$) whereas at high supersaturation primary homogeneous nucleation predominates ($f(v) = 1$) thereby facilitating discrimination of the crystallization mechanism (Söhnel and Mullin, 1988).

Figure 5.8 shows a typical plot of log (t_{ind}) versus (log S)$^{-1}$ for Cyanazine™ crystallized form aqueous ethanol solution.

The shape of the two part plot in Figure 5.8 can be attributed to homogeneous nucleation at high levels of supersaturation ($f(v) = 1$) for the steeper, linear curve, while the second part of the curve is consistent with heterogeneous nucleation at low supersaturation ($f(v) < 1$). The inferred mechanism

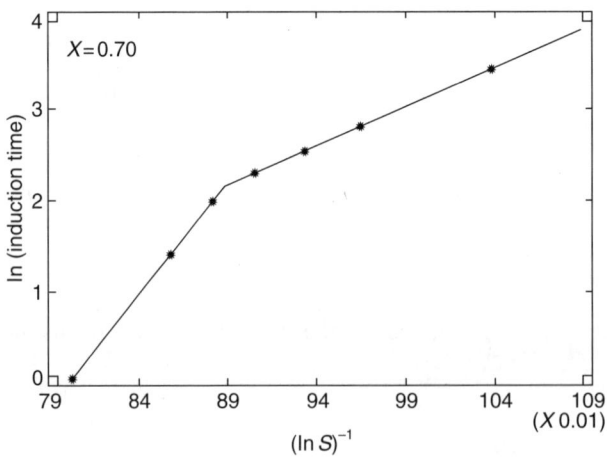

Figure 5.8 *Induction time-supersaturation plot for cyanazine in 70% w/w aqueous ethanol at 20°C (Hurley et al., 1985)*

controlling the induction period is thus consistent with primary nucleation followed either by polynuclear or diffusion controlled growth ($t_n \cong t_g$).

Interfacial tension

Induction period measurements can also be used to determine interfacial tensions. To validate the values inferred, however, it is necessary to compare the results with an independent source. Hurley et al. (1995) achieved this for Cyanazine™ using a dynamic contact angle analyser (Cahn DCA312). Solid–liquid interfacial tensions estimated from contact angle measurements were in the range 5–12 mJ/m² which showed closest agreement with values (4–20 mJ/m²) obtained from the log-log plots of induction time versus supersaturation based on the assumption of $t_n \cong t_g$.

Determination of crystal growth and nucleation kinetics

In addition to induction time measurements, several other methods have been proposed for determination of bulk crystallization kinetics since they are often considered appropriate for design purposes, either growth and nucleation separately or simultaneously, from both batch and continuous crystallization. Additionally, Mullin (2001) also describes methods for single crystal growth rate determination.

Batch methods

Several authors developed methods for the determination of growth kinetics of added seed crystals by measuring the change in solution concentration and crystal mass. This method assumes the absence of nucleation during an experiment with an estimated mean level of supersaturation within the metastable limit (e.g. Tanimoto et al., 1964; Bujac and Mullin, 1969; Jones and Mullin, 1973). Alternatively, initial derivatives of concentration and supersaturation may be employed (Garside et al., 1982). The overall mass growth rate, R_G, is related to the linear growth rate, G, and the change in concentration dc/dt by

$$R_G = k_g A \Delta c^g = \frac{1}{A}\frac{dM_c}{dt} = \frac{3f_v \rho_c}{f_a} G = -\frac{M_h}{A}\frac{dc}{dt} \tag{5.31}$$

Schierholtz and Stevens (1975), Noor and Mersmann (1993) and Chen et al. (1996) determined nucleation rates by integrating the total crystal number formed over a period and related it to an estimate of supersaturation in the precipitation of calcium carbonate, barium carbonate and barium sulphate respectively.

Several authors have presented methods for the simultaneous estimation of crystal growth and nucleation kinetics from batch crystallizations. In an early study, Bransom and Dunning (1949) derived a crystal population balance to analyse batch CSD for growth and nucleation kinetics. Misra and White (1971), Ness and White (1976) and McNeil et al. (1978) applied the population balance to obtain both nucleation and crystal growth rates from the measurement of crystal size distributions during a batch experiment. In a refinement, Tavare and

Garside (1986) applied the laplace transformation to the population balance. Qui and Rasmusson (1991) and Nývlt (1989), respectively, measured solution concentration during seeded batch cooling crystallizations and determined the crystal size distribution from which growth and nucleation rates were determined. Witkowski *et al.* (1990) used a non-linear parameter estimation technique to estimate nucleation and growth rates based on solution concentration and light obscuration measurements. Gutwald and Mersmann (1990) estimated growth and nucleation rates from constant supersaturation controlled batch crystallization. Qui and Rasmuson (1994) proposed a direct optimization method based on solution concentration and product CSD data from seeded batch experiments. Aoun *et al.* (1999) reviewed methods for determining precipitation kinetics and also presented a method for the simultaneous determination of growth and nucleation kinetics from batch experiments.

Continuous methods

Similarly, several authors have presented MSMPR methods for kinetics determination from continuous crystallizer operation (Chapter 3), which have become widely adopted. In an early study, Bransom *et al.* (1949) anticipated Randolph and Larson (1962) and derived a crystal population balance to analyse the CSD from the steady state continuous MSMPR crystallizer for growth and nucleation kinetics. Han (1968) proposed a method of kinetics determination from the moments of the CSD from a cascade of continuous crystallizers and assessed the effect of sample position. Timm and Larson (1968) suggested the use of the extra information present in transient response data to determine kinetics, followed by Sowul and Epstein (1981), Daudey and de Jong (1984) and Jager *et al.* (1991). Tavare (1986) applied the s-plane analysis to the precipitation of calcium oxalate, again assuming nucleation and growth only.

It is often observed, however, that crystals do not exist as discrete entities but comprise agglomerated particles which can confound simple MSMPR data for kinetic analysis determination (Budz *et al.*, 1987b; Hostomský and Jones, 1991). Again, broadly speaking, two types of crystal agglomeration can be distinguished viz. primary (as a result of 'mal-growth' of crystals), and secondary (as a consequence of crystal/crystal collisions; see Jones, 1989, 1993). A further complication occurs in the assessment of crystallization kinetic data if crystal attrition or breakage occurs. The processes of particle disruption and agglomeration and disruption are considered in detail in Chapter 6.

Particle breakage processes

Particle formation can also occur via breakage processes that start with existing particles and form new, smaller, ones of varying sizes. *Comminution* is the generic name for particle size reduction and generation whilst *secondary nucleation* occurs due to the presence of existing crystals in supersaturated solutions. Thus, the same process can occur within crystallizers giving rise to the more complex process of secondary nucleation.

Nature of comminution

Many crystalline products, including fine chemicals, foodstuffs and pharmaceuticals, require a final particle size that is significantly smaller than that produced during the crystallization or precipitation step. One way of achieving the required particle size is to employ a subsequent size-reduction step using some form of comminution device, frequently a mill.

Comminution can occur by several mechanisms including:

- crushing
- grinding
- milling.

to form new particles of small size from existing large particles.

The basic principle is simple. A solid particle is struck and the energy transferred causes the particle to fragment creating a size distribution of smaller particles (Figure 5.9).

There are thus two important aspects to comminution viz. the energy consumption by the feed and the size distribution of the product.

The process of comminution reduces particle size (μm), thereby

- increasing specific surface area (m^2/kg), but also
- creates fines – which can be a substantial dust nuisance.

The mechanisms of comminution are complex involving breakage along particle cracks and fissures etc., and depend on the hardness and structure of the feed particle. The Institution of Chemical Engineers (London) produced a major report on comminution (IChem^E, 1975), which was followed by reviews by Bemrose and Bridgwater (1987), Prior et al. (1990) and Jones (1997). These reviews included sections on both the fundamental and practical aspects of comminution and attrition in process equipment, test methods and an extensive list of references.

Single particle breakage

All large lumps or particles contain cracks, microcracks or lines of weakness. Any normal particle structure will contain imperfections, dislocations and

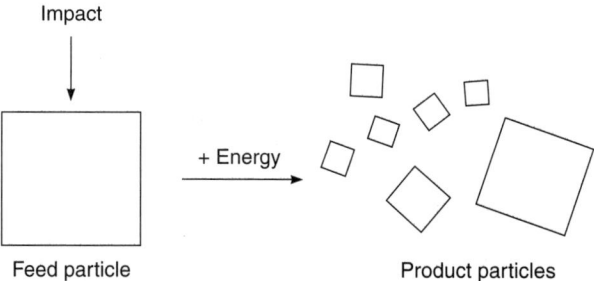

Figure 5.9 *Principle of comminution*

impurities. These defects give rise to non-uniform mechanical behaviour and hence to sources of breakage in a structure (Lawn, 1983). Particle deformation may be elastic and reversible according to Hooke's law, or irreversible normally resulting in eventual breakage. In slow crushing, as may occur in moving-bed systems or in equipment such as rock crushing mills, particle deformation can be considered isothermal, whereas with higher impact velocities the local conditions may include large temperature changes. During elastic deformation, the energy stored in a particle is available for the release of cracks (Lawn and Wilshaw, 1975; Schonert, 1972).

If a flaw exists in the particle, then when some critical value of the term

$$\frac{\sigma^2 a}{E} \tag{5.32}$$

is reached, the crack will be propagated (the Griffith criterion, Griffith, 1920, 1924). Here, σ denotes stress, a the crack length and E the Young modulus. Only if the critical value is achieved before the yield strength of the particle is reached will fracture, or crack propagation, occur. Otherwise, increasing the stress causes plastic deformation. The formula indicates that fracture requires higher stresses as the flaw length decreases, and also implies a limiting particle size below which fracture due to crack propagation is impossible. It is well known that the breakage strength of particles increases markedly as the diameter decreases because the size of flaws and microcracks becomes smaller at smaller particle sizes, and there is a change from brittle to plastic behaviour (Hess and Schonert, 1981; Schonert *et al.*, 1972).

Typical comminution process

A typical comminution process is depicted in Figure 5.10. A particular feature that comminution shares with granulation is the use of high solids recycle ratios to achieve the desired product particle size distribution.

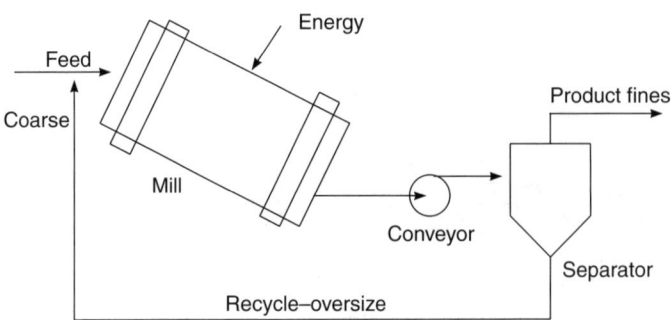

Figure 5.10 *Closed-circuit grinding mill*

Energy for size reduction

In comminution processes, energy consumption is often the most important design consideration. In fact, it has been estimated that 1 per cent of global energy consumption is used in comminution. Energy consumption is a function of the size and hardness of the material and the required degree of breakage or surface area formation. Empirically

$$\frac{dE}{dD} = -kD^n \qquad (5.33)$$

where E = energy [MJ/kg], D = particle size [mm], k = function of machine design and material of construction, $-1 > n > -2$.

Note that this relationship indicates that it is more difficult to break down fine than coarse material.

Three specific cases have been proposed:

Rittinger's law (von Rittinger, 1867)

This law assumes that the power required is proportional to the increase in specific surface area (i.e. energy $\propto 1/D$; $n = -2$)

$$E_r = k_r \left[\frac{1}{D_2} - \frac{1}{D_1} \right] \qquad (5.34)$$

where k_r = Rittinger's constant.

Kick's law (Kick, 1885)

Assumes that the power required is constant with size reduction ratio (e.g. the energy required for 1 mm $\rightarrow$ 0.5 mm $\equiv$ 10 cm $\rightarrow$ 5 cm; $n = -1$)

$$E_k = -k_k \ln \frac{D_2}{D_1} \qquad (5.35)$$

where k_k = Kick's constant.

Bond's law (Bond, 1952)

This is an intermediate case ($n = -3/2$).

$$E_b = k_b \left[\frac{1}{\sqrt{D_2}} - \frac{1}{\sqrt{D_1}} \right] \qquad (5.36)$$

where k_b = Bond's constant.

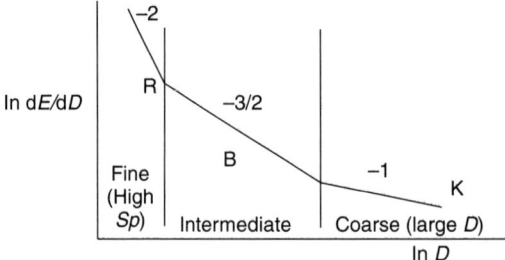

Figure 5.11 *Schematic representation of the three laws of comminution: B, Bond; K, Kick; R, Rittinger*

These three relationships are summarized graphically in Figure 5.11.

Example

A chemical works mills *Compound A* from a feed of Sauter mean size of 6350 μm to a material whose size analysis produced:

Weight (in %)	$\bar{L}(\mu m)$
5	273
80	137
15	82

If the energy consumption of milling the compound A is 100 MJ/tonne what would it be if the material fed to the mill were changed to *Compound B* of Sauter mean size of 6350 μm being milled to 82 μm?

Assume Crushing strength: Compound A, 350 bar; Compound B, 550 bar.

Solution

Compound A

$$\text{Sauter-mean size, } L_{sm} = \frac{1}{\sum_i \frac{x_i}{L_i}}$$

where x_i = wt/wt fraction and L_i = particle size.

5% by weight $\dfrac{x}{L} = \dfrac{0.05}{273} = 1.8315 \times 10^{-4}$

80% by weight $\dfrac{x}{L} = \dfrac{0.80}{137} = 58.3941 \times 10^{-4}$

15% by weight $\dfrac{x}{L} = \dfrac{0.15}{82} = 18.2926 \times 10^{-4}$

$\sum = 78.5182 \times 10^{-4}$

$$L_{sm} = \frac{1}{78.5182 \times 10^{-4}} = 127 \, \mu m.$$

Compound B

$L_{sm} = 127 \, \mu m$ product

$L_{sm} = 6350 \, \mu m$ feed

The particles are 'fine' in size therefore we can use Rittinger's law

$$E = K'_r \cdot f_c \left[\frac{1}{L_{sm2}} - \frac{1}{L_{sm1}} \right]$$

where f_c = the crushing strength

$$\frac{E_1}{E_2} = \frac{f_{c1} \left[\frac{1}{L_2} - \frac{1}{L_1} \right]}{f_{c2} \left[\frac{1}{L_3} - \frac{1}{L_4} \right]}$$

$$\frac{100}{E_2} = \frac{350 \left[\frac{1}{127} - \frac{1}{6350} \right]}{550 \left[\frac{1}{82} - \frac{1}{6350} \right]}$$

$$E_2 = \frac{100 \cdot 1100}{700} \times \frac{0.0120}{0.0077} = 245\,245\,\text{MJ/tonne}$$

Thus the energy requirement of the mill is more than doubled on changing to compound B.

Prediction of product particle size distribution

Population balance

The size distribution formed in the product of comminution can again be predicted using the population balance methodology (Randolph and Larson, 1988). For a breakage process the population balance becomes

$$\frac{\partial n}{\partial t} + \frac{\partial (nG)}{\partial L} + \frac{n - n_0}{\tau} = B_d - D_d \qquad (5.37)$$

For solution of the population balance equation, many forms exist for the particle disruption terms B_d and D_d respectively (Randolph and Larson, 1988; Petanate and Glatz, 1983) but a particularly simple form, which requires no integration of a fragment distribution, is the two-body equal-volume breakage function. It is assumed that each particle breaks into two smaller pieces, each of half the original volume from which it follows that

$$B_d(v) = 2D_d(2v) \qquad (5.38)$$

Randolph (1969) proposed a form for D that was proportional to the volume of the rupturing particle and the population density at that volume. The disruption functions then become

$$D_d = K_d n(v) \qquad (5.39)$$

where K_d is a disruption parameter, and

$$B_d = 4K_d v n(2v) \qquad (5.40)$$

For agitated suspensions, however, recent work suggests that micro-attrition of parent crystals can occur in agitated suspensions (Synowiec et al., 1993). Such breakage via micro-attrition will result in many fine crystals generated and a relatively unchanged parent particle (cf. secondary nucleation, see below) with a correspondingly more complex distribution function (Hill and Ng, 1995, 1996). These alternative attrition models are considered in more detail later.

Hydrodynamic determinants of attrition

Particle break-up may occur by two general modes:

1. *Collisional break-up* of crystals suspended in stirred vessels may occur as a result of collision between crystal–crystal, crystal–impeller or crystal–vessel, and has been described by many authors e.g. Ottens and de Jong (1973), Kuboi et al. (1984), Mazzarotta (1992).
2. *Fluid mechanical break-up* due to turbulent fluid flow has also been inferred in some studies e.g. Evans et al. (1974), with direct evidence being provided by Powers (1963), Sung et al. (1973) and Jagannathan et al. (1980) respectively.

The breakage process is determined by two opposing factors, namely

1. the mechanical strength of crystals, and
2. applied breaking forces, respectively.

Several potential mechanisms exist for the attrition process with the breaking energy of particles originating from either bulk circulation e.g. Ottens and de Jong (1973), Nienow and Conti (1978), Conti and Nienow (1980), Kuboi et al. (1984), Laufhütte and Mersmann (1987), Ploß and Mersmann (1989) or the turbulent motion of the fluid e.g. Evans et al. (1974), Glasgow and Luecke (1980), Jagannathan et al. (1980), or both e.g. Synowiec et al. (1993).

The most important stresses are considered to be:

1. impact-induced stresses
(a) crystal–crystal collisions
(b) crystal–wall collisions
(c) crystal–impeller collisions.
2. fluid-induced stresses
(a) shear stresses
(b) drag stresses
(c) pressure i.e. normal stresses.

Thus the total rate of fine particle generation is

$$\frac{dn_e}{dt}\bigg|_{tot} = \frac{dn_e}{dt}\bigg|_{imp} + \frac{dn_e}{dt}\bigg|_{turb} \qquad (5.41)$$

where $(dn_e/dt)|_{imp}$ describes rate of fine particle generated by means of impact and $(dn_e/dt)|_{turb}$ concerns the rate of fine fragment generation by turbulent fluid forces respectively.

Impact attrition

If it is assumed that the impact crystal–impeller attrition rate is a function of:

- the stirring intensity, f_{st},
- the impact energy applied per unit energy needed to produce one attrition fragment from the crystal surface (E_{ci}/e_{ci}),
- the target efficiency, ηT, which is similar to impact probability between a crystal of given size and an impeller (Lapple, 1950),
- the material properties of crystal and stirrer,
- the total number of parent particles in the vessel, n_c.

Then the *crystal–impeller* impact attrition rate may be described by Synowiec *et al.* (1993)

$$\frac{dn_e}{dt}\bigg|_{ci} = \frac{\pi^2}{2} k_v L_0^3 \rho_c \frac{Q_0}{P_0} \varepsilon \eta_T \left(\frac{1+\sigma_c}{1+\sigma_{st}} \cdot \frac{Y_{st}}{Y_c}\right) \frac{n_c}{e_{ci}} \quad (5.42)$$

implying a linear dependence of impact attrition rate on both power input and suspension density, and a fifth-order dependence on particle size ($\eta T \propto L^2$).

If it is assumed that the crystal–crystal attrition process is a function of:

- the crystal–crystal collision frequency, f_{cc} (Ottens and de Jong, 1973),
- the crystal–crystal impact energy per unit energy needed to produce one attrition fragment from the crystal surface (E_{cc}/e_{cc}),
- the total number of parent particles in the vessel, n_c.

Then the *crystal–crystal* attrition rate may be expressed as

$$\frac{dn_e}{dt}\bigg|_{cc} \approx K_2 L_0^8 \varepsilon^{\frac{3}{2}} n_c^2 \quad (5.43)$$

which implies an eighth-order dependence on crystal size and a second-order dependence on crystal number.

Nienow and Conti (1978) developed a model of particle abrasion at high solids concentration based on Rittinger's law of comminution. When tested experimentally using copper sulphate and nickel ammonium sulphate crystals in two non-solvent liquids, measured abrasion rates were consistent with a second-order dependence of concentration as predicted (Figure 5.12).

Turbulent attrition

For turbulent fluid-induced stresses acting on particles it is necessary to consider the structure and scale of turbulence in relation to particle motion in the flow field. There is as yet, however, no completely satisfactory theory of turbulent flow, but a great deal has been achieved based on the theory of isotropic turbulence (Kolmogorov, 1941).

For particles to break-up in a turbulent flow field, fluid eddies responsible for break-up have to be of both less than the critical size and also possess sufficient disruptive energy. Eddies that are larger than the critical size tend to entrain

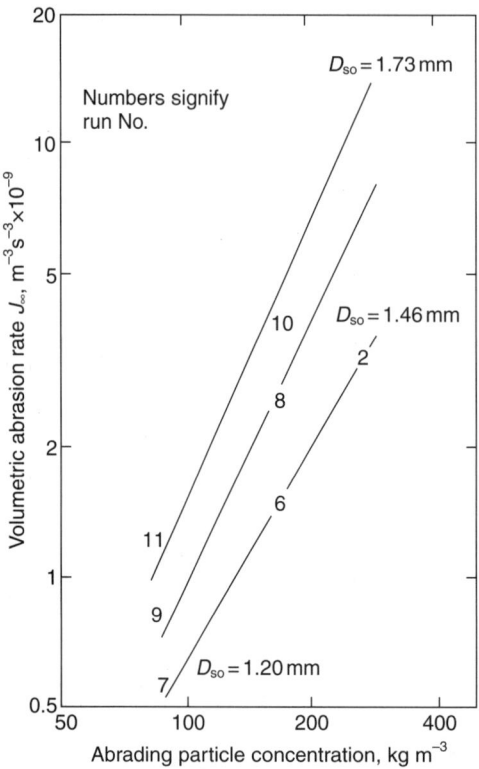

Figure 5.12 *The relationship between volumetric abrasion rate and particle concentration (after Nienow and Conti, 1978)*

particles thus causing little surface stress. Eddies smaller than the size of particle, however, will flow over the particles providing particle shear, surface drag and pressure forces.

Thus the turbulent break-up of parent crystals will depend on:

- the frequency of eddies, f_e (Levich, 1962),
- the relation between turbulent energy component and unit energy required to generate an attrition particle,
- the total number of parent crystals, n_c.

And the attrition rate due to turbulence is given by:

$$\frac{dn_e}{dt}\bigg|_{turb} = f_e \left(\frac{E_n}{e_n} + \frac{E_d}{e_d} + \frac{E_s}{e_s} \right) \tag{5.44}$$

where subscripts n, d, and s refer to disruption energy generated from pressure (normal), drag and shear forces respectively.

Particles $>200\,\mu m$ are predicted by Synowiec *et al.* (1993) to be mainly affected by the drag component of the turbulent forces. For particles $200\,\mu m$, shear forces become more significant in the process of particle break-up.

Pressure forces are estimated to be much smaller than both drag and shear forces (Synowiec et al., 1993).

Whence the *fluid turbulence* attrition rate is given by

$$\frac{dn_e}{dt}\bigg|_{turb} = 0.258 k_v L_0^3 \left[\frac{\varepsilon}{v}\right] \times \left(2.55 \times 10^{-2} L_0^2 \Delta \rho \left[\frac{\varepsilon}{v}\right]^{\frac{1}{2}} \frac{1}{e_d} + \frac{\mu}{\rho_s}\right) n_c \quad (5.45)$$

implying a 1–1.5 order dependence of turbulent attrition rate on power input, the maximum being interestingly similar to that (1.5) for crystal–crystal impacts (equation 5.43), but now with a fifth-order dependence on parent particle size and first-order dependence on solids concentration.

These models have been tested in a crystal attrition cell using alternately stainless steel and rubber coated impellers to determine impact and turbulent contributions respectively by Synowiec et al. (1993) as shown in Figure 5.13.

The respective dependencies of attrition rate on particle number, n_c, and size, L_0, and average unit power input, ε, for each mechanism considered are summarized in Table 5.1.

The experimental results are consistent with crystal attrition occurring via both crystal–impeller impacts and turbulent disruption with no significant effect of crystal–crystal collisions on the number of fine fragments produced in the dilute agitated suspensions.

The small size of daughter particles and only very small changes of the parent crystal size distribution during each run indicate that substantial gross particle fragmentation does not occur. Particle dispersion occurs mainly due to erosion of the crystal surface in the inert solutions.

A significant effect of impeller hardness on the average attrition rate was also detected supporting the observations of Evans et al. (1974) and Shah et al.

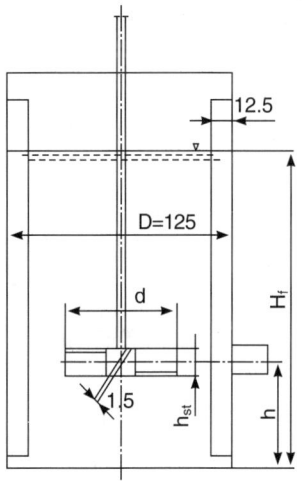
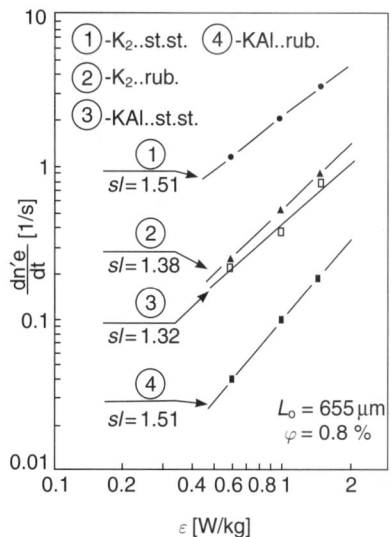

Figure 5.13 *Crystal attrition cell and effect of power input and agitator material on attrition rate (after Synowiec et al., 1993)*

Table 5.1 *Predicted and observed dependence of attrition rate on suspension variables (Synowiec et al., 1993)*

	Predicted and observed dependencies		
	Number n_c	Initial size L_0	Power input ε
Impact			
crystal–crystal	2	8	1.5
crystal–impeller	1	5	1
experimental data impact		5.04–5.11	1.02–1.20
Turbulence			
crystal–shear	1	3	1
crystal–drag	1	5	1.5
experimental data turbulence		5.10–5.37	1.38–1.51
Overall (impact + turbulence)			
exp. data total	0.97–1.04	5.20–5.28	1.13–1.32

(1973). A dependence of both crystal and impeller material properties as well as the probability of crystal–impeller collision on fine particle generation rate has also been demonstrated. Thus the relative effects of impact, drag and shear forces responsible for crystal attrition have been identified. The contribution of shear forces to the turbulent component is predicted to be most significant when the parent particle size is smaller than $\approx 200\,\mu m$ while drag forces mainly affect larger crystals, the latter being consistent with the observations of Synowiec *et al.* (1993).

The maximum contribution of turbulent attrition rate varies in the range 30–40 per cent of total fine numbers with the assumption that impact attrition fragments are not generated by the rubber coated turbine (i.e. similar to the 25 per cent estimate of Evans *et al.*, 1974).

Secondary nucleation

Secondary nucleation is an important particle formation process in industrial crystallizers. Secondary nucleation occurs because of the presence of existing crystals. In industrial crystallizers, existing crystals in suspension induce the formation of attrition-like smaller particles and effectively enhance the nucleation rate. This process has some similarity with attrition but differs in one important respect; it occurs in the presence of a supersaturated solution.

Several modes of secondary nucleation have been identified (see Garside and Davey, 1980 for a review):

- initial breeding
- needle breeding
- polycrystalline breeding
- shear nucleation
- contact nucleation (or collision breeding, or collision nucleation).

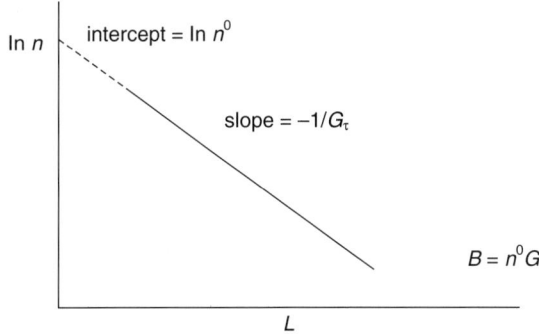

Figure 5.14 *MSMPR size distribution on log-linear co-ordinates*

MSMPR crystallizer studies

Evidence for secondary nucleation has come from the early continuous MSMPR studies. MSMPR crystallization kinetics are usually correlated with supersaturation using empirical expressions of the form

Nucleation rate

$$B = k_b \Delta c^b \tag{5.46a}$$

Growth rate

$$G = k_g \Delta c^g \tag{5.46b}$$

$$\Rightarrow B = K_R G^i \tag{5.46c}$$

where the crystallization kinetic index $i = b/g$.

Plots of log population density versus crystal size of the type shown in Figure 5.14 enable the crystallization kinetics to be determined. Some early literature data reporting such analyses are summarized in Table 5.2.

Table 5.2 *Crystallization kinetics obtained in some early MSMPR studies (after Garside and Shah, 1980)*

System	Author	g	b	$i = \dfrac{b}{g}$
Ammonium sulphate	Larson and Klekar (1973)	1.1	1.6	1.5
Citric acid	Sikdar and Randolph (1976)	0.65	0.54	0.83
Magnesium sulphate	Sikdar and Randolph (1976)	2.29	2.59	1.13
Potassium alum	Garside and Jančić (1979)	1.33	2.10	1.58
Potassium alum	Garside and Jančić (1979)	1.33	2.52	1.89
Potassium dichromate	Timm and Cooper (1971)	1.7	0.9	0.5
Potassium nitrate	Helt and Larson (1977)	1.0	1.6–1.9	1.6–1.9
Potassium sulphate	Randolph and Sikdar (1976)	1.29	0.67	0.52

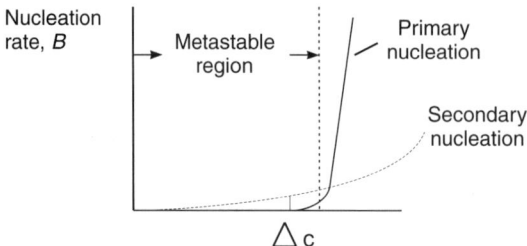

Figure 5.15 *Form of nucleation rate equations showing sensitivity to Δc*

These and other kinetic data in which only i and not Δc is measured in general give i values between 1 and 2. Since the maximum value of $g \approx 2$, then the maximum value of $b \approx 3\text{–}4$, with most values closer to 1. These data therefore imply that the dependence of nucleation in these studies is much lower than the corresponding exponential dependence predicted for primary nucleation, illustrated in Figure 5.15.

Similarly, the dependence j of nucleation rate B on magma density M_T and stirrer speed N in MSMPR crystallizers given by

$$B = K M_T^j N^k G^i \qquad (5.47)$$

Where most values of $i \approx 1$, within the range 0.14–1.07. Additionally, a dependence on stirrer speed k is observed being in the range $0 < k < 7.8$ (Garside and Shah, 1980). Thus the observed nucleation rates in MSMPR crystallizers depend on both magma density and stirrer speed. They are thus likely to have been due to secondary rather than primary nucleation. Further evidence suggests these effects result from crystal/solution interactions (see Chapter 5).

Mechanisms of secondary nucleation

Denk and Botsaris (1972) used left- and right-handed optical properties of sodium chlorate to discriminate between the crystal surface and the solution as sources of secondary nuclei. They found that at high supersaturation levels 50 per cent of nuclei are of each optical form, which implies the solution as the source, in other words, primary nucleation occurs. At lower supersaturation, however, all nuclei are of same form as the parent crystal, implying crystals as the source, i.e. secondary contact nucleation.

In other experiments using a device to strike crystals in solution within a cell under the microscope, Garside and Larson (1978) provided direct observation of crystal surface during secondary contact nucleation.

Secondary nucleation kinetics

The overall nucleation rate in a crystallizer is determined by the interaction of the secondary nucleation characteristics of the material being crystallized with the hydrodynamics of the crystal suspension. When crystallizing a given material, crystallizers of different size, agitation levels, flow patterns, etc. will

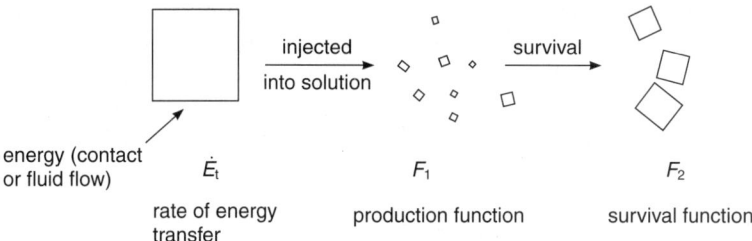

Figure 5.16 *Secondary nucleation in crystallizers*

produce different nucleation rates. In order to incorporate nucleation kinetics in design models these changes in nucleation rates should be predictable.

A number of authors have developed mechanistic descriptions of the processes causing secondary nucleation in agitated crystallizers (Ottens *et al.*, 1972; Ottens and de Jong, 1973; Bennett *et al.*, 1973; Evans *et al.*, 1974; Garside and Jančić, 1979; Synowiec *et al.*, 1993). The energy and frequency of crystal collisions are determined by the fluid mechanics of the crystallizer and crystal suspension. The numbers of nuclei formed by a given contact and those that proceed to survive can be represented by different functions.

These processes are represented schematically in Figure 5.16.
So, following Botsaris (1976), the rate of secondary nucleation is given by

$$B = \dot{E}_t F_1 F_2 \tag{5.48}$$

where $\dot{E}_t$ is the rate of energy transfer to the crystals. If these nuclei are produced by crystal collisions then

$$\dot{E}_t = \int_0^\infty E(L) f(L) n(L) \, dL \tag{5.49}$$

The product of the collision energy $E(L)$ and collision frequency $f(L)$ is integrated over all crystals in the distribution to obtain the total rate of energy transfer. Different approaches have been used to estimate $E(L)$ and $f(L)$, both for particle impacts and turbulent fluid induced attrition.

The term F_1 is a production function giving the number of particles generated per unit of transferred energy while F_2, a survival function, represents the fraction of these particles that survive to become nuclei and subsequently grow to populate the size distribution. F_1 depends on the particular crystallizing system and on the supersaturation, temperature and impurity level. F_2 is related to the size of the fragments and their growth characteristics. At present these two functions cannot be fully determined separately. In practice, they are combined and written as a simple power function of the concentration driving force, and the coefficients determined experimentally.

The precise nucleation mechanism occurring in any particular case is often a subject of debate, however, and in practice the data are normally correlated empirically by an expression including a dependence on solids hold up of the form:

$$\frac{dN}{dt} = B = k_b M_T^j \sigma^b \tag{5.50}$$

where σ is the absolute supersaturation ($S - 1$). It is generally observed that the order b of secondary nucleation is less than the corresponding value for primary nucleation. From the attrition theory discussed above, for crystal/crystallizer collisions the solids hold-up dependence $j = 1$, for crystal/crystal collisions $j = 2$ and for fluid shear $j = 1 - 2$ while in practice both modes may be operative (Synowiec et al., 1993).

Scale-up of secondary nucleation kinetics

The original development of Ottens et al. (1972) results in the expression

$$B = K_N \left[\frac{Q_0}{P_0}\right] \bar{\varepsilon} M_T \Delta c^b \tag{5.51}$$

where $\bar{\varepsilon}$ = power input per unit mass of magma, P_0 and Q_0 are the power and flow numbers respectively, and K_N incorporates the proportionalities introduced in deriving the equation, and so will depend on the crystallizing system, temperature, impeller material of construction, etc.

This, and other similar equations, show that scale-up of secondary nucleation kinetics on the basis of constant specific power input, $\bar{\varepsilon}$, should at most produce only a modest increase in nucleation rate. Constant tip speed or constant stirrer speed to just maintain the crystals in suspension should both result in a decrease in nucleation rate with increasing scale of operation (Garside and Davey, 1980).

Experimental data comparing results at different scales of operation with equations of the type quoted above are given in: Ottens and de Jong (1972), Bennett et al. (1973), Garside and Jančić (1979), Bourne and Hungerbuehler (1980), Ploß and Mersmann (1989), Synowiec et al. (1993).

In the analysis of Synowiec et al. (1993), while the turbulent attrition rate is predicted to be constant with increasing scale of operation at constant power input, the impact attrition rate is predicted to decline with the square of vessel size. In combination

$$\left[\frac{n_e}{n_{elab}}\right]_{total} \approx \left[B\left(\frac{d}{d_{lab}}\right)^a\right]_{imp} + [C]_{turb} \tag{5.52}$$

where empirically $a = -2$, $B = 0.7$, and $C = 0.3$, which is qualitatively consistent with the data of Ploß and Mersmann (1989). These data are summarized in Figure 5.17.

Scale-up of the apparatus is predicted to decrease the total number of attrition particles and reduce the proportion of fines contributed by impeller impact relative to turbulent fluid erosion of parent particles, but it was concluded more work is still necessary to quantify and experimentally confirm those aspects. The implication, however, is that mean crystal size increases with scale up – a fortuitous result often observed in practice.

Grootscholten et al. (1982) discuss the influence of different impeller types on secondary nucleation rates. There is some evidence (Scrutton et al. 1982) that secondary nucleation rates are particularly sensitive to small clearances between the stirrer and a draft-tube by which the stirrer is shrouded.

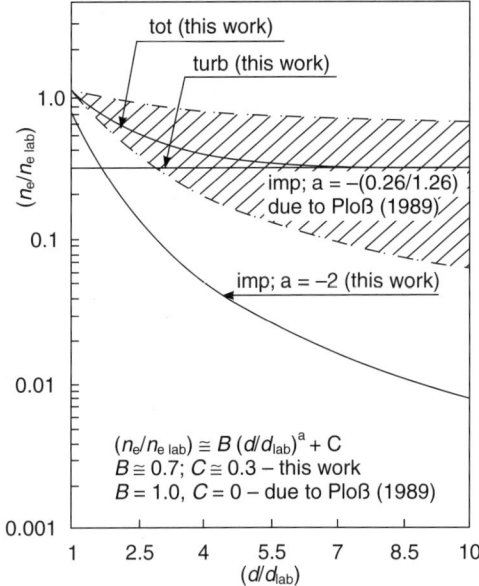

Figure 5.17 *Effect of scale-up factor* (d/d_{lab}) *on overall attrition rate for constant* ε *(Synowiec et al., 1993)*

Mazzarotta *et al.* (1996) demonstrate a time dependence of batch-wise crystal attrition, the rate of new particle formation declining rapidly after the initial high rate period.

Growth of secondary nuclei, small crystals and attrition fragments

Bujac (1976) observed that a batch of pentaerythritol fragments in the size range 5–40 μm did not grow until supersaturation exceeded 35 per cent. Similarly, van't Land and Wienk (1976) reported that NaCl attrition fragments < 40 μm did not grow whereas well-formed cubic NaCl crystals of all sizes grew. Garside and Shippey (1982) report the effect of such 'growth rate dispersion' on the product of a batch crystallizer (Figure 5.18).

Effect of supersaturation on secondary nucleation

Garside *et al.* (1979) measured size distributions of secondary nuclei and reported their variation with supersaturation. Significant increase of nuclei with supersaturation is observed. Thus the process is not simply an attrition event alone, but is also related to the level supersaturation at which parent crystal is growing. Jones *et al.* (1986) also observed anomalous growth of secondary nuclei in a study of the continuous MSMPR crystallization of potassium sulphate with consequences inferred for secondary nucleation rates. Girolami and Rousseau (1986) demonstrate the importance of initial breeding mechanism in seeded potash alum batch crystallization. The number of crystals

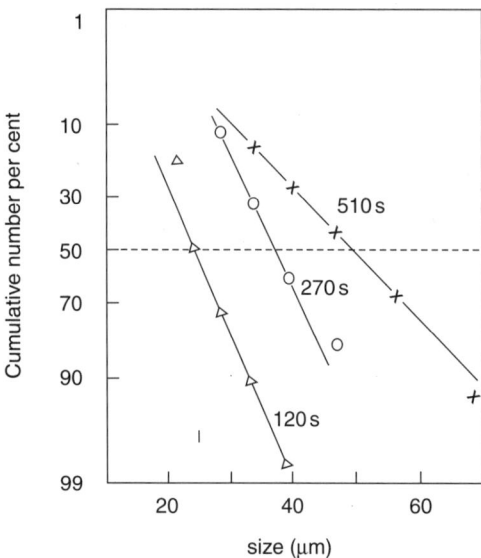

Figure 5.18 *Dispersion of Potash alum seeds during growth in a batch crystallizer (Garside and Shippey, 1982)*

formed was found to depend on the surface area of the seed crystals but not the prevailing supersaturation, but the latter did affect subsequent growth. An initially narrow distribution of *ca.* 5 μm crystals broadened substantially, attributed to growth rate dispersion. Chianese *et al.* (1993) studied the effect of secondary nucleation on the crystal size distribution from a seeded batch crystallizer. The crystal size distribution data were well predicted by attrition kinetics at low cooling rates, but indicate an excess of secondary nucleation at high cooling rates (Figure 5.19) consistent with a change in mechanism at high supersaturation levels.

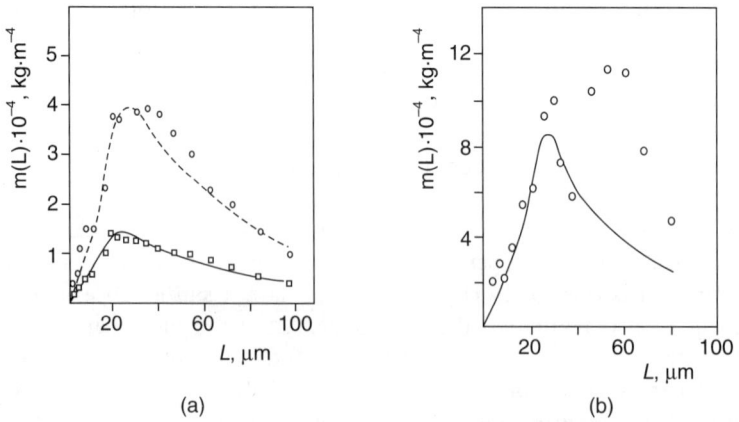

Figure 5.19 *Mass density distributions for crystallization at (a) high; and (b) low cooling rates (Chianese et al., 1993)*

These effects of supersaturation imply that contact secondary nucleation results from a more complicated process than straightforward attrition and many authors have speculated about the actual mechanism involved. Strickland–Constable (1979) suggested that the 'mosaic structure' of the crystal provides the basic fragments. During a collision plastic deformation of the crystal takes place, stresses build up at the surface and large numbers of particles corresponding to the mosaic structure but modified by the deformation are produced. The modulus of elasticity and the elastic limit of the material may now be seen as possible important parameters determining the nature and extent of plastic deformation and the initiation of brittle fracture.

Dislocations and solution inclusions may also play a role in secondary nucleation since they promote weak points in the crystal. Ristic *et al.* (1988, 1991) demonstrated that the increase in solubility due to crystal strain is a likely cause of crystal growth rate variations (see Ristic and Sherwood, 2001 for a comprehensive review of work done in this area). Van Der Heijden *et al.* (1994) presented a model of secondary nucleation incorporating formation, removal and growth mechanisms thereby giving a physical explanation of the supersaturation dependence generally found in empirical correlations of secondary nucleation rates. Gahn and Mersmann (1999) analysed the mechanics of brittle fracture in crystallization processes and developed a model to predict particle attrition rates and size distribution, and the observed size-dependence of crystal growth rates and crystallizer scale-up. Gertlauer *et al.* (2001) develop a two co-ordinate population balance model based on both particle size and lattice strain of the individual crystals and predict a strong dependence of the steady state mass density function on the different assumptions adopted for the relaxation of the internal lattice strain during crystal growth.

The consequences of growth rate dispersion are:

- Broadening of the crystal size distribution in batch crystallizers; thus the effect of seeding is particularly important.
- For continuous crystallizers the effect of size-dependent growth and growth rate dispersion are difficult to distinguish (Janse and de Jong, 1978; Randolph and White, 1977).

Use of 'growth rate diffusivity' (Randolph and White, 1977; Tavare and Garside, 1982) or 'size dependent growth' (Abegg *et al.*, 1968; Mydlarz and Jones, 1993) have both been proposed as alternative phenomenological means to describe the effect of growth dispersion on crystal size distributions; the latter being simpler mathematically than the former, but in all probability both mechanisms can occur.

Summary

Nucleation and crystal growth are the primary particle formation processes during crystallization and have a large effect in determining product crystal size distribution. Both processes depend on the degree of supersaturation within a solution, with nucleation having the stronger dependence. Classical theories of nucleation and crystal growth, however, are incapable of predicting rates for a particular substance under given conditions, which must therefore be determined experimentally. In particular, nucleation in industrial crystallizers is complicated by the presence of existing crystals giving rise to so-called secondary nucleation. These two kinetic quantities can, however, readily be determined simultaneously from the continuous MSMPR crystallizer operated at steady state.

Attrition of particulate materials can also have an important affect on the product size distribution from industrial crystallizers. Because crystals are usually polydisperse and all sizes are subject to breakage and wear, the concepts of attrition can be difficult to describe in simple terms. In assessing the attrition that is likely to occur in industrial crystallizers, the various mechanisms must be considered. Underlying theoretical insights from materials science should ultimately enable these mechanisms to be interrelated for isolated single particles and some progress has been made in understanding the relationship between mechanical and thermal fracture employing the concepts of the existence of flaws and stored elastic energy. A further complication arises, however, that since attrition can be due to particle motion within equipment it will be affected by scale-up, normally decreasing with increasing vessel size. Stresses in any narrow clearances in process equipment such as in pumps and shrouded impellers are particularly important in enhancing attrition. Within industrial crystallizers, attrition is closely related to secondary nucleation. The latter is also dependent, however, on solution supersaturation thereby adding to the complexity of the process.

6 Crystal agglomeration and disruption

Agglomeration is a particle size enlargement process by which fine particles, rather than ions or molecules, are joined in an assembly e.g. within a suspension crystallization or precipitation process, or alternatively by sticking dry powders together with a liquid binder, e.g. in rotating dishes, drums or pans, when the process is often known as *granulation*. These processes therefore result in relatively rapid size enlargement. In agitated crystallizers, particle disruption can also occur, thereby complicating the agglomeration process. In this chapter, the processes occurring during agglomerative crystallization are analysed and the determination of their kinetics is considered in detail. The processes of size enlargement by agglomeration and granulation have been reviewed by Sherrington and Oliver (1981), Stanley-Wood (1990) and Pietsch (1991).

Agglomerate particle form

The particle characteristics obtained in the product depend strongly on the mechanism of agglomeration and the processing conditions. Agglomerates typically have the form shown schematically in Figure 6.1, although the sizes of both primary particles and agglomerates can vary considerably.

For many applications, the main need is for agglomerated particles having an open particulate array which rapidly dissolves (as in some pharmaceuticals), while in others a more dense structure is required giving a slow release of its components (as in fertilizers).

Agglomerative crystallization

It is often observed that crystals do not exist as discrete entities but, especially when viewed under the optical microscope, appear to comprise agglomerated particles. Even a cursory glance at the literature quickly shows that the subject of agglomeration spans several disciplines including crystallography, materials

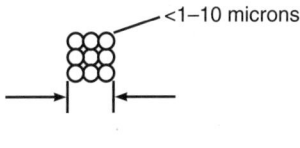

Figure 6.1 *Typical agglomerate or granule (actual dimensions vary)*

science, surface chemistry, colloid science, civil engineering and chemical engineering and perhaps in consequence, the terminology used is not consistent. Similarly, contrary definitions have appeared in Europe, the United States and elsewhere. That used in this chapter follows the usage as defined below.

There are two main types of agglomeration viz. primary and secondary. Firstly, a crystalline particle may undergo a form of mal-growth, related to its crystallography, and comprise individual crystals within a structure of parallel units, dendrites or twins. Such formation of composite crystals can be termed primary agglomeration. Secondly, crystals suspended in liquids may collide induced by the flow and join together i.e. aggregate to form a larger particulate entity which may subsequently be disrupted and redisperse or fuse to form a secondary agglomerate. A schematic diagram illustrating these processes is depicted in Figure 6.2. Both types of agglomeration may, of course, occur simultaneously. An agglomerate is thus a firmly cemented particle joined by solid crystalline bridges. Other terms commonly encountered are coagulate and flocculate which refer to the grouping of two or more particles loosely held together by weak cohesive forces while an aggregate is a weakly cemented crystal cluster. Coagulation flocculation and aggregation are reversible to a greater or lesser extent by the process of disruption and redispersion and precede agglomeration. Agglomerates can only be disrupted and redispersed by crystal breakage attrition or erosion. For convenience, agglomerate will be used as the generic term for a strongly bound poly-crystalline particle, unless otherwise specified locally in the text.

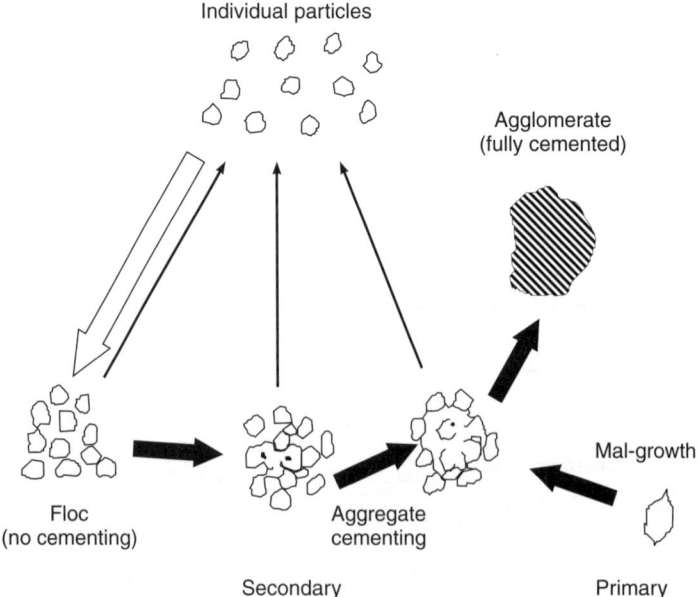

Figure 6.2 *Particle formation via aggregation and agglomeration (adapted from Low and White, 1975)*

Crystal agglomerates can arise by a variety of mechanisms and give rise to a wide variety of physical forms of differing particle strength. These range from raspberry to star-like assemblages with properties, which may be an advantage or a disadvantage depending on the circumstances. For example, such particles can deform, may be porous and retain liquor thereby giving rise to filtration and purity problems, break forming dust or simply be unacceptable in appearance. On the other hand, the formation of agglomerates can enhance the solid–liquid separation characteristics of fine precipitates and decrease the packing density of the product. The occurrence of agglomeration thus greatly complicates the analysis of crystallization and precipitation processes.

Agglomeration and its consequences is of considerable scientific and technological importance and has been the subject of several international conferences (Capes, 1961, 1977, 1981, 1985; Wyn-Jones and Gormally, 1983; Davis *et al.*, 1984; Family and Landau, 1984; Gregory, 1984a), selective reviews (Schubert, 1981; Gregory, 1984b; Jones *et al.*, 1996; Jones, 1988; Hounslow *et al.*, 2001), and a vast number of related articles in journals.

Primary agglomeration

As mentioned earlier, agglomerated forms of crystals may arise not only from the aggregation of small units originating separately but also by the mal-growth of crystals sometimes called composites i.e. polycrystals, dendrites and twins. While it is difficult to imagine that some agglomerate forms could have arisen via an aggregation process, it is in fact often difficult to discriminate by microscopic observation alone between the alternative origins. Primary agglomerate growth has been postulated to arise either as a consequence of impurity action or by diffusion field limitations especially at high growth rates (Walton, 1967) but there have been relatively few reports of primary agglomerate solution growth kinetics. The crystallographic analysis of primary agglomerates has been long known, however. Twinned crystals in particular appear to be composed of two intergrown individuals, similar in form, joined symmetrically about an axis (a twin axis) or a plane (a twin plane).

Dendrite growth

Crystallization often produces tree-like formations called dendrites, as illustrated in Figure 6.3 and will be considered here to be primary agglomerates. A seed stem first grows quite rapidly, and at a later stage primary branches grow out of the stem at a slower rate than the stem, often at right angles to it. In certain cases, small secondary branches may grow out of the primaries. Eventually branching ceases and the pattern becomes filled in with crystalline material. Dendrites form most frequently in melt crystallization, snow flakes and 'Jack Frost' being the commonest examples – thought to form due to heat transfer limitations – but dendrites can also form from solution or the vapour and may similarly be controlled by transport process limitations.

Davey and Rutti (1976) studied the effect of polycrystals growth on agglomeration of hexamethylene tetramine (HMT). Regular six member clusters were

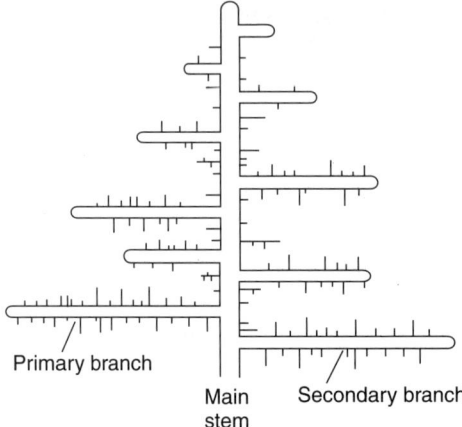

Figure 6.3 *Typical dendrite*

observed, the continued growth of which also resulted in occlusions. The creation of clusters was considered due to initial dendritic growth and a crystal packing model was proposed consistent with these observations.

Parallel growth

A second and perhaps simplest form of primary agglomerate results from the phenomenon known as parallel growth in which individual crystals of the same substance grow in a stack on top of one another in such a manner that all corresponding faces and edges of the individuals are parallel. The classic example

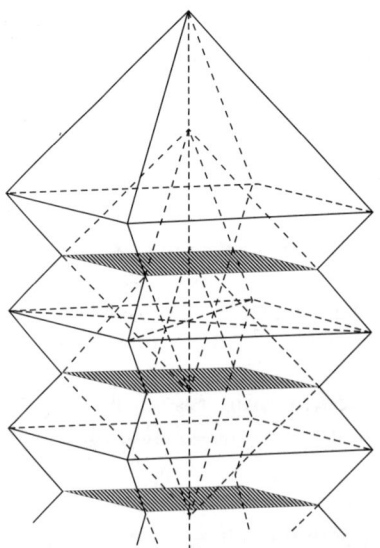

Figure 6.4 *Parallel growths of potash alum*

is given by the case of the alums as shown in Figure 6.4 in which cubic octahedra are often formed into pagoda-like columnar piles of parallel crystals. The planes of contact are parallel to the cube or (001) face, and the centres of successive octahedra, indicated by the dots, lie in the extension of the vertical axis. In some cases it is possible for two substances to exhibit parallel growth on one another. This is thought to occur where congruency of the space lattice exists.

Mydlarz and Jones (1990a,b) determined agglomeration of potash alum in continuous MSMPR crystallization from aqueous solution and its effect on slurry filterability (see Chapter 9).

Twinning

Another primary agglomerate crystal frequently observed is known as a twin or a macle composed of two or even three constituents united in a regular manner, other than parallelism, the particular mode of union being characteristic of the substance. The constituents are joined symmetrically about an axis (a twin axis, e.g. a crystal edge) or a plane (a twin plane, e.g. a crystal face). Many types of twins form into simple shapes such as a V, + and L are possible, or they may show interpenetration giving the appearance of one crystal having passed directly through the other either completely as shown for the cube in Figure 6.5a or partially as shown for the rhombohedra in Figure 6.5b. In some cases, a

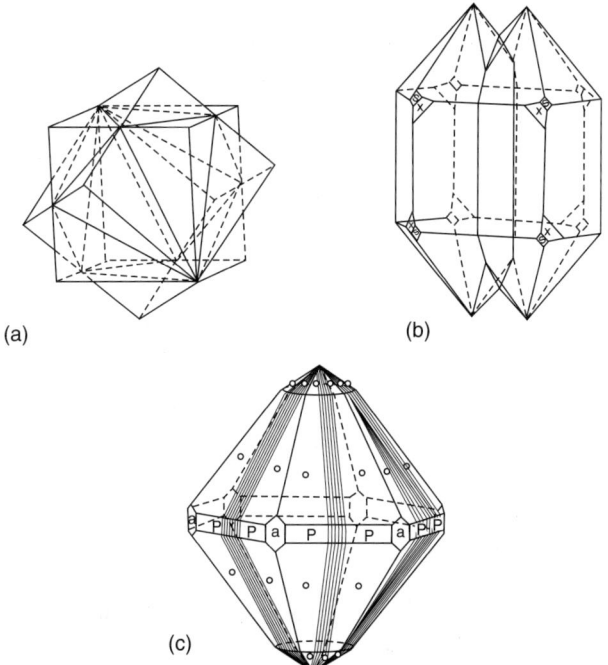

Figure 6.5 (a) *Cubic Twin of Fluor-Spar*; (b) *Rhombohedric of Quartz*; (c) *Hexagonal Bipyramid Triplet of Potassium Sulphate* ($a = (100)$; $o = (111)$; $p = (110)$)

twinned crystal may represent (mimic) the outward appearance of a form that processes a higher degree of symmetry (i.e. of higher crystal system) than that of the individuals, known as mimetic twinning. The best-known example of this being that of orthorhombic potassium sulphate, which can form a triplet looking almost identical to a hexagonal bipyramid (Figure 6.5c).

Mantovani *et al.* (1983) reported a crystallographic and kinetic study of the twinning of sucrose crystals and determined the enhancement of twin types due to a common impurity (raffinose). Aquilano and Franchini-Angela (1985) determined twin laws of calcium oxalate trihydrate (COT) in aqueous solutions and reasoned by geometrical arguments that these laws are defined by the twin axes (001), (010) and (001), respectively. Prasad (1985) observed cross-like twinning in palmitic acid crystals form an evaporated organic solvent (methyl alcohol–acetone mixture). A shift in the position was explained crystallographically. Aoki and Nakamuto (1984) determined factors affecting penetration twins of potassium chloride crystallized from pure and impure aqueous solution. Increasing levels of supersaturation and the concentration of lead ions enhanced the presence of interpenetrating octahedra and cuboctahedra. Budz *et al.* (1987b) observed crystal agglomeration in the continuous MSMPR crystallization of potassium sulphate from aqueous solution and determined its effect on inferred crystallization kinetics. Both primary and secondary agglomerates were observed under the microscope.

Thus, while much is known about crystal structure, relatively little is still known about primary agglomerate crystal growth. Apart from the existence of a compatible twin axis or plane, the main factors that often favour their formation are high levels of supersaturation, excessive seeding, poor agitation and the presence of certain impurities. Twin formation can still occur in agitated systems, however. Twinning most frequently (but not exclusively) occurs when the crystals belong to the orthorhombic and monoclinic systems.

Secondary agglomeration

While primary agglomeration can occur originating from a single crystal, a second form of agglomeration occurs because of the presence and motion of more than one crystal in a suspension, leading to secondary crystal aggregation. Two types of secondary agglomeration occur:

(a) Perikinetic due to Brownian motion of small ($<1\,\mu m$ say) particles, or
(b) Orthokinetic due to larger ($>1\,\mu m$ say) particles entrained in fluid velocity gradients.

Each process can give rise to collisions, hence has the potential of causing aggregation leading to secondary agglomeration. These processes are illustrated in Figures 6.6(a) and (b) respectively.

The irregular part of the motion comes from the apparently random bombardment of the particle by surrounding fluid molecules i.e. Brownian motion. The systematic part derives from the action of various external influences – mechanical, electrical and gravitational for example – the strength of which change in time and place.

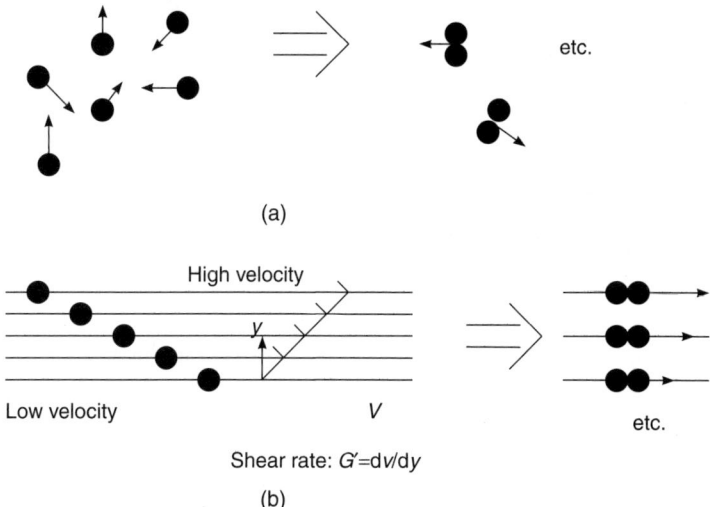

Figure 6.6 (*a*) *Perikinetic aggregation*; (*b*) *Orthokinetic aggregation* (*schematic*)

Perikinetic motion of small particles (known as 'colloids') in a liquid is easily observed under the optical microscope or in a shaft of sunlight through a dusty room – the particles moving in a somewhat jerky and chaotic manner known as 'the random walk' caused by particle bombardment by the fluid molecules reflecting their thermal energy. Einstein propounded the essential physics of perikinetic or Brownian motion (Furth, 1956). Brownian motion is stochastic in the sense that any earlier movements do not affect each successive displacement. This is thus a type of Markov process and the trajectory is an archetypal fractal object of dimension 2 (Mandlebroot, 1982).

Larger particles, however, are not so affected by molecular motion but are strongly influenced by the presence of velocity gradients within the fluid enabling them to catch up and collide with one another – orthokinetic motion.

Colloid and surface science

The term colloid generally refers to small particles of dimensions approximately in the size range say 1 nm – 1 μm. Such particles have a high surface: volume ratio for which particle interactions and surface properties including charge become important, even dominant, in determining their behaviour. Hence, the study of colloid and surface science which now occupies a vast literature with application in fields as divers as chemicals, detergency and water treatment, sols, elusions, oils, fats, paints etc. (Shaw, 1980; Parfitt, 1981; Lyklema, 1985; Russel, 1987; Wedlock, 1994; Williams, 1994). One characteristic of colloidal particles is their tendency to aggregate. Firstly, however, we need to consider the forces that small particles experience and how they affect their behaviour.

Particle forces

Briefly, there are four main classes of forces that affect the state and kinetics of aggregation (Dickinson, 1986):

1. Interparticle colloidal forces (attractive and/or repulsive) stemming from electrostatic stabilization (or destabilization).
2. Rapidly fluctuating stochastic forces associated with Brownian diffusion.
3. Direct systematic forces due to externally applied fields (e.g. electrostatic or gravitational).
4. Indirect systematic forces transmitted through the hydrodynamic medium by movement or other particles or externally applied flow fields (e.g. shear or extensional).

Fortunately, often one or more of the above classes can be ignored, thereby simplifying the problem substantially.

There are three main types of particle interaction (Lyklema, 1985): Van der Waals attraction (mainly London forces), electrostatic forces and steric effects. These interactions depend on the shape and size of the particle, surface charge, solution composition (pH, ionic strength etc.) and temperature, and – most importantly, the particle separation distance, h.

The Van der Waals forces always lead to a weak attraction of particles composed of similar materials described by an expression of the form

$$V_v = -A_{12} f(h) \tag{6.1}$$

where A_{12} is the Hamaker constant for the material. Van der Waals forces decay rapidly with increasing separation distance.

Surface charge

Electrical double layer

Particles dispersed in aqueous solutions frequently develop a surface charge (most commonly negative but can be positive) due to preferential adsorption of certain ionic species, dissociated surface groups or polymers etc. The principle of the electrostatic interaction is that these surface charges are compensated by an equal, but opposite, charge surrounding them such that an electrical double layer forms (see Figure 6.7). Although overall they are electroneutral, such double layers of equal sign repel each other with an energy given approximately by an expression of the form

$$V_r = B' r \exp(-kh) \tag{6.2}$$

where the coefficient B' is related to the surface charge and hence particle and solution properties. Electrokinetic behaviour of colloidal particles with thin ionic double layers has been analysed by Natarajan and Schechter (1987).

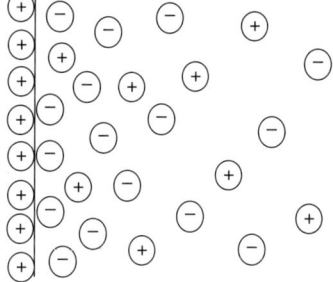

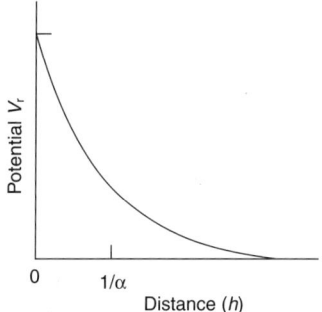

Figure 6.7 *Electrical double layer*

Particle interactions

As two particles approach in a liquid their charge fields may interact and form two minima as depicted in Figure 6.8. If the particles approach to a distance L_1, known as the primary minimum they aggregate to form a configuration with minimum energy – and rapid coagulation is said to take place. On the other hand, if the particles remain separated at a distance L_2, the secondary minimum, loose clusters form which do not touch. This is known as slow coagulation and is the more easily reversed.

Steric interaction

It has long been known that added polymers can have a drastic effect on particle interaction. Depending on conditions, polymers can either stabilize or destabilize colloidal systems. The former occurs by a protection mechanism whereby the charged adsorbed polymer layers repel each other. In the latter case, polymers can enhance aggregation in two ways: (1) by making the particles more susceptible to salts as above; or (2) by flocculating the system without the aid of electrolytes by forming polymer bridges. This technique is often used in water treatment (Gregory, 1984a). A necessary condition for the effectiveness of polymer flocculants in terms of polymer size and particle separation distance has been proposed (Kashiki and Suzuki, 1986).

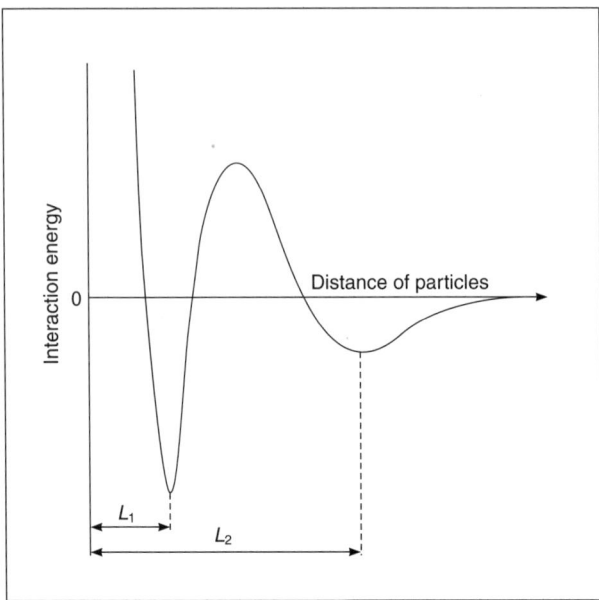

Figure 6.8 *Interaction of charged particles in a liquid*

Adsorption

An alternative analysis of colloidal interaction and stability based on classical solution thermodynamics has been proposed independently by Hall (1972) and Ash *et al.* (1973) and the practical implications with particular interest in the effects of polymers and surfactants has been discussed by Pethica (1986). In this theory, the forces between particles are governed in a straightforward way by the adsorptions of the components of the system and their dependence of particle separation and chemical potentials (which are defined by the composition) according to the equation

$$\left(\frac{\partial f}{\partial \mu_i}\right)_{\mu_i h} = -2\left(\frac{\partial \Gamma_i}{\partial h}\right)_{\mu_i} \tag{6.3}$$

where the Γ_i are the adsorption per particle. Equation 6.3 states that the interparticle force varies directly with chemical potential if its adsorption varies with interparticle separation distance at a given solution composition.

Stability

Prediction of the stability of colloidal suspensions has been analysed in two ways. The first sums the above energy changes due to London attraction and double layer repulsion respectively as particles approach each other with the net effect passing through a maximum (Figure 6.9). This is known as the DLVO (Derjaguin–Landau and Verwey–Overbeek) model and its development forms a major part of the statistical mechanical theory of colloid stability (Derjaguin

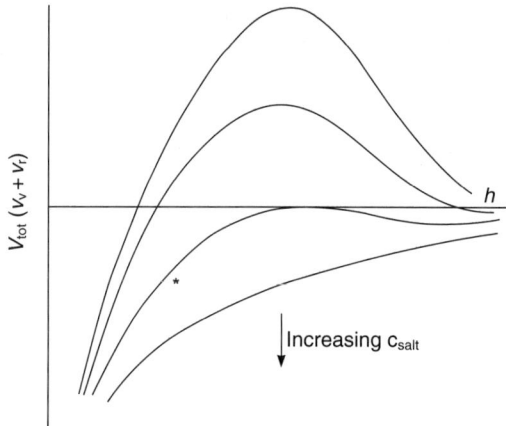

Figure 6.9 *Influence of electrolyte concentration on colloid stability (*denotes change from a stable to an aggregated state)*

and Landau, 1941; Vervey and Overbeek, 1948; Overbeek, 1949). One interesting prediction of the theory is that addition of a non-adsorbing salt can render the double layer more diffuse leading to greater aggregation and to a change from a stable to an aggregated state (denoted by the asterisk in Figure 6.9).

Similarly, it is predicted that aggregation generally increases with particle size (Figure 6.10) but in practice a limit is reached due to hydrodynamic forces (see later).

Electrostatic characterization of particles is commonly determined via their electrokinetic or zeta potential i.e. the potential of a slipping plane, notionally located slightly away from the particle surface approximately at the beginning of the diffuse part of the double layer using, for example, electrophoresis. In some cases, zeta potential can be used as a criterion for aggregation.

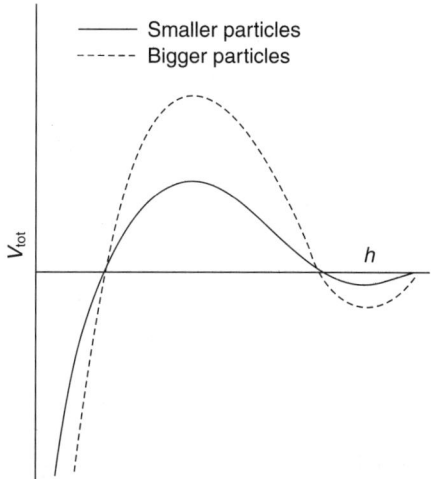

Figure 6.10 *Effect of particle radius on colloidal interaction*

In the thermodynamic theory, the time dependence of the variation of adsorption with separation distance determines the colloidal stability and hence aggregation and, although such data are not yet generally available, the theory can provide qualitative insight and has an advantage of being independent of particle characteristics.

More detailed predictions of the stability and aggregation behaviour of particulate suspensions require consideration of the kinetics of aggregation and their dependence on fluid-particle hydrodynamics.

Motion of crystals in suspended particulate systems gives rise to collisions, hence aggregation (a coming together of the particles) leading to secondary agglomeration, and these processes can be enhanced by agitation. At high agitation levels, however, particle disruption can also occur tending to disperse the particles and limit the maximum attainable particle size. Thus, the final form of secondary agglomerates is a result of the balance between these opposing rate processes.

Population balance

The population balance provides the mathematical framework incorporating expressions for the various crystal formation, aggregation and disruption mechanisms to predict the final particle size distribution. Note, however, that while particles are commonly characterized by a linear dimension the aggregation and particle disruption terms also require conservation of particle volume. It was shown in Chapter 2 that the population balance accounts for the number of particles at each size in a continuous distribution. The quantity conserved is thus the number (population) density and may be thought of as an extension of the more familiar mass balance. The population balance is given by (Randolph and Larson, 1988)

$$\frac{\delta n}{\delta t} + \frac{\delta}{\delta L}(nG) + \frac{n - n_0}{\tau} = B_a - D_a + B_d - D_d \tag{6.4}$$

where $n(L, t)$ is the population density defined on a crystal size basis, $G(=dL/dt)$ is the linear growth rate and τ is the mean residence time within the vessel. $B_a - D_a$ and $B_d - D_d$ represent the net formation of particles at size L by aggregation and disruption respectively.

Although it looks complicated, the population balance simply accounts for all the mechanisms by which particles can enter and leave a particular size interval – by nucleation, growth, inflow or outflow to the vessel, breakage and aggregation. This was illustrated for an idealized continuous MSMPR crystallizer in Chapter 3 but its use for agglomeration is less well developed than that for crystallization considered in Chapter 5. This arises partly historically but in part due to the fact that particle growth by agglomeration concerns at least three particles: two particles of arbitrary size undergoing (say) binary collision and the product particle which may itself subsequently be disrupted. Thus, the particle number accounting rules are a little more complex to implement. Nevertheless some useful solutions to simplified problems

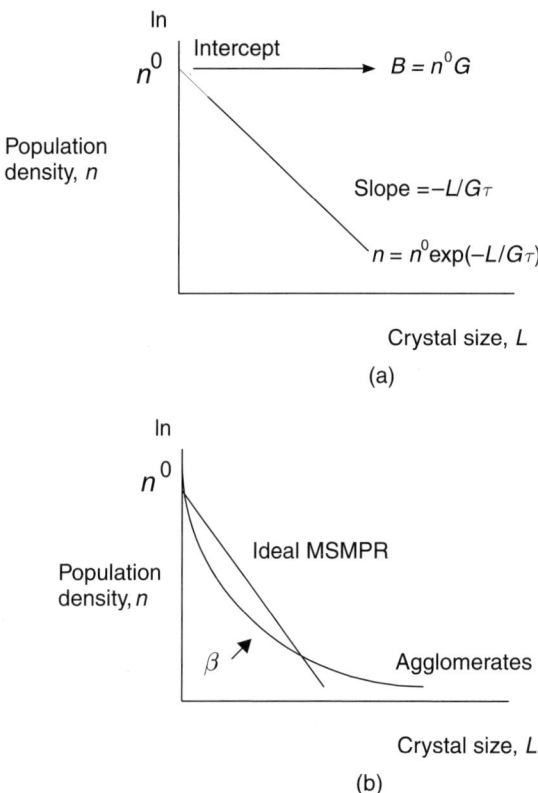

Figure 6.11 *Idealized MSMPR CSD*: (a) *Nucleation and growth only*; (b) *effect of agglomeration*

are available, together with numerical solutions for the more complicated cases.

It was shown in Chapter 3 that in the absence of crystal aggregation, the product crystal size distribution is log-linear, as illustrated in Figure 6.11(a).

In the presence of crystal agglomeration, however, the number of small particles is reduced as they stick together with a corresponding increase in the number of larger ones. This gives rise to concave curvature, as illustrated in Figure 6.11(b).

Consider the crystal size distribution in a model MSMPR crystallizer arising because of simultaneous nucleation, growth and agglomeration of crystalline particles. Let the number of particles with a characteristic size in the range L to $L + dL$ be $n(L)dL$. It is assumed that the frequency of successful binary collisions between particles (understood to include both single crystals and previously formed agglomerates) of size L' to $L' + dL'$ and L'' to $L'' + dL''$ is equal to $\beta n(L')n(L'')dL'dL''$. The number density $n(L)$ and the collision frequency factor β are related to some convenient volumetric basis, e.g. unit volume of suspension.

In the MSMPR crystallizer at steady state, the increase of particle number density brought about by particle growth and agglomeration is compensated by withdrawal of the product from the crystallizer.

The population balance can then be written as

$$\frac{L^2}{2}\int_0^L \frac{\beta n(L')n(L'')\mathrm{d}L}{L''^2} - n(L)\int_0^\infty \beta n(L')\mathrm{d}L' - G\frac{\mathrm{d}n(L)}{\mathrm{d}L} = \frac{n(L)}{\tau} \qquad (6.5)$$

where by conservation of volume $L'^3 + L''^3 = L^3$ with the boundary condition

$$n^0 = \frac{B^0}{G} \qquad (6.6)$$

In general formulations β appears as a size-dependent kernel in the integrals in equation 6.5 as does the crystal growth rate $G(L)$. Formulation of the aggregation kernel, $\beta(L', L'')$ is chosen to correspond to the particular mechanism of aggregation. It accounts for the physico-chemical forces that affect the aggregation process and is observed to depend on crystal size and solution supersaturation amongst other factors (see p. 171). Similar equations to include breakage or attrition terms can be included in the model.

The general population balance equation requires numerical methods for its solution and several have been proposed (e.g. Gelbard and Seinfeld, 1978; Hounslow, 1990a,b; Hounslow *et al.*, 1988, 1990), of which more later. Fortunately, however, some analytic solutions for simplified cases also exist.

Analytical approximations

Continuous MSMPR

Equation 6.5 can be solved in an analytical form for two limiting cases in which besides nucleation only either (1) crystal growth; or (2) particle agglomeration occurs.

Nucleation and crystal growth

Solution of equation 6.4 is the now familiar exponential distribution which is linear in a semi-logarithmic plot $\ln n(L)$ versus L (Figure 6.11a)

$$n(L) = n^0 \exp\left(\frac{-L}{G\tau}\right) \qquad (6.7a)$$

or

$$\ln n(L) = \ln\frac{B_0}{G} - \frac{L}{G\tau} \qquad (6.7b)$$

The effect of crystal agglomeration is generally to cause concave curvature on the log population density versus size plot (see Figure 6.11b).

Nucleation and agglomeration

For purely (size-independent) agglomerative process, equation 6.5 was solved by Hostomský (1987) using Laplace transformation which yields

$$\ln n(L) = \ln n(L_0) - \left(\frac{5}{2}\right) \ln \frac{L}{L_0} - \left[\left(\frac{L}{L_0}\right)^3 - 1\right] \ln \left[\frac{\alpha}{\left(\alpha + \frac{1}{2}\right)}\right] \quad (6.8)$$

where L_0 is the primary particle size and

$$\ln n(L_0) = \ln \left\{\frac{3}{2}\left[\frac{1+2\alpha}{\pi}\right]^{\frac{1}{2}} \frac{1}{\beta \tau}\right\} + \ln \frac{\alpha}{\alpha + \frac{1}{2}} - \ln L_0 \quad (6.9)$$

and the dimensionless parameter α is defined as

$$\alpha = \beta_0 B^0 \tau^2 \quad (6.10)$$

where β_0 is the collision frequency factor averaged over the whole size range. For high values of α

$$n(L) = \left(\frac{L}{L_0}\right)^{-\frac{5}{2}} \quad (6.11)$$

Thus, in the purely agglomerative process with a typical value of the parameter $\alpha \gg 1$, the number density distribution $n(L)$ according to equation 6.11 is linear when plotted in logarithmic co-ordinates, i.e. $\ln n(L)$ versus $\log L$, thus exhibits an initial slope equal to $-5/2$, as shown in Figure 6.12. (*Note*: If the population density is defined on a unit solid volume, rather than size, basis, then the corresponding slope is $-3/2$, see also Jones *et al.*, 1996.)

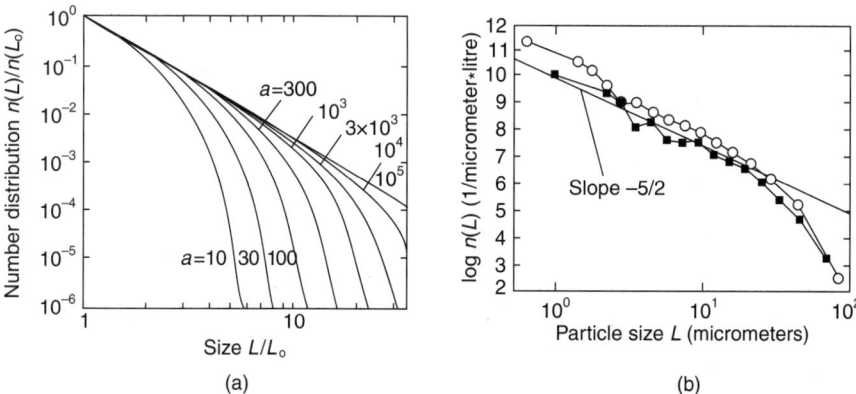

Figure 6.12 *Particle size distributions for agglomerative precipitation: (a) Theory according to equation 4.25; (b) Experimental results* ■ *$CaCO_3$,* ○ *$BaSO_4$ (Hostomský and Jones, 1993a,b)*

Well-mixed batch

Similarly, for pure agglomeration of an initial population of particles, the dynamic size distribution can be predicted as follows.

Perikinetic aggregation

By equating Fick's second law and the Stokes–Einstein equation for diffusivity, Smoluchowski (1916, 1917) showed that the collision frequency factor takes the form

$$\beta(L', L'') = \frac{2kT}{3\mu}[L' + L''][L'^{-1} + L''^{-1}] \tag{6.12}$$

For monodisperse particles, $[L' + L''][L'^{-1} + L''^{-1}] = 4$, the aggregation kernel becomes independent of particle volume and the aggregation rate is given by

$$-\frac{dN}{dt} = \frac{4kT}{3\mu}N^2 \tag{6.13}$$

and is thus a second order process.

Analytical solutions for the particle number distribution in the absence of growth and nucleation may be given as

$$N(t) = \frac{N_0}{\left(1 + \frac{4kT}{3\mu}N_0 t\right)} \tag{6.14}$$

Thus, a plot of $1/N$ against t is a straight line (Figure 6.13) and is independent of the agitation rate. Perikinetic motion is generally thought to apply to particles of less than ca. 1–10 µm, depending on the particle–fluid motion.

Orthokinetic aggregation

Smoluchowski also presented a simple theory of aggregation kinetics assuming collisions of perfect collection efficiency to predict spherical particle size distributions in a uniform liquid shear field of constant velocity gradient. The aggregation kernel is then expressed as

$$\beta(L', L'') = \frac{4}{3}G'[L' + L'']^3 \tag{6.15}$$

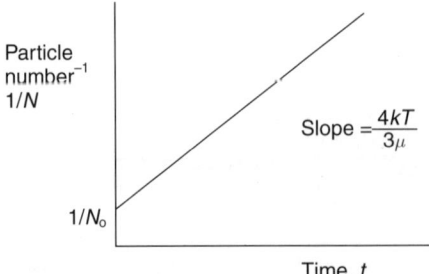

Figure 6.13 *Perikinetic particle aggregation*

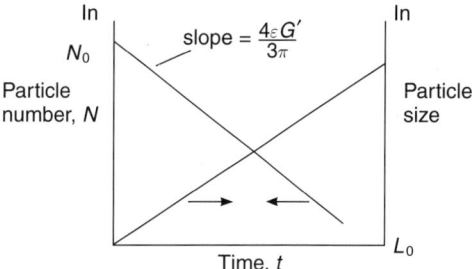

Figure 6.14 *Orthokinetic particle aggregation*

Thus, for monodisperse particles

$$-\frac{dN}{dt} = \frac{2}{3} G' L^3 N^2 \quad (6.16)$$

and is also a second order process.

If the particles are initially monodisperse and the total solids volume fraction ($\varepsilon = \pi L^3 N/6$) remains constant in the absence of crystal growth and nucleation then

$$N(t) = N_0 \exp\left(-\frac{4\varepsilon G' t}{3\pi}\right) \quad (6.17)$$

A plot of $\ln N$ against t thus yields a straight line (Figure 6.14) the slope of which depends on agitation rate.

Thus under ideal circumstances the modes of aggregation can be discriminated by such plots. Deviations below the expected slopes are usually attributed to 'collision inefficiency' leading to imperfect aggregation. In a crystallization or precipitation process, of course, deviations may also occur due to growth and nucleation unless properly accounted for.

Agglomeration and disruption during precipitation

Hartel *et al.* (1986a) and Hartel and Randolph (1986b) carried out an elegant study of the precipitation and subsequent aggregation of calcium oxalate dihydrate (CaOx) crystals in an MSMPR mini-nucleator/Couette aggregator sequence. Aggregation increased strongly with increasing CaOx supersaturation while disruption rates decreased indicating that a change in ionic conditions at the crystal surface enhanced the probability and strength of particle attachment on collision. Aggregation was only weakly dependent on rotation speed while rupture of the aggregates increased and the latter was presumed to be due to increasing turbulence.

The general form of the population balance including aggregation and rupture terms was solved numerically to model the experimental particle size distributions. While excellent agreement was obtained using semi-empirical two-particle aggregation and disruption models (see Figure 6.15), PSD predictions of theoretical models based on laminar and turbulent flow considerations

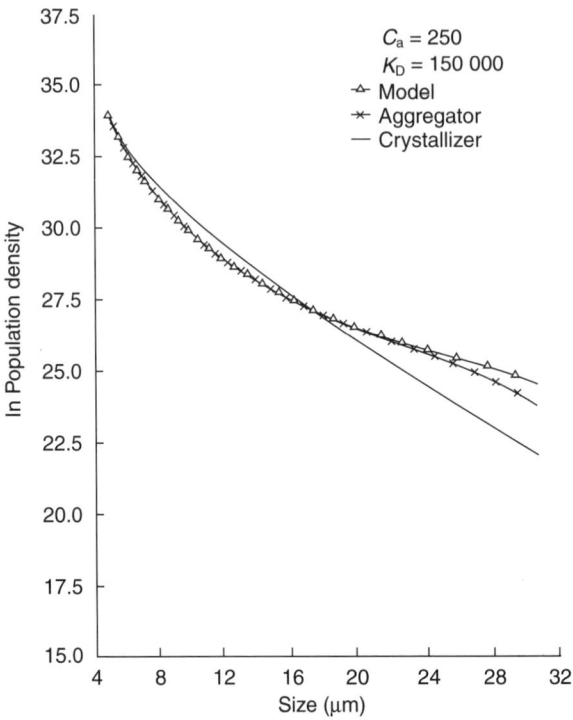

Figure 6.15 *Predicted and experimental calcium oxalate agglomerate size distributions (Hartel and Randolph, 1986)*

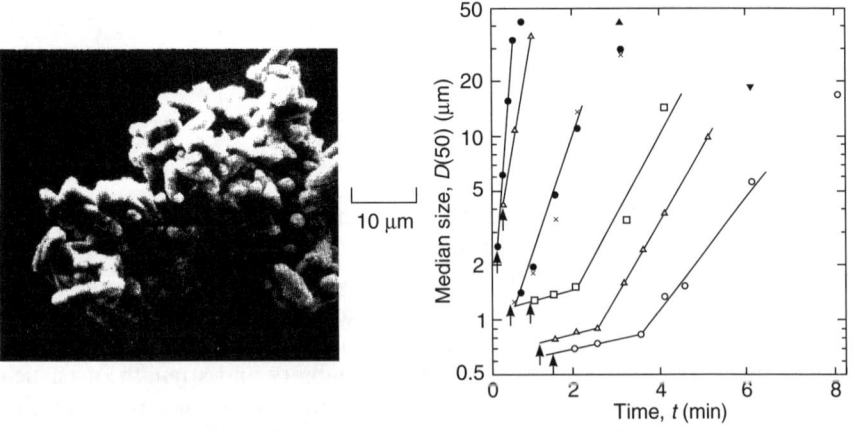

Figure 6.16 *Agglomerates and transient size of* $SrMoO_4$ *crystals (after Söhnel et al., 1988)*

deviated from the data indicating that more work is required to deduce the precise aggregation and disruption mechanisms.

Using an alternative approach to determine kinetics, induction time measurements were made in a recent study of the well-mixed batch precipitation of

Strontium molybdate (SrMoO$_4$) using laser light scattering (Söhnel et al., 1988). Bimodal distributions were observed and the presence of agglomerates was confirmed by microscopy (Figure 6.16).

Soon after the induction period, however, the small SrMoO$_4$ particles agglomerate exhibiting a log-linear transient particle size consistent with an orthokinetic mechanism (Figure 6.16) and develop into much larger agglomerates, the size of which depends on both the intensity of stirring and the initial supersaturation and ultimately stabilizes. Again, a supersaturation dependence of attachment efficiency was detected.

Collection efficiency

Orthokinetic collision due to shear is predicted to increase with increasing particle volume and shear rate but in practice the collection efficiency is relatively low and decreases with increasing particle size thereby imposing particle size limitations. This is thought to be due to not all collisions being successful, again due to fluid shear. Thus, Smoluchowski introduced a size dependent term in his model. Mumtaz et al. (1997) developed a mean field hydrodynamic model of aggregation to account for aggregation efficiency. They predict a size dependence of aggregation that passes through a maximum with increasing mean shear rate together with a correlating parameter for the aggregation efficiency, ψ, based on the ratio of aggregate strength: applied force between particles, M, Figure 6.17. Mersmann and Braun (2001) and Hounslow et al. (2001) provide detailed reviews of work in this area.

Hollander et al. (2001) report numerical simulations of orthokinetic aggregation in a turbulent channel flow and in a stirred tank, respectively. Using a

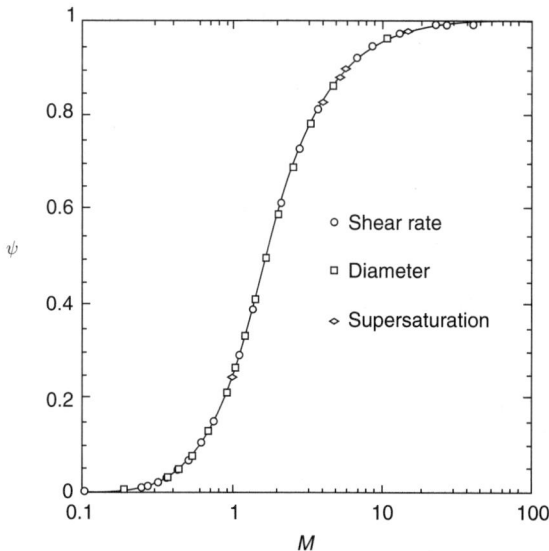

Figure 6.17 *Simulation results of Mumtaz et al. (1997) plotted as efficiency against the correlating parameter M (Hounslow et al., 2001)*

lattice Boltzmann scheme to simulate the turbulent flow field, and a Monte Carlo algorithm to solve for the particle size distribution, the dependence of the agglomeration rate on the shear rate and turbulent transport properties was investigated. It was found that local flow information is needed to model agglomeration and that a description in terms of a volume-averaged agglomeration rate constant β_0 is inadequate for both geometries. Furthermore, turbulent transport of particles is predicted to be an important factor determining the overall agglomeration rate. Subsequent disruption and break-up of a formed aggregate may also occur if the applied forces via hydrodynamic motion are sufficient to overcome its particulate shear strength (see also Chapter 8).

Agglomerate strength

Wójcik and Jones (1997, 1998b) report a decrease in the disruption rate of calcium oxalate and of calcium carbonate, respectively, with increasing growth rate. When agglomeration takes place during precipitation aggregates are only recently attached, so the crystalline bridges between the primary particles are not yet completely desolvated and cemented. The bonds of some aggregate fragments (not necessarily of the same coordination number as that originally attached) are thus broken relatively easily subsequently by agitation. Presumably, primary particles would usually (though not necessarily invariably) be the easiest to remove by fluid shear or collision. On the other hand, previously aged and dried crystal agglomerates, as often for seeds, are mechanically stronger and should therefore be more difficult to break so their attrition occurs at a lower rate. The role of liquid bridges during agglomeration is reviewed by Schubert (1981) and Pepin *et al.* (2001). Pratola *et al.* (2000) describe a novel microscopic method for determining the strength of agglomerating crystals in supersaturated solution (Figure 6.18) based on the MFB (Micro Force Balance) technique of Simons and Fairbrother (2000).

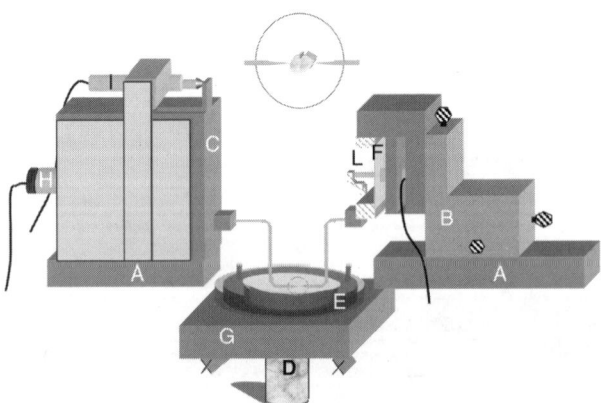

Figure 6.18 *Micro Force Balance (A) stage, (B) and (C) micromanipulators, (D) objective, (E) temperature controller device, (F) flexible blade, (G) travelling platform, (H) travel piezo adjuster, (I) and (L) LVDT (after Pratola et al., 2000)*

Determination of precipitation kinetics

Reliable kinetic data are of paramount importance for successful modelling and scale-up of precipitation processes. Many data found in the literature have been determined assuming MSMPR conditions, analogous to the CSTR model in reaction engineering. Here, a method developed by Zauner and Jones (2000a) is outlined.

The aim of the parameter estimation is to deduce the growth rate G, nucleation rate B^0, agglomeration kernel β_{aggl} and disruption kernel β_{disr} from the experimental CSD. The CSD is described mathematically by the population balance (Randolph and Larson, 1988)

$$\frac{\partial n}{\partial t} + \frac{\partial (Gn)}{\partial L} + n\frac{d(\log V)}{dt} = B - D - \sum_k \frac{Q_k n_k}{V} \tag{6.18}$$

With the assumption that the growth rate G is not a function of the particle size (McCabe's ΔL-law), for a constant crystallizer volume and for a crystal-free inlet flow (no seeding) the population balance becomes

$$\frac{\partial n}{\partial t} + G\frac{\partial n}{\partial L} = B_{aggl} + B_{disr} - D_{aggl} - D_{disr} - \frac{n}{\tau} \tag{6.19}$$

with the birth terms for agglomeration and disruption

$$B_{aggl} + B_{disr} = \frac{L^2}{2} \int_0^L \frac{K_{aggl} n(L_u) n(L_v) dL_u}{L_v^2} + \int_{L_v}^{\infty} K_{disr} S'(L_u, L_v) n(L_u) n(L_v) dL_u \tag{6.20}$$

and the death terms for agglomeration and disruption

$$D_{aggl} + D_{disr} = n(L) \int_0^{\infty} K_{aggl} n(L_u) dL_u + K_{disr} n(L) \tag{6.21}$$

and the mean crystal residence time

$$\tau = \frac{V}{Q} \tag{6.22}$$

Kumar and Ramkrishna (1996a,b) present a solution to population balance problems of agglomeration processes without restricting the choice of the grid of the discretized length scale. Two arbitrarily chosen properties, for example total number and mass of particles, can be preserved.

$$\frac{dN_i}{dt} = \left(\frac{dN_i}{dt}\right)_{nucl} + \left(\frac{dN_i}{dt}\right)_{growth} + \left(\frac{dN_i}{dt}\right)_{aggl} + \left(\frac{dN_i}{dt}\right)_{disr} - \frac{N_i - N_{i,in}}{\tau} \tag{6.23}$$

Nucleation and growth can be modelled using the discretization due to Hounslow (1990a):

$$\left(\frac{dN_i}{dt}\right)_{nucl} = \begin{cases} B^0 & i=1 \\ 0 & i \neq 1 \end{cases} \qquad (6.24)$$

$$\left(\frac{dN_1}{dt}\right)_{growth} = \frac{G}{L_1}[(b+cr)N_1 + cN_2] \qquad (6.25)$$

$$\left(\frac{dN_i}{dt}\right)_{growth} = \frac{G}{L_i}(aN_{i-1} + bN_i + cN_{i+1}) \qquad (6.26)$$

In the case of preservation of the zeroth and third moments (number and mass respectively) the equations for the agglomeration term can be written in the following form

$$\left(\frac{dN_i}{dt}\right)_{aggl} = \sum_{\substack{j,k \\ x_{i-1} \leq (x_j+x_k) \leq x_{i+1}}}^{j \geq k} \left(1 - \frac{1}{2}\delta_{j,k}\right)\eta K_{aggl} N_j N_k - N_i \sum_{k=1}^{i_{max}} K_{aggl} N_k \qquad (6.27)$$

with

$$\eta = \begin{cases} \dfrac{x_{i+1} - v}{x_{i+1} - x_i} & x_i \leq v \leq x_{i+1} \\ \dfrac{v - x_{i-1}}{x_i - x_{i-1}} & x_{i-1} \leq v \leq x_i \end{cases} \qquad (6.28)$$

and x_i as the representative volume of the ith size range and v as the particle volume.

Furthermore, according to Kumar and Ramkrishna (1996a,b), disruption can be accounted for with

$$\left(\frac{dN_i}{dt}\right)_{disr} = \sum_{k=i}^{i_{max}} n_{i,k} K_{disr} N_k - K_{disr} N_i \qquad (6.29)$$

with

$$n_{i,k} = \int_{x_i}^{x_{i+1}} \frac{x_{i+1} - v}{x_{i+1} - x_i} b(v, x_k) dv + \int_{x_{i-1}}^{x_i} \frac{v - x_{i-1}}{x_i - x_{i-1}} b(v, x_k) dv \qquad (6.30)$$

For binary breakage the breakage function $b(v, x_k)$ becomes

$$b(v, x_k) = \frac{2}{x_k} \qquad (6.31)$$

Use of the parabolic attrition function

$$b(v, x_k) = \frac{6}{x_k}\left[4\left(\frac{v}{x_k}\right)^2 - 4\left(\frac{v}{x_k}\right) + 1\right] \qquad (6.32)$$

was proposed by Hill and Ng (1995), and is based on an empirical form suggested by Austin et al. (1976) and Klimpel and Austin (1984).

Both breakage functions can be checked for consistency using the relation

$$\int_0^{x_k} b(v, x_k) \mathrm{d}v = 2 \tag{6.33}$$

for the number of particles formed per (binary) breakage event.

The sets of non-linear ordinary differential equations obtained by discretization were solved by Zauner and Jones (2000a). They used FORTRAN 90 programming including the NAG subroutine D02EAF, which is particularly suitable for stiff systems of first-order ordinary differential equations (ODEs), as here. This variable-order, variable-step method implements the Backward Differentiation Formulae (BDF) and is of an explicit type. The ODEs are integrated over a time range of ten residence times, assuming that after that time steady state has been achieved. Subroutines calculated values for the change of population due to nucleation, growth, agglomeration and disruption. The population density of each class is calculated from the number of particles in each size class.

It is notoriously difficult to solve the so-called inverse problem and extract kinetic data using the population balance. Muralidar and Ramkrishna (1986) describe a procedure to obtain agglomeration frequencies from measured size distributions without the kinetic processes of nucleation, growth and disruption. The authors point out that even if the experimental data are very accurate, it is not always possible to estimate the aggregation frequency satisfactorily and to distinguish between different mechanisms.

It is even more difficult to estimate not only one but four parameters (nucleation rate, growth rate, agglomeration kernel and disruption kernel) simultaneously from a particle size distribution. The errors are likely to be unacceptably high and it might be impossible to distinguish between the mechanisms involved. Therefore, an alternative sequential technique has been developed to obtain the kinetic parameters nucleation rate, growth rate, and agglomeration and disruption kernels from experimental precipitation data.

Parameter estimation

The following equation of moments (Randolph and Larson, 1988) for an ideal MSMPR crystallizer

$$\frac{\mathrm{d}\mu_3}{\mathrm{d}t} + m_3 \frac{\mathrm{d}(\log V)}{\mathrm{d}t} = 3G\mu_2 + \left(-\frac{Q\mu_3}{V}\right) \tag{6.34}$$

can be manipulated in such a way that the growth and nucleation kinetics can be determined without estimation or approximation.

Growth and nucleation

Assuming steady state is reached, size-independent growth and that the third moment (mass) is conserved in the agglomeration and disruption process, i.e.

$$\frac{\mathrm{d}\mu_3}{\mathrm{d}t} + \mu_3 \frac{\mathrm{d}(\log V)}{\mathrm{d}t} = 0 \tag{6.35}$$

the following equation is obtained

$$0 = 3G\mu_2 - \frac{\mu_3}{\tau} \tag{6.36}$$

Therefore, knowing the moments of the distribution it is possible to calculate the growth rate G from

$$G = \frac{\mu_3}{3\mu_2 \tau} \tag{6.37}$$

Furthermore, from the relationship

$$n^0 = \frac{B^0}{G} \tag{6.38}$$

the nucleation rate B^0 can be calculated as

$$B^0 = n^0 G \tag{6.39}$$

with n^0 obtained from the intercept of the experimental distribution with the abscissa at $L = 0$ (or $L = L_0$).

Agglomeration kernel

The next kinetic parameter to be determined, the agglomeration kernel β_{aggl}, is calculated using the kinetic data already known. Various expressions for the agglomeration kernel are suggested in the literature with many of these kernels having a sound theoretical basis and others being purely empirical. Some of the kernels important for precipitation in an agitated vessel are listed as follows.

Smoluchowski kernel (Smoluchowski, 1916)

As early as 1916, Smoluchowski showed that aggregation of spherical particles in a laminar shear field can be expressed as

$$K_{aggl}(L_u, L_v) = \frac{1}{6}\dot{\gamma}[L_u + L_v]^3 = \beta_{aggl}[L_u + L_v]^3 \tag{6.40}$$

with the shear rate $\dot{\gamma}$ (velocity gradient) and the particle sizes L_u and L_v of the aggregating particles. As the agglomeration rate increases with the volume of the particle, it is more likely that a large particle is involved in an agglomeration event than a small particle. In similar form, agglomeration due to diffusion of particles in eddies in a turbulent flow regime can be expressed as

$$K_{aggl}(L_u, L_v) = kU[L_u + L_v]^3 = \beta_{aggl}[L_u + L_v]^3 \tag{6.41}$$

with U as the velocity gradient of the fluid in turbulent motion being proportional to $\varepsilon^{1/2}$ (Low, 1975).

Thompson kernel (Thompson, 1968)

This empirical formulation of a kernel is given as

$$K_{\text{aggl}}(L_u, L_v) = C_A E_i \frac{(u-v)^2}{u+v} = \beta_{\text{aggl}} \frac{[L_u^3 - L_v^3]^2}{L_u^3 + L_v^3} \qquad (6.42)$$

with the collection efficiency E_i and the aggregation parameter C_A. Hartel *et al.* (1986) and Hartel and Randolph (1986) successfully modelled the agglomeration kinetics of calcium oxalate using this kernel.

Invariant agglomeration kernel

Although the theoretical consideration above imply a positive size-dependence of aggregation, the particle size distributions observed can often be approximated by a size-independent kernel.

$$K_{\text{aggl}} = \beta_{\text{aggl}} \neq f(L_u, L_v) \qquad (6.43)$$

Both aggregation inefficiency (Adler, 1981) and particle disruption (Hartel and Randolph, 1986) increase with particle size. These dispersive processes can counteract the positive effect of aggregation thereby imposing agglomerate particle size limitations and may give rise to apparent size-independence.

Thus in a mixed system, as e.g. in a stirred tank, the rate of agglomeration additionally depends on the shear field and therefore on the energy dissipation ε in the vessel. Furthermore, in precipitation systems solution supersaturation plays an important role, as the higher the supersaturation, the 'stickier' the particles and the easier they agglomerate (Mullin, 2001). This leads to a general formulation of the agglomeration rate

$$K_{\text{aggl}}(L_u, L_v, \varepsilon, S) = \beta_{\text{aggl}} \varepsilon^p S^q f(L_u, L_v) \qquad (6.44)$$

where both the energy dissipation and the level of supersaturation are accounted for using a power law function. The number density distribution of the population balance model was fitted to the number density distributions of the continuous experiments and the sum of least squares of the population balance model and the experimental data minimized.

Disruption kernel

The function most often used in modelling breakage and disruption processes has the form

$$K_{\text{disr}}(L_u, L_v) = \beta_{\text{disr}}[L_u + L_v]^3 \qquad (6.45)$$

With this kernel, the disruption rate is proportional to the particle volume. This theoretical assumption of the disruption rate dependence on particle volume was validated by Synowiec *et al.* (1993), by demonstrating that a third-order dependence on the particle size (and therefore a proportionality on particle volume) can be explained by the dominant disruption mechanism of turbulent crystal-shear forces.

Additionally, the disruption rate is also a function of the degree of supersaturation prevailing in the reactor. High supersaturation results in high growth rates and strong agglomerates and thus reduces breakage (Wójcik and Jones, 1997, 1998b). In addition, the rate of disruption increases with increased power input. Therefore, the disruption rate function becomes

$$K_{\text{disr}}(L_u, L_v, \varepsilon, S) = \beta_{\text{disr}} \varepsilon^r S^s f(L_u, L_v) \tag{6.46}$$

A further complication in the breakage process is that depending on the breakage mechanism a breakage event can lead to

- two daughter particles of the same or nearly the same size (particle splitting)
- two daughter particles of very different sizes (attrition or abrasion)
- a number of daughter particles (multiple breakage or micro-attrition).

In order to describe these different mechanisms, various breakage functions have been proposed (Hill and Ng, 1995, 1996). For precipitation processes, a breakage function of the form given in equation (6.32) with $b(v, x_k)$ being the discretized number fraction of particles broken from size v into size interval x_k, seems particularly suitable as both attrition – with a high probability – and particle splitting – with a low probability – are accounted for.

Zauner and Jones (2000a) determined the evaluation of the disruption kernel using seeded batch disruption experiments, where an initial particle size distribution was stirred for a certain time and the change in the distribution analysed. The solution was saturated so that negligible growth, agglomeration and nucleation occurred and therefore all changes in the particle size distribution could be attributed to attrition and breakage. By fitting the population balance model for disruption to the experimental disruption data, the disruption kernel was estimated. The experiments were carried out at different stirrer speeds, and therefore the dependence of the disruption kernel on the specific power input could be determined.

Experimental method

Zauner and Jones (2000a) describe an experimental set-up for determination of precipitation kinetics, as shown in Figure 6.19. Briefly, the jacket glass reactor (1) (300 ml, $d = 65$ mm) is equipped with a polyethylene draft tube and four baffles. The contents are stirred using a three-blade marine-type propeller (5) with motor (Haake), which pumps the suspension upwards in the annulus and downwards inside the draft tube. Measured power inputs ranged from 3.3×10^{-3} to 1.686 W/kg.

The product withdrawal pipe is located in the annulus halfway up the draft tube where the vector of the circulation velocity of the suspension and that of the withdrawal velocity of the product point in the same direction. Using this arrangement ensures low particle classification effects. Furthermore, the feed tubes and withdrawal tube can be exactly positioned in the reactor and scaled

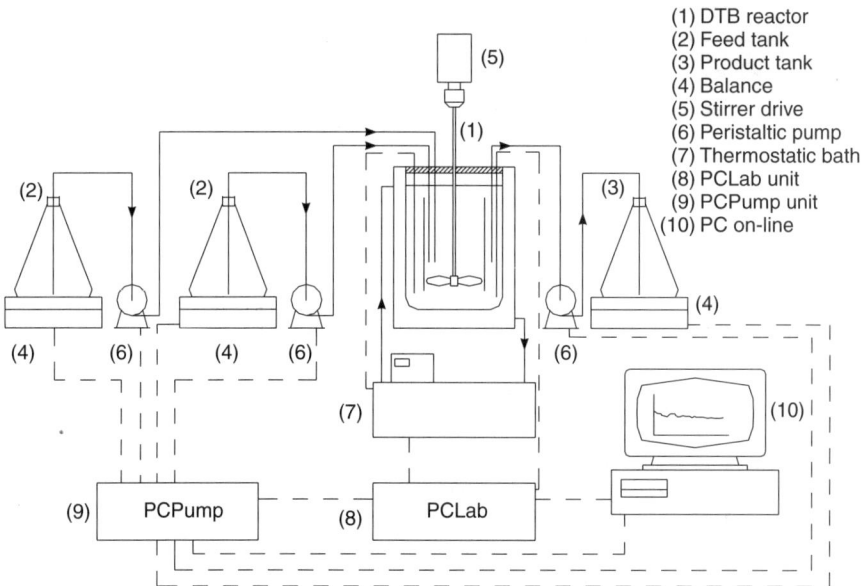

Figure 6.19 *Experimental set-up: continuous MSMPR reaction-crystallizer (Zauner and Jones, 2000a)*

up with high accuracy which would appear to be of fundamental importance for the supersaturation profile of the reaction plumes and subsequently for nucleation and growth.

The particle size distribution is determined with a Coulter Counter. In order to avoid secondary changes of the precipitate during the particle sizing, a suitable electrolyte has to be found which on the one hand fulfils the conductivity criteria necessary to detect particles as they pass through the orifice and on the other does not dissolve particles. A saturated solution of calcium oxalate with 3 per cent by weight sodium chloride proved to be suitable. The electrolyte is filtered twice through 0.1 µm Whatman cellulose nitrate membrane filters in order to achieve low background counts of particles already present in the electrolyte. A dispersant especially designed for a Coulter Counter measurement is added to avoid secondary aggregation of the particles during the measurement.

Crystal growth rates

The crystal growth rates can be directly determined from the second and third moment as described above. The calculated rates for calcium oxalate here are in the range 0.75×10^{-8} to 4.7×10^{-8} m/s. Literature values for the growth rate of calcium oxalate monohydrate vary considerably: 1.08×10^{-8} m/s (Kavanagh, 1992), 3.4×10^{-9} to 5.0×10^{-8} m/s (Garside *et al.*, 1982) and 2.8×10^{-10} to 1.11×10^{-8} m/s (Nielsen and Toft, 1984). The values obtained from the experiments are therefore within the range of the literature data. It should be borne in

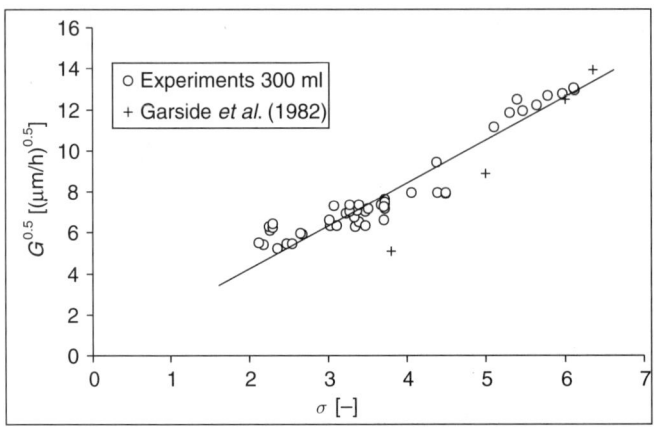

Figure 6.20 *Growth rate of calcium oxalate versus supersaturation at 37°C (Zauner and Jones, 2000a)*

mind, however, that direct comparison of kinetic data is not always possible. The growth rate of calcium oxalate crystals is correlated with the supersaturation σ in Figure 6.20.

The linear dependence of $\sqrt{G}$ on σ suggests a power law relationship of the form

$$G = k_g \sigma^2 \qquad (6.47)$$

A kinetic order of two was also observed by Nielsen and Toft (1984) and by Nancollas and Gardner (1974).

The second-order dependence of the growth rate on the supersaturation can be explained by a number of growth theories. The most convincing, however, is that of Burton *et al.* (1951). In their BCF theory about the screw dislocation centred surface spiral step, it is assumed that growth units enter at kinks with a rate proportional to σ and that the kink density is also proportional to σ which gives the factor σ^2 in the rate expression.

Nucleation rates

Nucleation rates, which are calculated from the growth rates and the population densities at $L \to 0$ cover a range from 5.6×10^8 to 2.8×10^{11} m^{-3}s^{-1}. Literature values for nucleation rates of calcium oxalate monohydrate are rare; Garside *et al.* (1982) found nucleation rates from 2.78×10^7 to 8.33×10^7 m^{-3}s^{-1} at 37°C and 0.001 M feed concentration, and those calculated from the data of Brown *et al.* (1990) are between 2.78×10^4 and 2.7×10^{10} m^{-3}s^{-1}, also at 37°C.

The dependence of the nucleation rate on stirrer speed is strongly influenced by the feed point position and the feed concentration. The small influence of the stirrer speed and hence the power input on the nucleation rate is observed for the feed point positions in a zone of small turbulence (od). For the feed point inside the draft tube (id) and a residence time of 660 s, B^0 reaches a maximum at

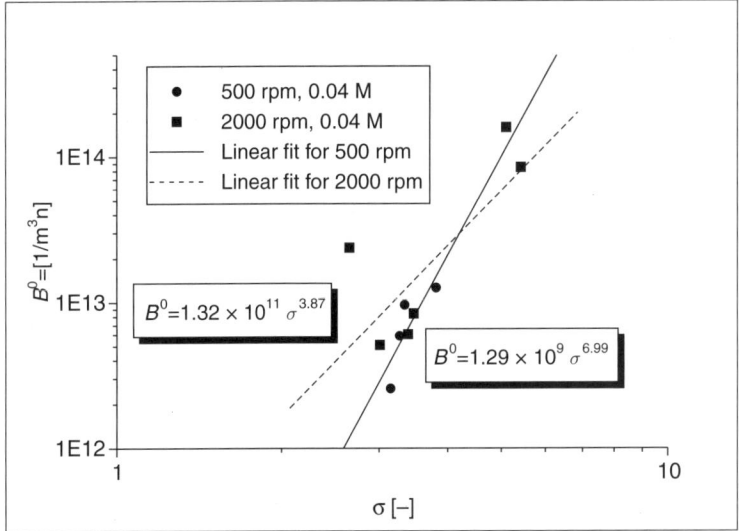

Figure 6.21 *Nucleation rate of calcium oxalate versus supersaturation at* 37 °C *(Zauner and Jones, 2000a)*

25 Hz and decreases at 33.4 Hz. By comparing data of different feed point positions, it was found that the nucleation rates for the experiments with the feed point located near the impeller were higher than those for feed points outside the draft tube.

The nucleation rate is plotted versus the supersaturation for different stirrer speeds in a log-log diagram (Figure 6.21). The kinetic order n in the correlating equation

$$B^0 = k_n \sigma^n \tag{6.48}$$

varies between 3.9 and 7.0 depending on the stirrer speed. As the experimental data scatter substantially, the correlation (equation 6.48) seems rather more arbitrary than satisfying. This might be an early indication that the nucleation kinetics is mixing-limited. Different stirrer speeds lead to different distributions of the supersaturation and therefore to inhomogeneities and steep local gradients in the nucleation rates. Therefore, the approach described above is not suitable for determining fundamental nucleation kinetic data. Söhnel and Garside (1992) describe other methods based on e.g. turbidity and induction time measurements. For calcium oxalate, data published by Brown *et al.* (1991) can be used to model primary nucleation kinetics. The authors measured the change in turbidity and related the nucleation kinetics to supersaturation ratio leading to the kinetic expression

$$B^0 = 9.38 \times 10^{11} \exp^{-\frac{52.09}{\ln^2 S}} \tag{6.49}$$

indicating a high-order dependence of nucleation rate on the degree of supersaturation.

Disruption kernel

As mentioned above, the disruption kernels were determined from batch experiments under zero supersaturation $\sigma = 0$ ($S = 1$). Figure 6.22 shows the initial particle size distribution and the distribution after 2 h of stirring. Due to attrition and breakage of particles, the distribution gradually shifts to smaller sizes as large particles are reduced in size by breakage or attrition. Furthermore, the measured total number of particles increases as each breakage event leads to a number of daughter particles.

The expected increase of the disruption rate with the stirrer speed (power input) was confirmed for all the experiments (Figure 6.23).

A nearly linear dependence of the disruption kernel on the power input was observed leading to the relation

$$\beta_{\text{disr}} \propto \varepsilon \qquad (6.50)$$

The disruption experiments were carried out at $\sigma = 0$ ($S = 1$) and therefore did not account for any effects of the supersaturation on the disruption process. Hartel and Randolph (1986b) and Wójcik and Jones (1997) reported a decrease in the disruption rate of calcium oxalate and of calcium carbonate, respectively, with increasing growth rate. Based on these findings, a linear decrease of the disruption kernel with the growth rate was assumed giving

$$\beta_{\text{disr}} \propto G^{-1} \quad \text{or} \quad \beta_{\text{disr}} \propto \sigma^{-2} \quad \text{or} \quad \beta_{\text{disr}} \propto S^{-2.15} \qquad (6.51)$$

The final relationship proposed for the disruption kernel becomes therefore

$$\beta_{\text{disr}} = 6.25 \times 10^{-5} \varepsilon S^{-2.15} \qquad (6.52)$$

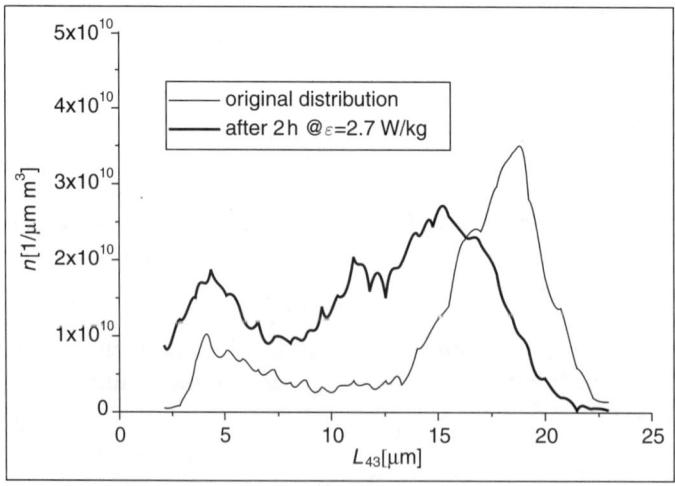

Figure 6.22 *Particle size distribution for disruption experiment (calcium oxalate, $\varepsilon = 2.7$ W/kg) (Zauner and Jones, 2000a)*

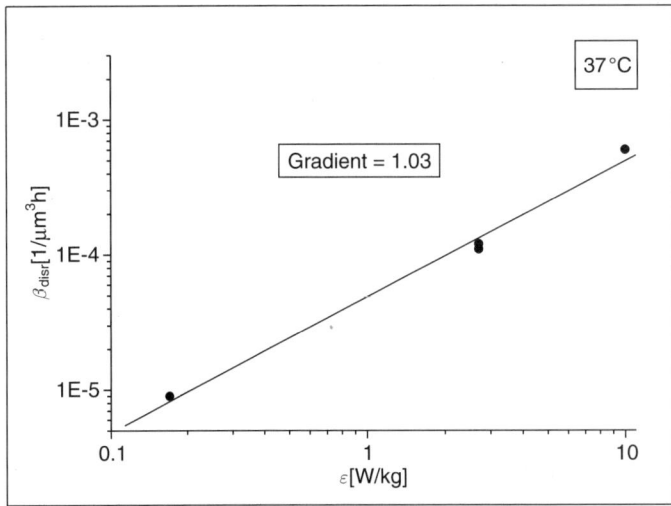

Figure 6.23 *Disruption kernel of calcium oxalate versus power input (Zauner and Jones, 2000a)*

Agglomeration kernel

To extract the agglomeration kernels from PSD data, the 'inverse problem' mentioned above has to be solved. The population balance is therefore solved for different values of the agglomeration kernel, the results are compared with the experimental distributions and the sums of non-linear least squares are calculated. The calculated distribution with the minimum sum of least squares fits the experimental distribution best.

Parameter estimation was performed for different agglomeration kernels and the results for one run are compared in Figure 6.24.

The size-dependent agglomeration kernels suggested by both Smoluchowski and Thompson fit the experimental data very well. For the case of a size-independent agglomeration kernel and the estimation without disruption (only nucleation, growth and agglomeration), the least square fits substantially deviate from the experimental data (not shown). For this reason, further investigations are carried out with the theoretically based size-dependent kernel suggested by Smoluchowski, which fitted the data best:

$$K_{\text{aggl}}(L_u, L_v) = \frac{1}{6}\dot{\gamma}[L_u + L_v]^3 = \beta_{\text{aggl}}[L_u + L_v]^3 \quad (6.53)$$

For stirrer speeds of 4.2, 8.4, 16.7, 25 and 33.4 Hz, agglomeration kernels obtained in this study vary from 0.01 to 183 s^{-1}. Unfortunately, no other measured data for agglomeration of calcium oxalate analysed using Smoluchowski's kernel were found in the literature. The corresponding values reported by Wójcik and Jones (1997) for calcium carbonate, however, cover a range from 0.4 to 16.8 s^{-1}.

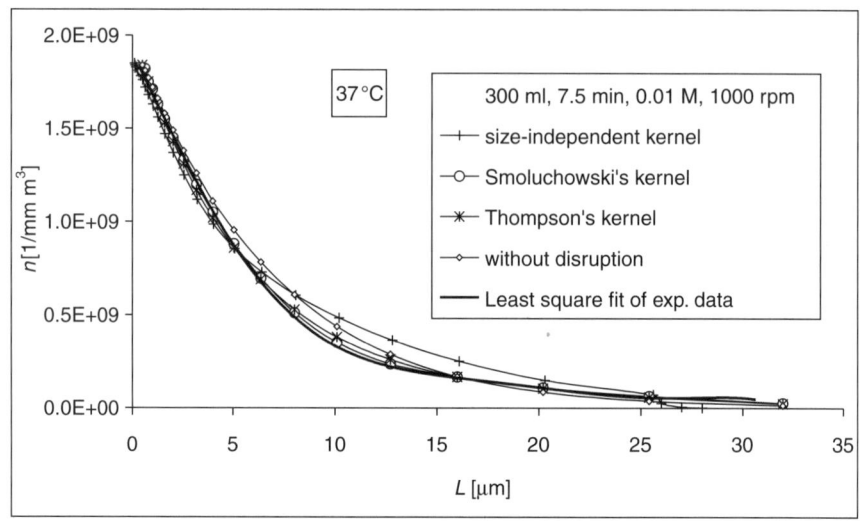

Figure 6.24 *Comparison of various agglomeration kernels (calcium oxalate) (Zauner and Jones, 2000a)*

According to Smoluchowski's theory (equation 6.53), the agglomeration rate increases proportional with the fluid shear rate $\dot{\gamma}$

$$\beta_{aggl} \propto \dot{\gamma} \tag{6.54}$$

Camp and Stein (1943) originally proposed for flocculation that the mean fluid shear rate is proportional to the square root of the mean energy dissipation

$$\bar{\dot{\gamma}} = \sqrt{\frac{\bar{\varepsilon}}{\mu}} \tag{6.55}$$

therefore

$$\beta_{aggl} \propto \sqrt{\bar{\varepsilon}} \tag{6.56}$$

Based on these considerations, and assuming an aqueous solution of density $1000\,\text{kg/m}^3$ and viscosity $1 \times 10^{-3}\,\text{Ns/m}^2$ (not measured), corresponding estimated Smoluchowki aggregation kernels (one sixth mean shear rate) range from 9.5 to $216\,\text{s}^{-1}$, being somewhat higher than those inferred experimentally, above. Figure 6.25 shows the dependency of the aggregation rate on the shear rate as obtained from the experiments.

Recent work has, however, cast doubt on the precise relationship between agglomeration rates and power input as indicated by equation 6.58 (see Rielly and Marquis, 2001). Nevertheless, at low power inputs the suspension is poorly mixed with wide variations in local energy dissipation (Zauner, 1999), and agglomeration is then very inefficient. At maximum occurs in the agglomeration rate at medium power inputs. This can be explained by the proposition that due to very high power inputs the particles have insufficient time to 'stick' together and form agglomerates. Mumtaz et al. (1997) found a similar beha-

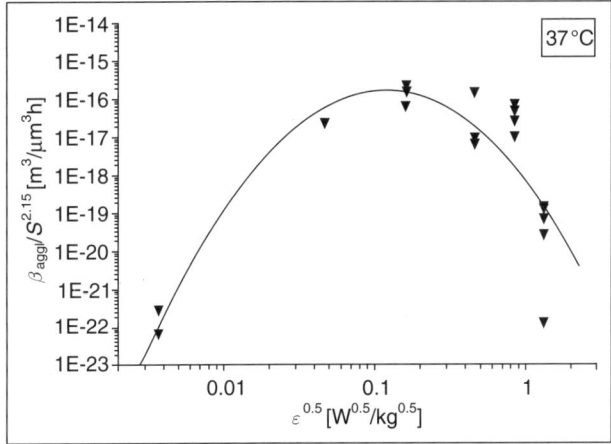

Figure 6.25 *Dependence of calcium oxalate aggregation rates on power input (Zauner and Jones, 2000a)*

viour in their agglomeration experiments with calcium oxalate monohydrate and attributed theoretically the decrease in the agglomeration rate after a maximum at a certain shear rate to decreased agglomeration efficiency. Collier and Hounslow (1999) drew a similar conclusion.

Agglomeration rates also depend on the level of supersaturation in the reactor and on the power input. Wójcik and Jones (1997) found a linear increase of the agglomeration kernel with the growth rate. Therefore, the level of supersaturation was accounted for by Zauner and Jones (2000a) using the relation

$$\beta_{aggl} \propto G \quad \text{or} \quad \beta_{aggl} \propto \sigma^2 \quad \text{or} \quad \beta_{aggl} \propto S^{2.15} \tag{6.57}$$

The final modelling equation proposed for the agglomeration kernel is

$$\beta_{aggl} = 5.431 \times 10^{-17}(1 + 2.296\varepsilon^{\frac{1}{2}} - 2.429\varepsilon)S^{2.15} \tag{6.58}$$

In this equation, the first term relates to agglomeration effects independent of power input and the second term in brackets to agglomeration due to shear as described in Smoluchowski's theory. The third term describes the reduced agglomeration efficiency found experimentally at higher shear rates and is consistent with the findings of Mumtaz *et al.* (1997).

Troubleshooting agglomeration

It is plain from the above review of the literature that agglomeration can be a very complicated phenomenon in which contrary behaviour is often observed, and despite much work the theoretical basis is yet to be fully developed, particular problems tending to require specific solutions. Nevertheless, faced with a given problem there is some guidance that can be offered for systematic study, *caveat emptor*.

The main factors that appear to control agglomeration in many systems are:

1. Level of supersaturation
2. Suspension density
3. Particle size
4. Degree of agitation
5. Ionic strength
6. Presence of impurities.

In all such laboratory studies, plant conditions and compositions should be employed as far as possible. Agglomeration rates tend to increase with the level of supersaturation, suspension density and particle size (each of which will, of course, be related but the effects may exhibit maxima). Thus, agglomeration may often be reduced by operation at low levels of supersaturation e.g. by controlled operation of a batch crystallization or precipitation, and the prudent use of seeding. Agglomeration is generally more predominant in precipitation in which supersaturation levels are often very high rather than in crystallization in which the supersaturation levels are comparatively low.

The occurrence of perikinetic or orthokinetic agglomeration can be deduced by the effect of agitation. Perikinetic is unaffected by flow and occur to particles in the nm size range. Particles >1–$10\,\mu m$ may be controlled by modifying flow conditions. Agitation rates should be decreased with caution, however, consistent with achieving adequate particle suspension and liquid mixing to avoid local peaks in supersaturation levels. On the other hand, very high shear rates tend to break-up and diminish agglomerates, of course, despite increasing collision rates.

To reduce or avoid agglomeration removal of certain impurities such as charged polymers could be effective – once identified, while to enhance agglomeration certain additives could be used as are commonly employed as flocculating agents to enhance solid–liquid separation in the water industry for example. These effects have to be determined empirically with care, however, since they can be pH and supersaturation dependent.

Such studies described above simply entail effect experiments in which the data may be correlated empirically over the practical range of interest. More rigorous fundamental studies require the determination of agglomeration kinetics and mechanisms. These are best carried out under defined conditions of composition, temperature and degree of agitation. If certain simplifying assumptions can be made, the analysis of agglomeration can be quite straightforward but simultaneous crystallization and agglomeration kinetic studies require the full formulation of the population balance models.

Finally, it should be emphasized that in all these studies more than one particle characteristic should be determined and this will usually entail more than one method of measurement. These can include various sizing techniques and the value of microscopy should not be ignored. Often, however, end-use tests provide the most meaningful data e.g. sedimentation rates or filtrability.

Summary

Crystal agglomeration is a particle formation process that leads to a rapid increase in particle size, particularly during precipitation processes where the degree of supersaturation is very high. Particle movement affects these growth processes, but that selfsame process also induces particle disruption. Thus, crystal agglomeration is the net result of two opposing processes – aggregation and dispersion. Agglomeration tends to occur more in precipitation than in simple crystallization processes since the supersaturation and hence numbers of primary particles are by far the higher, as is their propensity for inter-growth. A full population balance analysis of nucleation, crystal growth and particle agglomeration is available but its solution is complex. If crystal growth is negligible as a size enlargement mechanism in comparison with agglomerative particle growth, however, then the population balance can be integrated to yield a limiting agglomerate size distribution from both batch and continuous MSMPR crystallizers.

Precipitation kinetic rates can also be extracted from the product particle size distribution from an MSMPR crystallizer. For crystal growth, it is possible to relate inferred rates to the measured level of supersaturation in the reactor. For nucleation, however, a mixing limitation of the kinetics is found due to the instantaneous reaction of the feed solutions, resulting in high local variations in the nucleation rate. It is therefore not possible to correlate the nucleation kinetics to the mean level of supersaturation and specific power input with precision. Measurement techniques based on turbidity or induction times, for example, might then be more suitable. The size-dependent agglomeration kernel exhibited a maximum with increasing power input, suggesting decreasing agglomeration efficiency at higher power inputs and therefore a deviation from Smoluchowski's theory. In addition, however, breakage occurs as a size-reducing kinetic process, linearly increasing with power input. Thus both mechanisms viz. aggregation inefficiency and particle disruption reduce the number of agglomerates observed at higher stirrer speeds.

Unfortunately, present theoretical knowledge does not permit prediction crystallization and precipitation kinetics for a particular substance *a priori*. Experimental determination is essential but this is often a slow process unless efficient experimental design is employed. Additionally, there is no single or universal method that is suitable for kinetics determination. Indeed, it is often advisable to use more than one method to determine a good understanding of crystallization kinetics in a given case – batch crystallization, single crystal growth, seeded growth, induction time measurements, attrition determination, continuous MSMPR studies, microscopic observation etc.

7 Crystallizer design and performance

Tailoring of the particle size of the crystals from industrial crystallizers is of significant importance for both product quality and downstream processing performance. The scientific design and operation of industrial crystallizers depends on a combination of thermodynamics – which determines *whether* crystals will form, particle formation kinetics – which determines *how fast* particle size distributions develop, and residence time distribution, which determines the *capacity* of the equipment used. Each of these aspects has been presented in Chapters 2, 3, 5 and 6. This chapter will show how they can be combined for application to the design and performance prediction of both batch and continuous crystallization.

Further details can be found in several texts including those on the theory of particulate processes (Randolph and Larson, 1988), crystallization (e.g. Van Hook, 1961; Bamforth, 1965; Nývlt, 1970; Jančić and Grootscholten, 1984; Garside *et al.*, 1991; Nývlt, 1992; Tavare, 1995; Mersmann, 2001; Myerson, 2001; Mullin, 2001).

Design and performance of batch crystallizers

Batch crystallizers are widely used in the chemical and allied industries, solar saltpans of ancient China being perhaps the earliest recorded examples. Nowadays, they still comprise relatively simple vessels, but are usually (though not always) provided with some means of agitation and often have artificial aids to heat exchange or evaporation. Batch crystallizers are generally quite labour intensive so are preferred for production rates of up to say 10 000 tonnes per year, above which continuous operation often becomes more favourable. Nevertheless, batch crystallizers are very commonly the vessel of choice or availability in such duties as the manufacture of fine chemicals, pharmaceutical components and speciality products.

Techniques

It was shown in Chapter 3 that supersaturation, or concentration driving force, is essential for any crystallization. In a batch crystallizer supersaturation can be generated in several ways, either solely or in combination:

1. cooling
2. evaporation
3. precipitation.

Cooling is the simplest method. It can be achieved simply by use of cooling medium (usually water) passing through a coil or jacket attached to the vessel.

Evaporation can be achieved simply by blowing air across a solution free surface, an effect that is enhanced by operation under vacuum during which cooling may also take place. Precipitation can occur in several ways, either by the addition of a reagent producing a substance that is relatively insoluble, by 'salting out' with a common ion, or by changing the power of the solvent e.g. by 'drowning-out' with a less powerful 'co-solvent' (often called an 'anti-solvent', or 'diluent'). In many cases a batch of nascent liquor is also seeded in order to provide sites for crystal growth. Jones, 1984 and Tavare (1987) provide reviews.

Equipment

As mentioned above batch crystallizers are usually simple vessels provided with some means of mechanical agitation or particulate fluidization. These have the effect of reducing temperature and concentration gradients, and maintain crystals in suspension. Baffles may be added to improve mixing and heat exchange or vacuum systems may be added, as appropriate. Various design combinations are available and some are illustrated in Figure 7.1.

Fluid mixing and particle suspension in unbaffled vessels (Figure 7.1(a)) may be poor, experiencing vortexing but this can be improved by the addition of wall

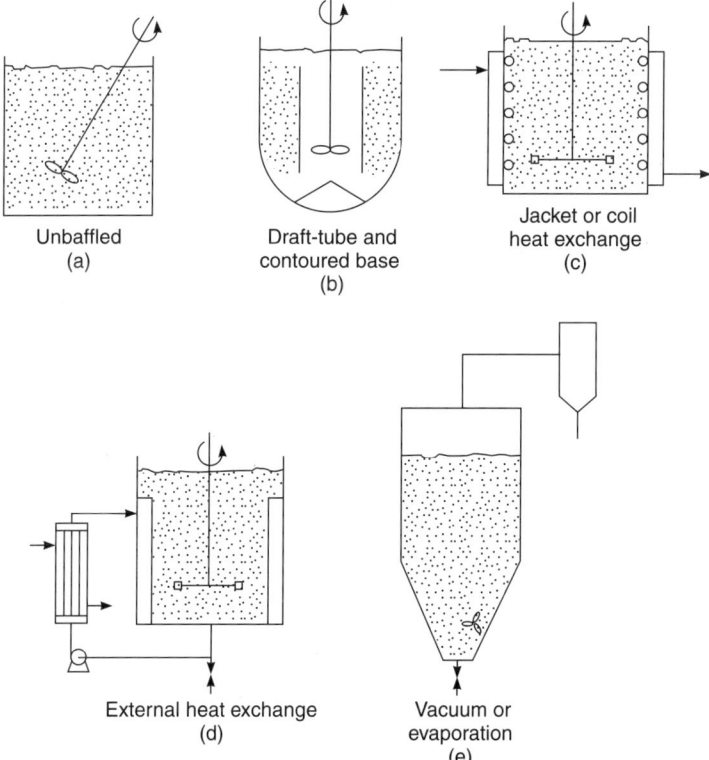

Figure 7.1 *Some agitated batch crystallizers: (a) Simple unbaffled; (b) draft-tube baffled; (c) jacket or coil heat exchange; (d) external heat exchange; and (e) vacuum or evaporation*

baffles (Figure 7.1(d)). These have the action of breaking up the swirling flow pattern and inducing turbulence. The draft-tube (Figure 7.1(b)) is essentially a shroud surrounding the impeller. It directs the flow in a vertical direction (up or down) and ensures good blending of the vessel contents from top to base. Heat exchange coils or jackets may be added for cooling (Figure 7.1(c)). A side stream can be drawn from the vessel through an external loop by a pump and be passed through a heat exchanger prior to return to the vessel in a more supersaturated condition. Vacuum or evaporation may be applied (Figure 1.1(e)).

Batch crystallizer design and operation

Design of the batch crystallizer has three main important aspects:

Vessel sizing

- Crystallizer capacity, which is determined by throughput and batch time.
- Agitator or pump speed and power, which are determined by particle size and suspension density.

Operating policy

- Specification of the appropriate cooling curve, evaporation rate, etc.
- A policy for seeding (mass and size) may be needed.
- The batch time must be selected to ensure the final particle size.

Performance measures

Performance measures are normally based on some aspect of product 'quality' (expressed in terms of the corresponding CSD and/or crystal purity) and crystal yield on a mass basis.

These design aspects may be subject to constraints including maximum supersaturation level (which may affect scaling or encrustation), etc. Thus for any given duty, the size of the vessel, the operating policy and performance of the crystallizer are interrelated.

Heat and mass balances

As for other mass transfer operations in chemical engineering, several authors have proposed equations for the calculation of heat and mass balances used for the estimation of crystal yield, heat load, and evaporation duty in batch crystallizations, e.g. (Mullin, 2001)

(i) Overall mass balance

$$M_{lo} = M_{lf} + M_{cf} + M_g + V_{\rho_1} \tag{7.1}$$

(ii) Anhydrous solute mass balance

$$M_{ho}c_o + M_{hf}c_f + \frac{M_{anh}}{M_{hyd}} M_{cf} \qquad (7.2)$$

(iii) Overall heat balance (assuming constant specific heat)

$$M_g h_{lg} = Q + M_c h_c + C_{po}(\theta_o - \theta_f) M_{lo} \qquad (7.3)$$

Thus for batch crystallizations the mass yields and heat loads are given by:

(a) Cooling crystallization

$$M_{cf} = \frac{M_h R(c_o - c_f)}{1 - c_f(R - 1)} \qquad (7.4)$$

$$-\frac{Q}{V\rho_l} = \frac{M_{cf}}{V\rho_l} h_c + C_{po}(\theta_o - \theta_f) \qquad (7.5)$$

(b) Evaporative crystallization

$$M_{cf} = \frac{M_{ho} R(c_o - c_f(1 - F))}{1 - c_f(R - 1)} \qquad (7.6)$$

$$-\frac{Q}{V\rho_l} = \frac{M_g}{V\rho_l} h_{lg} - \frac{M_{cf}}{V\rho_l} h_c + C_{po}(\theta_o - \theta_f) \qquad (7.7)$$

$$M_g = \frac{h_c R(c_o - c_f) + C_{po}(\theta_o - \theta_f)(1 + c_o)\{1 - c_f(R - 1)\}}{h_{lg}\{1 - c_f(R - 1)\} - h_c R c_f} \qquad (7.8)$$

Therefore, if the solubility data for a substance are known, it is a simple matter to calculate the potential yield of pure crystals that could be obtained from batch crystallization (equations 7.4 and 7.6). Conversely, the degree of evaporation to produce a specified yield may be estimated (equation 7.8).

In practice, there will be some residual supersaturation (i.e. $c_f > c_f^*$) so the potential yield will be somewhat less, the precise amount can only be calculated via the coupled population and mass balances, as will be seen later (pp. 194, 195).

Prediction of CSD

A theoretical analysis of an idealized seeded batch crystallization by McCabe (1929a) lead to what is now known as the 'ΔL law'. The analysis was based on the following assumptions: (a) all crystals have the same shape; (b) they grown invariantly, i.e. the growth rate is independent of crystal size; (c) supersaturation is constant throughout the crystallizer; (d) no nucleation occurs; (e) no size classification occurs; and (f) the relative velocity between crystals and liquor remains constant.

Thus in all size ranges the number of seeds dN_s of size L_s is equal to the number of product crystals after growth, dN_p to size L_p i.e.

$$\frac{dM_s}{f_v \rho_c L_s^3} = \frac{dM_p}{f_v \rho_c L_p^3} = \frac{dM_p}{\alpha \rho_c (L_s + \Delta L)^3} \tag{7.9}$$

where ΔL is the growth increment. Therefore

$$dM_p = \left(\frac{L_s + \Delta L}{L_s}\right)^3 dM_s \tag{7.10}$$

Integrating over all sizes the final product crystal mass is given by

$$M_p = \int_0^{M_s} \left(1 + \frac{\Delta L}{L_s}\right)^3 dM_s \tag{7.11}$$

Thus, if M_p is known, from the initial mass of seeds plus the change in solubility over the temperature range, then the growth increment ΔL and consequent product CSD could be evaluated. McCabe (1929b) verified the method in experimental work. It was observed, however, that nucleation occurred despite operation at low supersaturation levels – subsequently referred to as 'secondary' nucleation i.e. nucleation in the presence of crystals (Chapter 5).

A fuller review of the ΔL law together with an illustration of the method is given by Mullin (2001).

Incidentally, if the seeds are 'monodisperse' then equation 7.11 reduces to

$$L_p = L_{so} + \Delta L = L_{so}\left(\frac{M_{so} + M_{cf}}{M_{so}}\right)^{\frac{1}{3}} \tag{7.12}$$

where M_{cf} is the crystal yield (equations 7.4 and 7.6).

While this approach gives a reasonable first estimate of product crystal size in some cases, a general approach clearly has to account for the occurrence of nucleation. The resultant increase in crystal numbers over which the solute mass crystallizes, and consequent reduction in mean crystal size is again accounted for using the 'population balance'.

Population balance

The population balance concept enables the calculation of CSD to be made from basic kinetic data of crystal growth and nucleation and the development of this has been expounded by Randolph and Larson (1988), as summarized in Chapters 2 and 3. Batch operation is, of course, inherently in the unsteady-state so the dynamic form of the equations must be used. For a well-mixed batch crystallizer in which crystal breakage and agglomeration may be neglected, application of the population balance leads to the partial differential equation (Bransom and Dunning, 1949)

$$\frac{\delta n}{\delta t} + \frac{\delta(Gn)}{\delta L} = 0 \tag{7.13}$$

which describes the crystal size distribution $n(L, t)$ as a function of both crystal size and time, where G is the overall linear growth rate (dL/dt). In these methods, however, it is necessary to define expressions for the crystallization kinetics and solubility (Chapters 3, 5 and 6).

For example, if the crystal growth rate is invariant with size, i.e. $G \neq G(L)$, then in the absence of particle breakage and agglomeration

$$\frac{d\mu_o}{dt} = B \tag{7.14}$$

$$\frac{d\mu_1}{dt} = \mu_0 G \tag{7.15}$$

$$\frac{d\mu_2}{dt} = \mu_1 G \tag{7.16}$$

$$\frac{d\mu_2}{dt} = \mu_2 G \tag{7.17}$$

where B is the nucleation rate (dN/dt).

The third moment of the CSD is related to total particle volume. It thus couples with the mass balance thereby providing closure

$$-\frac{dc}{dt} = 3f_v \rho_c \frac{d\mu 3}{dt} \tag{7.18}$$

In general, both nucleation and crystal growth depend on supersaturation and to lesser extent temperature and magma characteristics. Such data must therefore be collected to gain maximum benefit from the population balance approach (Jones and Mullin, 1974; Jones, 1974). Further simplifications to the describing equations are also possible, however (as follows).

Control of supersaturation

Two important considerations in the design of the batch crystallizer are the transient level of supersaturation, which determines scale formation on heat transfer surfaces, the relative rates of nucleation and crystal growth, and the consequent product crystal size distribution. Use of both crystal seeding and control of the operating conditions, e.g. cooling rate, were recognized by Griffiths (1925) as key factors to consider in batch crystallization (Figure 7.2).

Operation of the batch crystallizer under natural cooling (Figure 7.2(a)) rapidly increases the level of supersaturation with the liquor. It thereby exceeds the 'metastable limit' in the early stages of the operation leading to high or 'crash' nucleation rates and, since there is only a limited amount of solute to go round, excessive fine crystals 'fines' in the product. In Figure 7.2(b) the effect of combining seeding with what Griffiths called 'controlled cooling' is shown. In this technique the cooling rate is manipulated such that the level of supersaturation remains within the metastable zone at all times. An improved CSD of large mean size and narrow spread with the absence of fouling may thus be achieved.

These qualitative considerations of what is now known as 'programmed cooling' crystallization were subsequently described mathematically by Mullin and Nývlt

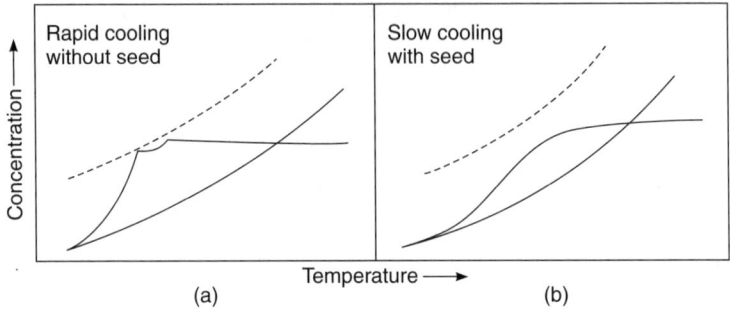

Figure 7.2 *Batch cooling crystallization: (a) Uncontrolled rapid cooling without seed crystals; (b) Controlled cooling with seed crystals added (after Griffiths, 1928)*

(1971) in order to determine quantitatively the controlled cooling curve needed to maintain the level of supersaturation at within the 'metastable limit' all times thereby avoiding excessive nucleation and hence improving the product CSD.

The supersaturation balance is given by

$$-\frac{d\Delta c}{dt} = \frac{dc^*}{dt} + k_g A \Delta c^g + k_m \Delta c^b \tag{7.19}$$

where the first term of the RHS represents the creation of supersaturation due to cooling and the second and third terms the desupersaturation rate due to growth on existing crystals and fresh nucleation respectively.

Now, if nucleation can be considered negligible by operation at a low level of supersaturation then the cooling curve needed to maintain the supersaturation within the 'metastable limit' may be expressed by (Mullin and Nývlt, 1971)

$$-\frac{d\theta}{dt} = \frac{3M_{so}}{\left[\frac{dc^*}{d\theta} + \frac{d\Delta c}{d\theta}\right]} \frac{G}{L_{so}} \left[L_{so} + \sum_{i=0}^{t-1} \frac{G}{L_{so}}\right]^2 \tag{7.20}$$

This equation was then solved numerically with known values of crystal growth rate (G) and solubility ($dc^*/d\theta$) for seeded solutions. It was further assumed that if $G \neq G(L, \theta)$, $dc^*/d\theta$ = constant and $d\Delta c/d\theta = 0$ then the simplified equation for calculating transient temperature results (Mullin and Nývlt, 1971)

$$\theta(t) = \theta_o - \phi Y Z \left(1 + YZ + \frac{1}{3} Y^2 Z^2\right) \tag{7.21}$$

where

$$\phi = \frac{3M_{so}}{\frac{dc^*}{d\theta}}$$

$$Y = \frac{(L_p - L_{so})}{L_{so}}$$

$$Z = \frac{t}{\tau}$$

These describing expressions are thus independent of knowledge of the crystallization kinetics. They imply that a quadratic convex form of controlled temperature profile is preferable to the more normal concave natural cooling curve. Further simplifications lead to a cubic cooling curve (Mullin and Nývlt, 1971; Mayrhofer and Nývlt, 1988).

Larson and Garside (1973) extended the controlled cooling approach to provide simplified equations to predict programmed evaporative crystallization policies based on the population balance.

Thus for controlled evaporation, the diminishing solvent mass is

$$\psi(t) = \psi_0 - YZ\left(1 + YZ + \frac{1}{3}Y^2Z^2\right) \tag{7.22}$$

where $\psi_0 = \left|\dfrac{M_h c}{3M_{so}}\right|$, $Y = \dfrac{(L_p - L_{so})}{L_{so}}$, $Z = \dfrac{t}{\tau}$

Tavare et al. (1980) further considered controlled 'dilution' crystallization, i.e. a type of precipitation

$$\frac{c_d - c_{do}}{c_{df} - c_{do}} = Z^4 \tag{7.23}$$

where c_d is the diluent concentration.

Programmed cooling crystallization

Recognizing that secondary nucleation can occur even within the metastable zone, the coupled population and mass balance equations for batch cooling crystallization in moment form

$$-\frac{d\Delta c}{dt} = \frac{dc}{d\theta} \cdot \frac{d\theta}{dt} + 3f_v \rho_c \frac{d\mu_3}{dt} \tag{7.24}$$

were solved to determine controlled cooling curves $\theta(t)$ for potassium sulphate solutions by Jones and Mullin (1974). As shown in Figure 7.3, the cooling curves predicted to control the level of supersaturation within the 'metastable limit' are all convex – the reverse of natural cooling.

Optimal operation

Whilst programmed cooling (i.e. operation at constant nucleation rate within the metastable zone) increases the mean product crystal size cf. natural cooling, is it the 'optimum' in producing the largest possible crystals? The problem is to find the maximum of the integral of crystal growth over the batch time. Thus because batch operation is by definition transient, a functional has to be maximized over time rather than just a function at some point in time. Jones (1972, 1974) addressed this problem by application of a particular result in

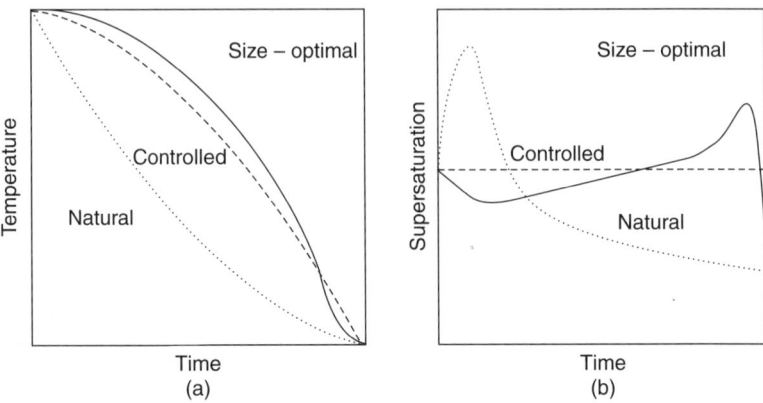

Figure 7.3 *Predicted: (a) cooling curves; and (b) transient supersaturation for potassium sulphate solutions (after Jones, 1994)*

optimal control theory known as the 'Continuous Maximum Principle' due to Pontryagin et al. (1962), which may be stated as

$$H(x,u) = H(\tau) = \text{constant} \tag{7.25}$$

is true for all $t \in [0, \tau]$ where x and u are vectors in the state and control variables respectively and H is the Hamiltonian function

$$H(x,u,t) = \lambda^t f \tag{7.26}$$

where λ^t is the adjoint variable vector being specified by the equation

$$\frac{d\lambda}{dt} = -\frac{\partial H}{\partial x} = -\lambda^t \left[\frac{\partial f}{\partial x}\right]^t \tag{7.27}$$

In the present case the *state* variables are most conveniently chosen as crystal size, moments of the distribution and solution concentration (which, as shown above, give rise to ordinary rather than partial differential equations, equations 7.14–7.18) while the *control* is solution temperature. The performance measure adopted is to maximize the terminal size of the 'S-crystals' (i.e. those originating as added seeds; those from nuclei being 'N-crystals')

$$F = L_s(\tau) = \int_0^\tau G_s(t)dt \tag{7.28}$$

The resulting numerical prediction for the 'size-optimal' cooling curve is shown in Figure 7.3. It predicts that in order to maximize the final sizes of the S-crystals, the temperature should be held constant for a period at both the start and end of the operation with a convex curve in between. This has the result of reducing both the early and terminal supersaturation levels and so maximizes solute deposition on the S-crystals and their growth rather than that of the N-crystals. Thus, programmed cooling is strictly 'sub-optimal', but nevertheless remarkably close to the optimum result in this case to be a practical alternative (Jones, 1974).

The predicted transient supersaturation levels (and corresponding nucleation rates) are also shown in Figure 7.3. These considerations predict that the high levels in the early period of natural cooling can be avoided by controlled cooling.

Example

An initial seed mass of 2.36×10^{-3} kg/kg H_2O of mean size 550 μm is added to an aqueous solution of potassium sulphate which is cooled from 60 °C to 25 °C in 3 h.

Estimate the maximum product crystal size and determine a simple controlled cooling curve (Jones, 1972, 1974; Mullin and Jones, 1974).

Solubility data

$$c^* = c_0 + c_1\theta + c_2\theta^2 \tag{3.1}$$

where $c_0 = 6.66 \times 10^{-2}$, $c_1 = 2.3 \times 10^{-3}$, $c_2 = -6.0 \times 10^{-6}$

$$L_{p\,max} \cong 0.550 \left[\frac{(2.36 \times 10^{-3} + 62.6 \times 10^{-3})}{2.36 \times 10^{-3}}\right]^{\frac{1}{3}} = 1.66\,\text{mm}$$

Controlled cooling curve

$$\theta(t) = \theta_o - \phi YZ\left(1 + YZ + \frac{1}{3}Y^2Z^2\right) \tag{7.21}$$

$$\phi = \frac{3M_{so}}{\frac{dc^*}{d\theta}}, \quad Y = \frac{(L_p - L_{so})}{L_{so}}, \quad Z = \frac{t}{\tau}$$

Thus

$$\phi = \frac{3 \times 2.36 \times 10^{-3}}{1.82 \times 10^{-3}} = 3.89$$

$$Y = \frac{(1660 - 550)}{550} = 2.02$$

$$Z = \frac{t}{180}, \quad YZ = 0.011t$$

where the temperature dependence of solubility is assumed constant $\cong 1.82 \times 10^{-3}$ over the temperature range 25–60 °C and t is expressed in minutes.

Thus the controlled temperature v time profile is given by

$$\theta(t) = 60 - 3.89 \times 0.011t\left(1 + 0.011t + \frac{1}{3}(0.11t)^2\right)$$

Programmed cooling crystallizer

An experimental implementation of these ideas of controlled cooling is illustrated in Figure 7.4. In this equipment the temperature of an agitated batch

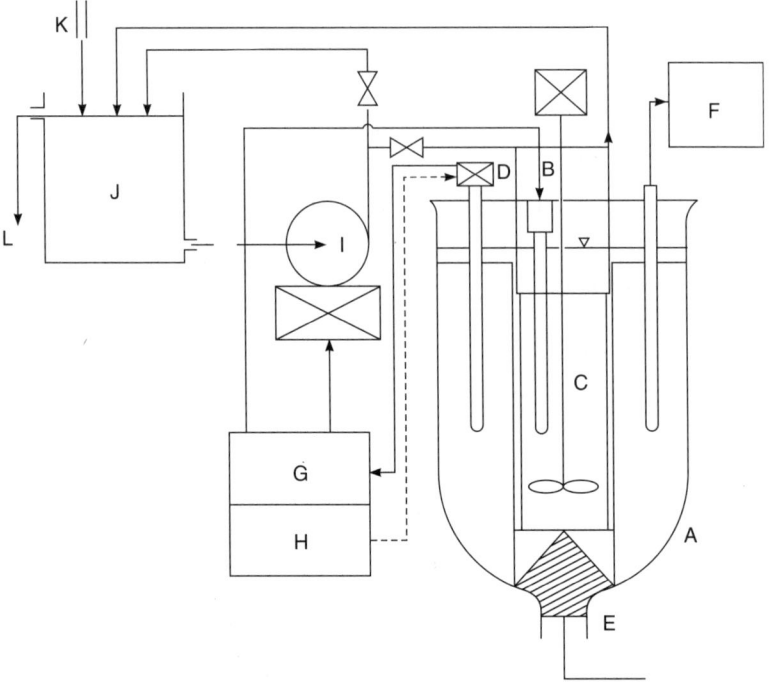

Figure 7.4 *Microcomputer programming of a batch cooling crystallizer: A, crystallization vessel; B, control heater; C, control cooler (surrounding the draft-tube); D, contact thermometer; E, discharge plug (and conical baffle); F, recorder; G, relay; H, temperature programmer; I, cooling water pump; J, cooling water reservoir; K, water inflow; L, water outflow (after Jones and Mullin, 1974)*

crystallizer is controlled by a combination of an immersion heater and circulating cooling water through a hollow walled draft-tube and temperature controller whose set point is determined via a microcomputer, which is programmed to follow any desired cooling curve.

Experimental results shown in Figure 7.5(a) clearly demonstrate the benefits of controlled cooling in increasing the median crystal size over that obtained by natural cooling – by about a factor of two in this case – as a result of keeping the transient supersaturation low (Figure 7.5(b)). Much of the crystal mass is below the maximum crystal size predicted, due to the occurrence of unwanted nucleation the effect of which can only be computed using the population balance approach employing additional, kinetic, data.

Computation of such optimum operating policies assumes that the seed parameters (mass and size) are already fixed. Doki *et al.* (1999) emphasize that with sufficient seed loading over a critical seed concentration, unimodal product crystals of grown seeds were obtained in seeded batch cooling crystallization using natural cooling and no scale-up effect on CSD was observed. A simple method was proposed for determining the critical seed amount of seed by a combination of calculation and experiment. Chung *et al.* (1999) reported a comprehensive theoretical investigation of the effect of seeding as a further variable. Again, based on the

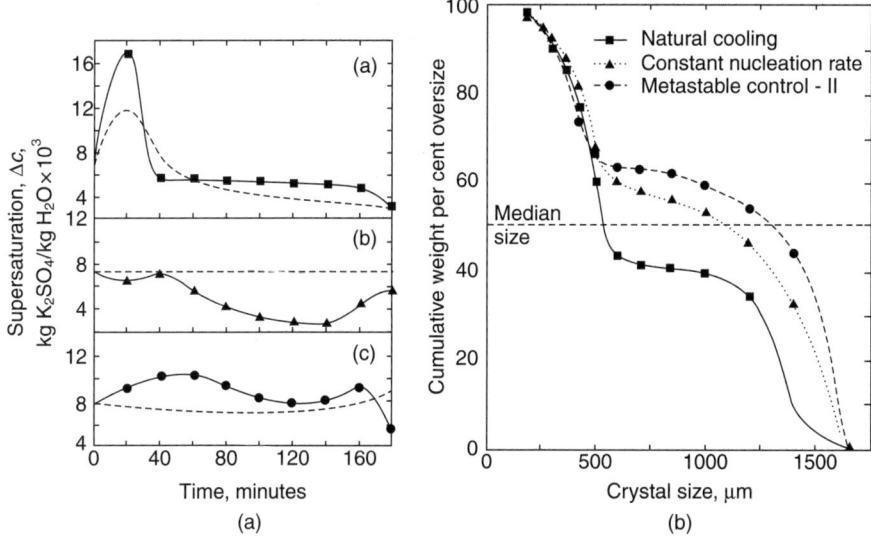

Figure 7.5 *Experimental*: (*a*) *transient supersaturation; and* (*b*) *consequent product crystal size distributions from batch cooling crystallizations* (*after Jones and Mullin, 1974*)

population balance moment equations, a dynamic programming formulation optimizes both the supersaturation profile and the seed distribution, and demonstrates the magnitude of variation in crystallizer performance.

Fines destruction during batch crystallization

Although programmed cooling crystallization clearly results in a larger mean crystal size than that from natural cooling it is also evident that some 'fines' i.e. small crystals are also present in the product. Since the solution was seeded these fine crystals must clearly have arisen from crystal attrition or secondary nucleation (see Chapter 5).

One method for improving the size distribution still further is to employ fines destruction (Jones *et al.*, 1984; Jones and Chianese, 1988). In this technique, a classified stream of suspension, i.e. containing only fine not coarse crystals, is removed from the crystallizer in an elutriation leg (Figure 7.6(a)).

The fines slurry is then heated to dissolve the crystals prior to cooling and return to the vessel. In this way, the total mass of solute remains unchanged but the 'effective' nucleation rate is reduced. The remaining crystals share the extra solute and grow slightly larger and have noticeably tighter size distribution (Figure 7.6(b)).

Programmed precipitation

The concept of programmed operation can also be applied to other types of batch crystallization e.g. precipitation via drowning-out with miscible solvents (Jones and Teodossiev, 1988).

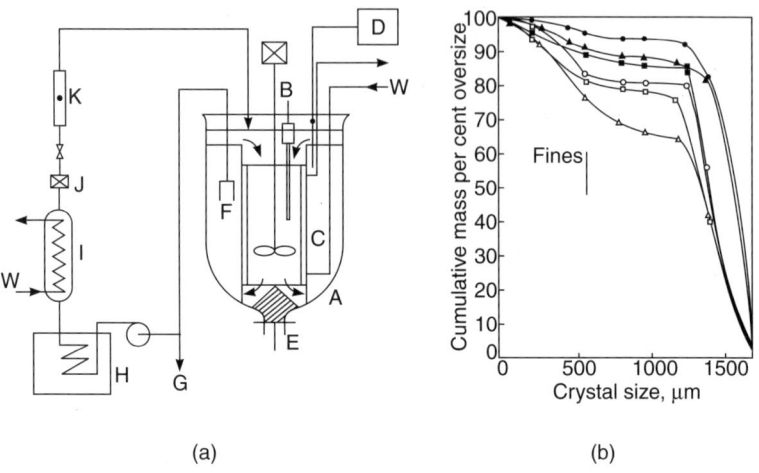

Figure 7.6 *Fines destruction during batch crystallization (after Jones et al., 1984)*

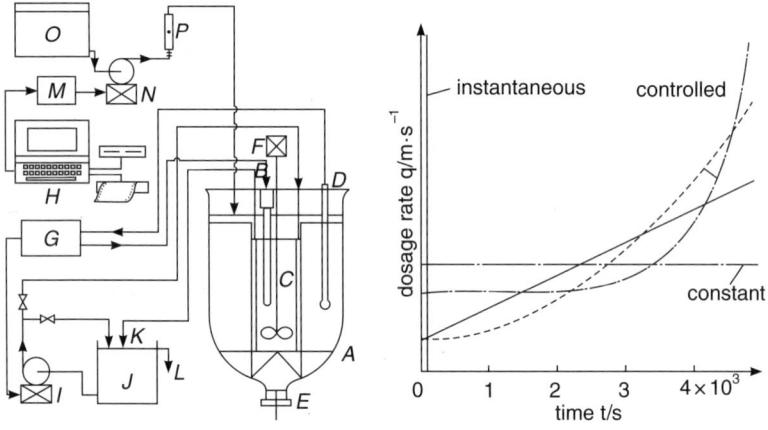

Figure 7.7 *Microcomputer programming of dosage rate during batch precipitation. Dosage rate curves: linear, $q = 0.3 + 3.15 \times 10^{-4} t$; parabolic, $q = 0.3 + 9.8 \times 10^{-8} t$; exponential, $q = 0.73 + 0.00153 \exp(0.00153 t)$ where q is the volumetric rate (mL/s) and t is the time (s) (after Jones and Teodossiev, 1988)*

In this case, the co-solvent dosage rate is programmed in order to control the transient level of supersaturation in an effort to improve on the product crystal size distribution from simply dumping in all the solvent at the start of the batch. An experimental crystallizer within which a programmed microcomputer determines the set point of a variable speed-dosing pump is shown in Figure 7.7. Controlled co-solvent dosing improves the product crystal size, with a consequent increase in the filterability of the product. These process concepts are developed further in Chapter 9.

Design and performance of continuous crystallizers

It was shown in Chapter 3 that the ideal continuous MSMPR crystallizer could be analysed using the population balance approach coupled with mass balances and crystallization kinetics to yield equations describing crystallizer performance in terms of the crystal size distribution, solids hold up etc. These concepts will now developed further to yield methods for continuous crystallizer design. Firstly, however, it is useful to consider how crystallization kinetics and crystallizer performance interact.

MSMPR interactions

It was shown above that the total crystal number, surface area and the mass mean size are affected by the mean residence time and the rates of nucleation and crystal growth respectively. Since both these kinetic processes depend upon the working level of supersaturation which will itself depend on the amount of surface area available and crystal mass deposited, the question arises 'what will be the effect of a change in residence time on crystallizer performance?'

Consider the idealized MSMPR crystallizer depicted in Figure 7.8.

Mass balance

The rate at which solute is removed from solution is balanced by the rate of mass crystallization. Thus

$$Q(c_{IN} - c_{OUT}) = QM_T \qquad (7.29)$$

Rate of change   Rate of mass
of mass in   crystallization
solution

Mass deposition rate

The mass deposition rate is also equal to the total flux of solute adding to the crystal surface, i.e.

$$QM_T = R_G A_T V \qquad (7.30)$$

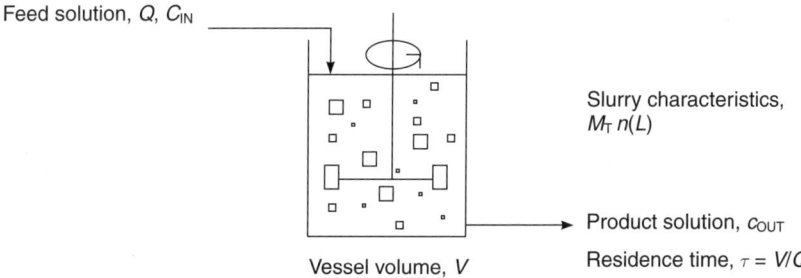

Figure 7.8 *Idealized MSMPR crystallizer*

where $R_G (= dM_T/dt)$ is the mass flux of solute from the solution to the crystal surface which is related to the (linear) crystal growth $(G = dL/dt)$ rate by

$$R_G = \frac{3\rho_c}{F} G \tag{7.31}$$

where F is the overall crystal shape factor of a crystal of density ρ_c, so

$$\frac{3\rho_c}{F} A_T V = Q(c_{IN} - c_{OUT}) \tag{7.32}$$

where the total crystal surface area in the slurry, A_T can be determined from the second moment of the crystal size distribution viz:

$$A_T = f_s \int_0^\infty nL^2 dL = 2f_s n^0 (G\tau)^3 \tag{7.33}$$

thus the linear crystal growth rate is given by

$$G = \frac{F(c_{IN} - c_{OUT})}{3f_s \rho_c \tau \int_0^\infty nL^2 dL} \tag{7.34}$$

i.e.

$$G = \frac{K(c_{IN} - c_{OUT})}{\tau A_T} \tag{7.35}$$

Supersaturation

The level of supersaturation (S) also changes. If it is assumed, for simplicity, that the crystal growth rate exhibits linear kinetics i.e.

$$G = k_g S \tag{7.36}$$

Then solving for S between equations 7.35 and 7.36 gives

$$S = \frac{K'(c_{in} - c_{OUT})}{\tau A_T} \tag{7.37}$$

Thus as $\tau \uparrow$, $S \downarrow$. In other words a change in crystallizer residence time has a corresponding but opposite effect on the level of supersaturation.

The nucleation rate is given by

$$B = k_b (M_T) S^b \tag{7.38}$$

Thus as $\tau \uparrow$, $G \downarrow$ and $B \downarrow$ but how does this affect the mean crystal size? Since the order of nucleation is usually more than that of growth, the nucleation rate decreases relatively more with an increase in τ. Thus, the mean size increases despite the lower growth rate (recall that the residence time is longer). Methods to estimate to what extent will be considered later.

Design equations for MSMPR crystallizers

From Chapters 2, 3 and 5, the basic describing equations for the MSMPR crystallizer at steady state are as follows

Total crystal mass per unit crystallizer volume

$$M_T = 6f_v\rho_c n^0 (G\tau)^4 \tag{7.39}$$

Solids production rate

$$P = M_T Q \tag{7.40}$$

Dominant product crystal size

$$L_D = 3G\tau \tag{7.41}$$

Crystallization kinetics

$$\text{Nucleation} \quad B = k_b M_T^j \Delta c^b \tag{7.42}$$
$$\text{Growth} \quad G = k_g \Delta c^g \tag{7.43}$$

Combining equations 7.42 and 7.43 to eliminate supersaturation gives

$$B = k_R M_T^j G^i \tag{7.44}$$

where the relative kinetic order i is given by

$$i = \frac{b}{g} \tag{7.45}$$

and

$$k_R = \frac{k_b}{k_g^{\frac{b}{g}}} \tag{7.46}$$

or, in terms of population density

$$n^0 = \frac{B}{G} = k_R M_T^j G^{i-1} \tag{7.47}$$

Combining the equations for mass balance, mean size and kinetics for the solids hold up gives (for $j = 1$)

$$M_T = 6f_v\rho_c[k_R M_T G^{i-1}]\left(\frac{L_D}{3}\right)^4 \tag{7.48}$$

and the growth rate required to achieve a desired dominant size is given by

$$G = \left[\frac{27}{2f_v\rho_c k_R L_D^4}\right]^{\frac{1}{i-1}} \tag{7.49}$$

and since $\tau = L_D/3G$, the corresponding residence time is

$$\tau = \frac{L_D}{3}\left[\frac{2f_v\rho_c k_R L_D^4}{27}\right]^{\frac{1}{i-1}} \tag{7.50}$$

and the crystallizer volume (capacity) is given by

$$V = \tau Q \tag{7.51}$$

Example

Calculate the residence time and volume of an MSMPR crystallizer required to produce 1000 kg/h of potash alum having a dominant crystal size of 600 μm using a slurry density of 250 kg crystals/m³ slurry.

Data

The relative nucleation: growth kinetics for this system are given by the relation (adapted from Garside and Jančić, 1979)

$$B^0 = 1.23 \times 10^{28} M_T G^{3.2}$$

where M_T (kg crystal/m³ slurry) is the magma density, B (crystals/s m³ slurry) is the nucleation rate and G is the crystal growth rate.

Solution

Thus, to recapitulate the data

Crystallizer performance: $L_D = 600\,\mu m$ (6×10^{-4} m); $P = 1000\,kg/h$; $M_T = 250\,kg/m^3$ slurry
Crystallization kinetics: $K_R = 1.23 \times 10^{28}\,no/s\,m^3\,kg/m^3\,(m/s)^{3.2}$; $i = 3.2$
Crystal characteristics: $f_v = 0.47$; $\rho = 1770\,kg/m^3$
The unknowns are: τ, V

The growth rate required to achieve the required dominant product crystal size is given by equation 7.49. Noting that

$$L_D^4 = 1.296 \times 10^{-13}\,m^4 \text{ and } \frac{i}{(i-1)} = \frac{1}{2.2} = 0.455$$

Then the required growth rate is given by

$$G = \left[\frac{27}{2 \times 0.47 \times 1770 \times 1.23 \times 10^{28} \times 1.296 \times 10^{-13}}\right]^{0.455} = (1.018 \times 10^{-17})^{0.455}$$

i.e.

$$G = 1.86 \times 10^{-8}\,m/s$$

Now, from equation 3.28, the dominant crystal size $L_D = 3G\tau$, hence the required residence time is given by

$$\tau = \frac{L_D}{3G} = \frac{6 \times 10^{-4}}{3 \times 1.86 \times 10^{-8}} 10750\,s = 2.99\,h$$

The volumetric flowrate through the vessel is given by $Q = P/M_T = 1000/250 = 4\,m^3$ slurry/h. Thus the vessel volume is given by equation 7.51 as

$$V = 4 \times 2.99 = 12.0\,m^3 \text{ slurry}$$

Typically, this would correspond to a standard diameter = 2 m and height = 3.8 m, together with a disengaging space above the slurry surface.

The design equations may also be used to infer nucleation kinetics from continuous crystallizer performance data.

Example

The CSD from an MSMPR crystallizer with a working volume of $10\,\text{m}^3$ operated with a magma density of $250\,\text{kg crystals/m}^3$ slurry and a production rate of $62\,500\,\text{kg crystals/h}$ has a mass mean size of $480\,\mu\text{m}$. Calculate:
(a) The overall linear crystal growth rate,
(b) The population density of nuclei, and
(c) The nucleation rate.
Take the crystal volume shape factor to be 0.4 and the crystal density to be $2600\,\text{kg/m}^3$.

Solution

Recall the design equations for the MSMPR
Slurry density: $M_T = 6 f_v \rho_c n^0 (G\tau)^4$
Mass mean size: $L = 4G\tau$
Production rate: $P = Q M_T$
Working volume: $V = Q\tau$

Calculation

(a) Crystal growth rate, $G = L/4\tau$

Mean residence time $\tau = M_T \times V/P = 250 \times 10/62\,500 = 0.04\,\text{h} = 144\,\text{s}$

$$G = \frac{480}{4} \times 0.04$$
$$= 3000\,\mu\text{m/h}$$
$$G = 8.33 \times 10^{-8}\,\text{m/s}$$

(b) Solids holdup $M_T = 6 f_v \rho_c n^0 (G\tau)^4$
Nuclei density (%),

$$n^0 = 250/[6 \times 0.4 \times 2600 \times (8.33 \times 10^{-8} \times 144)^4]$$
$$n^0 = 1.94 \times 10^{18}\,\text{nuclei/s m}^3$$

(c) Nucleation rate

$$B = n^0 G$$
$$= 1.94 \times 10^{18} \times 8.33 \times 10^{-8}$$
$$B = 1.6 \times 10^{11}\,\text{nuclei/m}^3\text{s}$$

Effect of residence time on mean crystal size

It was shown earlier that the dominant crystal size increases with mean residence time. From equation 7.50, the relationship is given by

$$\tau \propto L_D \cdot L_D^{\frac{4}{i-1}} \tag{7.52}$$

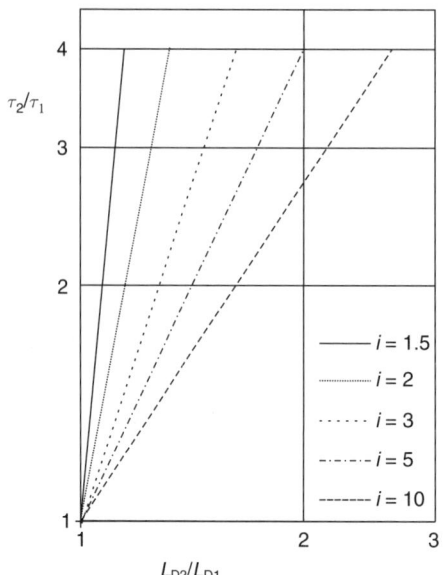

Figure 7.9 *Effect of mean residence time on mean crystal size (parameter $i = b/g$, the relative nucleation to growth order index)*

i.e.

$$\tau \propto L_D^{\frac{i+3}{i-1}} \tag{7.53}$$

Now if the special case of $i = 1$ (i.e. $b = g$) is assumed

$$L_D = \left[\frac{27}{2f_v \rho_c k_R}\right]^{\frac{1}{4}} \tag{7.54}$$

and the dominant crystal size is seen to be independent of mean crystallizer residence time. But in general $i > 1$, so generally the dominant size, L_D, increases with mean residence time, τ, but usually only weakly (Figure 7.9).

Example

The residence time of the crystallizer in previous example is doubled. If the slurry suspension density is kept constant, calculate the effect of this change on:

1. the crystal growth rate
2. the nucleation rate
3. the dominant crystal size.

Solution

Slurry suspension density is kept constant. This quantity is given by

$$M_T = 6f_v \rho n^0 (G\tau)^4$$

On changing τ, the supersaturation changes resulting in corresponding changes in G and B.

As M_T = constant, the above equation indicates that

$$n^0(G\tau)^4 = \text{constant} \tag{1}$$

The kinetics of the system indicate that
$$B \propto G^{3.2} \text{ for constant } M_T$$

i.e. $n^0 = \dfrac{B}{G} \alpha G^{2.2}$

$$\dfrac{n^0}{G^{2.2}} = \text{constant} \tag{2}$$

From (1) and (2) above

$$\dfrac{n_2^0}{n_1^0} = \left(\dfrac{G_1 \tau_1}{G_2 \tau_2}\right)^4 = \left(\dfrac{G_2}{G_1}\right)^{2.2}$$

Therefore

$$\left(\dfrac{\tau_1}{\tau_2}\right)^4 = \left(\dfrac{G_2}{G_1}\right)^{6.2}$$

or

$$G_2 = G_1 \left(\dfrac{\tau_1}{\tau_2}\right)^{0.65}$$

For $\tau_1/\tau_2 = 1/2$, $G_2 = 0.64 \times G_1$
i.e. an increase in $\tau \rightarrow$ a decrease in G.

$$\dfrac{B_2}{B_1} = \left(\dfrac{G_2}{G_1}\right)^{3.2} = \left[\left(\dfrac{\tau_1}{\tau_2}\right)^{0.65}\right]^{3.2}$$

or

$$B_2 = B_1 \left(\dfrac{\tau_1}{\tau_2}\right)^{2.08}$$

Thus for $\tau_1/\tau_2 = 1/2$, $B_2 = 0.24 B_1$
i.e. an increase in $\tau \rightarrow$ a decrease in B.
Now,

$$L_D \propto G\tau, \text{ so}$$

$$\dfrac{L_{D1}}{L_{D2}} = \dfrac{G_1 \tau_1}{G_2 \tau_2} = \left(\dfrac{\tau_2}{\tau_1}\right)^{0.65} \dfrac{\tau_1}{\tau_2}$$

or

$$L_{D2} = L_{D1} \left(\dfrac{\tau_2}{\tau_1}\right)^{0.35}$$

For $\tau_1/\tau_2 = 1/2$, $L_{D2} = 1.27 L_{D1}$
i.e. an increase in $\tau \rightarrow$ increase in L_D

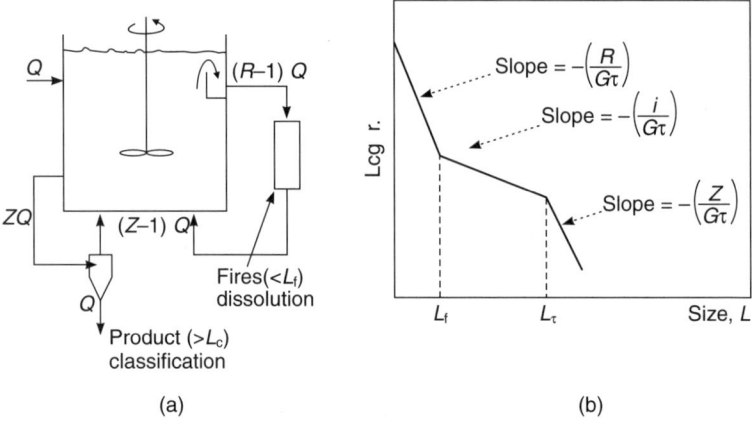

Figure 7.10 *Fines destruction and classified product removal during continuous mixed suspension crystallization*

Crystal size-dependent product removal crystallizers

It has been shown that varying the operating conditions, such as mean residence time, of the MSMPR crystallizer has only a rather weak effect on the mean size of the product crystals and the size distribution is of the same general shape of CV = 50%. Greater control of the mean size and shape of the distribution can be effected by modifying the product removal rates such that the residence time of a given size range of crystals differs from that of the liquid phase. This is normally achieved in one or both of two ways; fine destruction and classified product removal.

Fines destruction

As its name implies, fines destruction involves selectively withdrawing a slurry stream containing fine crystals (Figure 7.4). These tiny crystals are then redissolved, e.g. by heating or addition of fresh solvent, and the solute returned to the crystallizer. In this way, the working level of supersaturation is increased giving rise to increased growth rates but lower 'effective' nucleation rates. Thus, fewer crystals share the available solute, which leads to an increase in mean particle size and a tightening of the size distribution in the small crystal region (Figure 7.10). This technique is employed industrially and has been studied experimentally by Jazaszek and Larson (1977) and Randolph *et al.* (1977), respectively.

If L_f is the 'cut size' below which the fines are selectively removed, the residence times of fine crystals, τ_F, and product crystals, τ_P, respectively are given by

$$\tau_F = \frac{V}{Q_F + Q_P} \quad (L < L_F) \tag{7.55}$$

and
$$\tau_P = \frac{V}{Q_P} \quad (L < L_F) \tag{7.56}$$

The crystal size distribution then forms two parts (Figure 7.10) given by

$$n(L) = n^0 \exp\left(-\frac{L}{G\tau_F}\right) \quad (L < L_F) \tag{7.57}$$

and

$$n(L) = n^0 \exp\left(-\frac{L}{G\tau_P}\right) \quad (L < L_F) \tag{7.58}$$

The residence time required to produce a distribution of dominant crystal size L_D becomes:

$$\tau = \frac{fL_D}{3} \left[\frac{2f_v \rho K_R L_D^4}{27}\right]^{\frac{1}{(i-1)}} \tag{7.59}$$

where the fines removal factor f is given by

$$f = \left[\exp\left(-3\frac{\tau_P L_F}{\tau_F L_D}\right)\right]^{\frac{1}{(i-1)}} \tag{7.60}$$

Typical example calculations are given in Randolph and Larson (1988) and Larson and Garside (1973).

Example

A fines removal system is installed on the crystallizer designed in the first example. Assuming that the 'cut' size for the fines removal system is 50 µm and the ratio of mean residence times for product and fines, $\tau_P/\tau_F(=\gamma)$, is 10, calculate the mean product residence time now required to produce the same dominant size of 600 µm at the same production rate and suspension density.

Solution

With fines removal the system growth rate is given by

$$G = \left[\frac{27}{2f_v p K_R L_D^4 \exp\left(-3\gamma \frac{L_C}{L_D}\right)}\right]^{\frac{1}{i-1}}$$

$$\frac{3\gamma L_C}{L_D} = 3 \times 10 \times \frac{50}{600} = 2.5$$

whence

$$\exp(-2.5) = 0.0821$$

$$G = \left[\frac{1.018 \times 10^{-17}}{0.0821}\right]^{0.455} = (1.24 \times 10^{-16})^{0.455}$$

from previous calculation, i.e.

$$G = 5.79 \times 10^{-8} \, \text{m/s}$$

Thus, because less surface area is present in this case, the supersaturation increases and hence the growth rate also increases.

The residence time to achieve a given L_D is thus reduced:

$$\tau = \frac{L_D}{3G} = \frac{6 \times 10^{-4}}{3 \times 5.79 \times 10^{-8}} \times 3600 = 0.96 \, \text{h}$$

(cf. 2.99 h i.e. 1/3 the time)
$Q_p = 4 \, \text{m}^3$ slurry/h, as before
Therefore

$$V_u = 4 \times 0.96 = 3.84 \, \text{m}^3 \text{ slurry}$$

Now $\tau_F = \tau_p/\gamma + 0.096 \, \text{h}$, so

$$\tau_F = 0.096 = \frac{V_0 + V_u}{Q_F + Q_p} = \frac{V_0 + 3.84}{Q_F + 4}$$

Note that V_0 and Q_F are also related via the hydrodynamics of the settling region. This will give a second relation between V_0 and Q_F, so enabling both to be calculated.

Classified product crystal withdrawal

Attention can also be paid to the other end of the size distribution – that of the large crystals. By allowing the largest crystals to remain in the vessel longer than average, the mean product crystal size is similarly enhanced, but may lead to unstable crystallizer operation (Randolph and Larson, 1988).

Crystallizer dynamics

It has been shown that an increase in crystallizer residence time, or decrease in feed concentration, reduces the working level of supersaturation. This decrease in supersaturation results in a decrease in both nucleation and crystal growth. This in turn leads to a decrease in crystal surface area. By mass balance, this then causes an increase in the working solute concentration and hence an *increase* in the working level of supersaturation and so on. There is thus a complex feedback loop within a continuous crystallizer, illustrated in Figure 7.11.

One useful practical result of the study crystallizer dynamics is that it can take about eight residence times for the CSD to achieve steady state after start up or return to it following a disturbance, as illustrated in Figure 7.12.

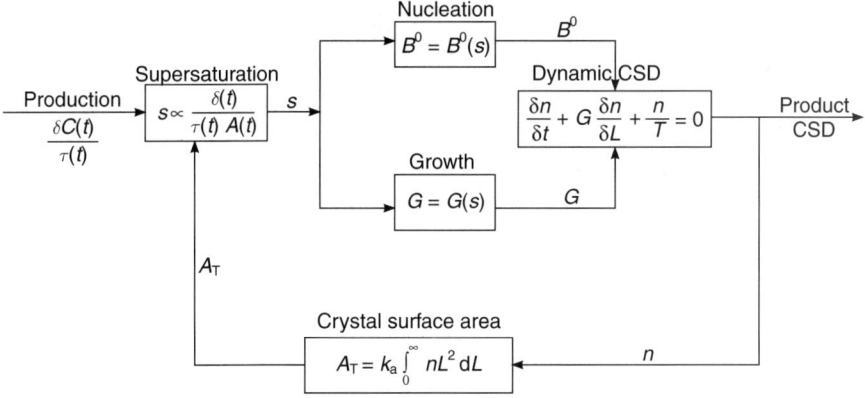

Figure 7.11 *Information flow in an MSMPR crystallizer (Randolph and Larson, 1988)*

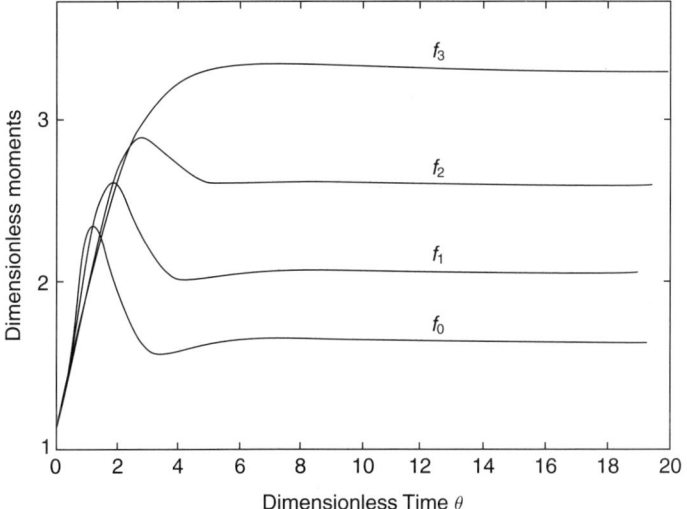

Figure 7.12 *Continuous MSMPR dynamics (Randolph and Larson, 1988)*

Summary

In this chapter the basic concept of the batch crystallizer has been described and the equations describing its operation and performance developed for cooling, evaporation and precipitation. Performance prediction starts with the idealized McCabe ΔL law which predicts product crystal size starting from the invariant growth of seed crystals while the effect of nucleation is accounted for using the population balance technique employing crystallization kinetics. It is seen that the performance of the batch crystallizer can be controlled via use of seeding and implementation of programmed operating conditions. The largest crystal sizes being obtained by controlling the level of supersaturation within the 'metastable limit' e.g. by convex cooling profiles or slowly increasing precipitant dose rates. Further improvements may be obtained by application of 'fines destruction' to tighten up the product crystal size distribution. Similar improvements in CSD can be made in controlled batch precipitation. The consequential effect of such controlled crystallization and precipitation on subsequent solid–liquid separation will be considered in Chapter 9.

Continuous crystallizer design is highly developed using the population balance approach. Given the crystallization kinetics, design equations are available to calculate the crystallizer residence time and vessel volume required to achieve a given mean product crystal size. The CSD from the continuous MSMPR crystallizer, however, is relatively insensitive to operating conditions; changes that are more substantial can be effected by use of crystal fines destruction and classified product removal techniques. The complex feedback loop for the MSMPR crystallizer gives rise to interesting CSD dynamics and the possibility of unstable behaviour following disturbances to process operating conditions.

The assumption of perfect mixing applied in the batch and continuous MSMPR crystallizer models does not always apply, however, especially as vessel size increases. Methods for accounting for imperfect mixing and scale-up are considered in the next chapter.

8 Crystallizer mixing and scale-up

Crystallization is widely used at varying scales of operation in the chemical and pharmaceutical industry. These processes often involve cooling, evaporation, rapid reaction or antisolvent addition. Successful prediction of product particle characteristics under differing mixing conditions is not straightforward and is frequently difficult. Local supersaturation levels and consequent crystallizer performance is often particularly sensitive to mixing conditions, particularly when the supersaturation generation is fast and the vessel size is large. Such local variations in supersaturation can give rise to varying rates of all particle formation kinetic processes, in addition to the fluid mechanical effects on secondary nucleation during crystallization that were considered in Chapter 5. Publications concerning tailor-made particle properties and scale-up of crystallization are increasing the literature, whilst their importance in the chemical and pharmaceutical industry has been emphasized (Krei and Buschmann, 1998). This chapter examines the effects of mixing and scale-up of crystallization and precipitation systems using firstly liquid reagents followed by an analysis of gas–liquid systems.

The fundamentals of crystallization and precipitation are reviewed in the books of e.g. Söhnel and Garside (1992), Mersmann (2001), Mullin (2001) and Myerson (2001), respectively, whilst Baldyga and Bourne (1999) provide a comprehensive account of turbulent mixing and chemical reactions.

Liquid–liquid crystallization systems

Crystallization systems frequently exhibit high levels of supersaturation around the points where it is generated, such as at colling surfaces, evaporation interfaces and where two or more liquid reactants are brought into contact. Attainment of uniform conditions throughout the reactor volume therefore becomes difficult and this mixing problem becomes magnified as the scale of operation increases, and can be particularly pronounced in fast precipitation systems. Attempts have therefore been made both to predict the effects of such imperfect mixing on crystallizer performance and to engineer alternative reactor configurations in order to minimize the problem.

Mahajan and Kirwan (1996) investigated micromixing (mixing on the molecular scale) and scale effects of the precipitation of Lovastatin in a two-impinging-jets (TIJ) precipitator. Due to the geometry of the TIJ mixer, it is possible to achieve very short micromixing times in the range of the induction time and therefore ensure homogeneous supersaturation throughout the reactor before nucleation starts. A scale-up criterion based on the Damköhler number (ratio of characteristic micromixing time constant to characteristic reaction time constant) was proposed. Houcine *et al.* (1997) investigated the influence of mixing conditions on the precipitation of calcium oxalate on a 20 l pilot scale.

Secker *et al.* (1995) and Wei and Garside (1997) studied precipitation using CFD assuming perfect micromixing within each cell. It was shown, however, the computational demand often increases enormously even for very simple reactor geometries. Using a different approach, van Leeuwen (1998) developed a compartmental mixing model for modelling precipitation processes based on the engulfment theory developed by Baldyga and Bourne (1984a–c). Mixing parameters between the feed and the bulk zone from the flow characteristics in the reactor are obtained and subsequently moments and mean sizes of the precipitate calculated.

Zauner and Jones (2000b) presented a computational methodology for predicting precipitate particle properties and mean size based on a hybrid SFM coupled with the population balance that can simultaneously account for micro-, meso- and macromixing effects, and by way of illustration will be described here in detail.

Mixing model: Segregated Feed Model (SFM)

The SFM is based on physically meaningful mixing parameters diffusive micromixing time and convective mesomixing time, and was therefore selected to model the mixing effects during precipitation mentioned above. The SFM model was used by Villermaux (1989) to investigate micromixing effects of consecutive–competitive semibatch reactions. Marcant (1996) subsequently applied it to predict the effects of mixing on the semibatch precipitation of

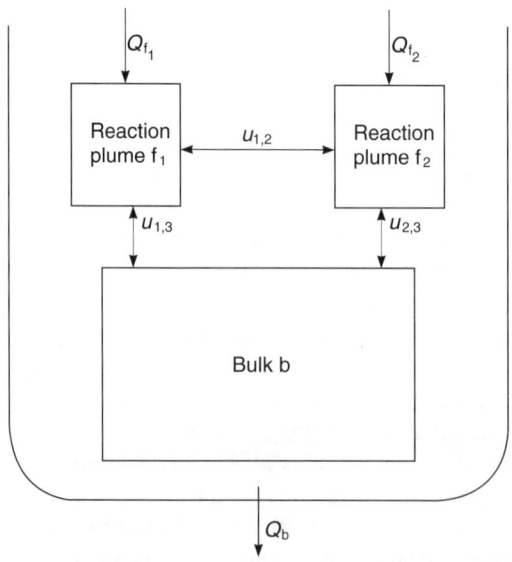

Figure 8.1 *Segregated Feed Model (SFM) (after Zauner and Jones, 2000b)*

barium sulphate without accounting for agglomeration and disruption. The SFM was found to be particularly suitable for modelling mixing effects, as it combines the advantages of both a compartmental model and a physical model.

In the SFM the reactor is divided into three zones: two feed zones f_1 and f_2 and the bulk b (Figure 8.1). The feed zones exchange mass with each other and with the bulk as depicted with the flow rates $u_{1,2}$, $u_{1,3}$ and $u_{2,3}$ respectively, according to the time constants characteristic for micromixing and mesomixing. As imperfect mixing leads to gradients of the concentrations in the reactor, different supersaturation levels in different compartments govern the precipitation rates, especially the rapid nucleation process.

Using the SFM, the influence of micromixing and mesomixing on the precipitation process and properties of the precipitate can be investigated. Mass and population balances can be applied to the individual compartments and to the overall reactor accounting for different levels of supersaturation in different zones of the reactor.

The individual volumes of the feed plume 1 (f_1) and feed plume 2 (f_2) and the total volume of the precipitation reactor can be obtained from

$$\frac{dV_{f_1}}{dt} = Q_{f_1} - \frac{V_{f_1}}{t_{meso1}} \tag{8.1}$$

$$\frac{dV_{f_2}}{dt} = Q_{f_2} - \frac{V_{f_2}}{t_{meso2}} \tag{8.2}$$

$$\frac{dV_{tot}}{dt} = Q_{f_1} + Q_{f_2} - Q_b \tag{8.3}$$

with

$$V_{tot} = V_{f_1} + V_{f_2} + V_b \tag{8.4}$$

Mass balances for a species j in the three zones give

$$\frac{d(V_{f_i} c_{j,f_i})}{dt} = r_{j,f_i} V_{f_i} + Q_{f_i} c^0_{j,f_i} - u_{j,f_{ib}} - u_{j,f_1 f_2} \tag{8.5}$$

$$\frac{d(V_b c_{j,b})}{dt} = r_{j,b} V_b - Q_b c_{j,b} + u_{j,f_{1b}} + u_{j,f_{2b}} \tag{8.6}$$

with the exchange flows between the compartments

$$u_{j,f_{ib}} = \frac{V_{f_i} c_{j,f_i}}{t_{mesoi}} + \frac{V_{f_i}(c_{j,f_i} - c_{j,b})}{t_{microi}} \tag{8.7}$$

$$u_{j,f_1 f_2} = \frac{(V_{f_1} + V_{f_2})(c_{j,f_1} - c_{j,f_2})}{t_{12}} \tag{8.8}$$

SFM applied to continuous precipitation

Using the following assumptions, the SFM will be applied to continuous precipitation (Zauner and Jones, 2000b):

- Instantaneous reaction: $A + B \xrightarrow{k_r} P$. The reaction between the two feed solutions occurs instantaneously as soon as they are mixed.

- Homogeneous conditions occur within the feed zones and the bulk. Within each compartment, there are no gradients in the field of supersaturation.
- Nucleation is the only kinetic process occurring in the feed plumes. Nucleation, growth, agglomeration and disruption take place in the bulk zone.
- Direct diffusional mass exchange between the feed plumes is negligible as the feed points are too far apart and diffusion is too slow to play a significant role in the diffusional mass exchange between the feed streams, leading to $t_{12} \to \infty$.

Volumes of compartments

$$V_{f_i} = Q_{f_i} t_{mesoi} \tag{8.9}$$

$$V_b = V_{tot} - (V_{f_1} + V_{f_2}) \tag{8.10}$$

Mass balances e.g. for Species A

$$\frac{d(V_{f_1} c_{A,f_1})}{dt} = Q_{f_1} c^0_{A,f_1} - \frac{V_{f_1} c_{A,f_1}}{t_{meso1}} - \frac{V_{f_1}(c_{A,f_1} - c_{A,b})}{t_{micro1}}$$

$$- B^0_{f_1} k_v L^3_0 \frac{\rho_c}{MW_c} V_{f_1} = 0 \tag{8.11}$$

$$\frac{d(V_{f_2} c_{A,f_2})}{dt} = -\frac{V_{f_2} c_{A,f_2}}{t_{meso2}} - \frac{V_{f_2}(c_{A,f_2} - c_{A,b})}{t_{micro2}}$$

$$- B^0_{f_2} k_v L^3_0 \frac{\rho_c}{MW_c} V_{f_2} = 0 \tag{8.12}$$

$$\frac{d(V_b c_{A,b})}{dt} = -Q_b c_{A,b} + \frac{V_{f_1} c_{A,f_1}}{t_{meso1}} + \frac{V_{f_1}(c_{A,f_1} - c_{A,b})}{t_{micro1}}$$

$$+ \frac{V_{f_2} c_{A,f_2}}{t_{meso2}} + \frac{V_{f_2}(c_{A,f_2} - c_{A,b})}{t_{micro2}} - B^0_b k_v L^3_0 \frac{\rho_c}{MW_c} V_b \tag{8.13}$$

$$- \frac{1}{2} G_b k_a m_2 \frac{\rho_c}{MW_c} V_b = 0$$

$$t_{micro} = 17.3 \times \left(\frac{\nu}{\varepsilon_{loc}}\right)^{\frac{1}{2}} \tag{8.14}$$

$$t_{meso} = \Lambda \frac{\varepsilon_{avg}}{\varepsilon_{loc}} \frac{Q^{\frac{1}{3}}}{N^{\frac{4}{3}} d_s} \tag{8.15}$$

As mentioned above, the inverse of the time constants t_{microi} (micromixing) and t_{mesoi} (mesomixing) can be interpreted as transfer coefficients for mass transfer by diffusion and convection, respectively. The former are given by equations 8.14 and 8.15 (Baldyga et al., 1995; Baldyga et al., 1997) in which the coefficients 17.3 and Λ are obtained from the literature and present experiments respectively. One of the advantages of using a compartmental model is that the 'full' population balance, including terms for size-dependent agglomeration and disruption etc. can be implemented in the model. Therefore, the population balance for the bulk becomes

$$\frac{\partial n_{P,b}}{\partial t} = -G_b \frac{\partial n_{P,b}}{\partial L} + B_b^0 + B_{aggl} + B_{disr} - D_{aggl} - D_{disr} - \frac{n_{P,b}Q_b}{V_{tot}} = 0 \quad (8.16)$$

with the birth terms for agglomeration and disruption

$$B_{aggl} + B_{disr} = \frac{L^2}{2} \int_0^L \frac{K_{aggl} n(L_u) n(L_v) dL_u}{L_v^2} + \int_{L_v}^\infty K_{disr} S'(L_u, L_v) n(L_u) n(L_v) dL_u \quad (8.17)$$

and the death terms for agglomeration and disruption

$$D_{aggl} + D_{disr} = n(L) \int_0^\infty K_{aggl} n(L_u) dL_u + K_{disr} n(L) \quad (8.18)$$

The expression for the nucleation rate B_j^0 in the compartment j is derived from the theory of primary nucleation and found to be (Mullin, 2001)

$$B_j^0 = A \exp\left[-\frac{16\pi\gamma^3 v^2}{3k^3 T^3 \ln^2 S_j}\right] \quad (8.19)$$

with the level of supersaturation defined as

$$S_j = \left(\frac{c_{A,j}^{\nu_A} c_{B,j}^{\nu_B}}{K_{sp}}\right)^{\frac{1}{\nu_A + \nu_B}} = \sigma_j + 1 \quad (8.20)$$

The overall nucleation rate in the reactor becomes

$$B_{tot}^0 = \frac{B_b^0 V_b + B_{f_1}^0 V_{f_1} + B_{f_2}^0 V_{f_2}}{V_{tot}} \quad (8.21)$$

The dependence of the growth rate on supersaturation is modelled using the power law expression

$$G_b = k_g \sigma_b^2 \quad (8.22)$$

Furthermore, the agglomeration and disruption kernels are also assumed to depend on the supersaturation in power law form (Zauner and Jones, 2000a)

$$K_{aggl} = \beta_{aggl} f(\varepsilon) \sigma_b^2 \quad (8.23)$$

$$K_{disr} = \beta_{disr} g(\varepsilon) \sigma_b^{-2} \quad (8.24)$$

The second moment of the particle size distribution used in the mass balances is obtained from

$$m_{2b} = \int_0^\infty n_{P,b}(L) L^2 dL \quad (8.25)$$

Determination of mixing times using Computational Fluid Dynamics (CFD)

As discussed in Chapter 2, agitated tanks are the most common form of chemical reactors. Nevertheless, due to high local gradients of the energy dissipation, the fluid dynamics are not well understood and depend to a large extent on the geometric details of the vessel. Different forms of impellers, baffles and draft-tubes can produce very different flow fields. As a stirred tank contains a moving impeller, the fluid cells surrounding the impeller are modelled as rotating blocks in CFD (Bakker *et al.*, 1997). A sliding mesh technique was chosen to account for the movement of the rotating impeller grid relative to the surrounding motionless tank cells. Xu and McGrath (1996) compared the sliding mesh simulation results for a stirred tank with experimental LDA data and found that the data corresponded very well.

In the studies by Zauner and Jones (2000b,c), the stirred tank is modelled as a single-phase isothermal system, i.e. only the hydrodynamics of the vessel are simulated. In the model equations of the turbulence the k-α model was used, assuming that turbulence is isotropic. The k-ε model offers a good compromise between computational economy and accuracy of the solution. Ranade (1997) has used it successfully to model stirred tanks under turbulent conditions. Manninen and Syrjänen (1998) modelled turbulent flow in stirred tanks, and tested and compared different turbulence models. They found that the standard k-ε model predicted the experimentally measured flow pattern best. One of the great advantages of CFD is that local data for the fluid velocity and energy dissipation can be obtained. As the local energy dissipation is a measure of the degree of local micromixing in the reactor, the micromixing time can be calculated directly from this parameter using equation 8.14.

In order to account for the rotation of the impeller, the zone surrounding the impeller was modelled using a sliding mesh approach. Grid-independence was established for all the problems solved in this research by stepwise and/or local refining of the grid until the solution for the energy dissipation of a grid cell close to the impeller no longer changed by more than 1 per cent. Three different types of impeller, a Rushton turbine and a propeller were defined as groups and implemented in the geometry.

In order to find the micromixing times in the feed zones and the mesomixing time characteristic of blending of the feed solutions with the bulk in the SFM, it is essential to determine the local distribution of the specific energy dissipation in the reactor. It is possible to obtain this local distribution using CFD, and find the diffusive and convective exchange parameters. Because the local energy dissipation is different on different scales of operation with the same mean specific power input, the micromixing (diffusion) and mesomixing (convection) times on different scales differ. Consequently, different diffusive and convective mass transfer between the feed zones and the bulk zone leads to different levels of supersaturation on different scales and therefore to different precipitation kinetic rates and mean crystal sizes with scale-up.

Model solutions

In order to solve the model equations 8.1–8.25 as a system of differential and algebraic equations, the population balance partial differential equation (equation 8.16) was discretized in the particle size domain. The discretization method of Kumar and Ramkrishna (1996a,b) was chosen for this purpose as it presents a solution to population balance problems without restricting the choice of the grid of the discretized length scale. The set of differential equations of the population balance was solved using the NAG® subroutine D02EAF which is particularly suitable for stiff systems of first-order ordinary differential equations. This variable-order, variable-step method implements the BDF and is of an explicit type (Schuler, 1996). The ODEs are integrated over a time range of ten residence times (continuous mode of operation), assuming that after that time steady state has been achieved, or over the feeding time (semi-batch mode of operation). Subroutines calculate values for the change of population due to nucleation, growth, agglomeration and disruption. Another NAG® subroutine solves simultaneously the coupled algebraic equations for the level of supersaturation in the different compartments and the exchange of mass and particles by micromixing and mesomixing. The population density of each class in the different compartments is calculated from the number of particles in each size class.

Model validation

In order to validate the predictions of the theoretical analysis based on the SFM, Zauner and Jones (2000b) studied the effect of reactor scale (capacity) on the precipitation of calcium oxalate obtained from reacting supersaturated solutions of calcium chloride $CaCl_2$ and sodium oxalate $Na_2C_2O_4$. The geometries of the 4.3 l and 12 l precipitation reactors are shown in Figure 8.2 with the experimental set-up shown in Figure 8.3.

The draft tubes of the reactor are hollow in order to allow cooling or heating fluid to circulate. The 300-ml laboratory-scale reactor (1) is geometrically similar to the larger reactors. A marine-type impeller (propeller) provides a smooth and even flow field throughout the reactor.

The specific power input was varied between 0.0024 and 8.09 W/kg in order to investigate the influence of mixing on scale-up.

Continuous mode of operation

In order to determine the mesomixing time, a least square fit of the 300 ml continuous calcium oxalate (CaOx) precipitation results for the number mean size and nucleation rate was performed. From these calculations, the factor Λ in equation 8.15 was obtained as 17.7. Using the kinetic parameters determined from the laboratory-scale continuous experiments (Zauner, 1999), the large-scale experiments were simulated with the SFM and compared with the experimental findings.

222 *Crystallization Process Systems*

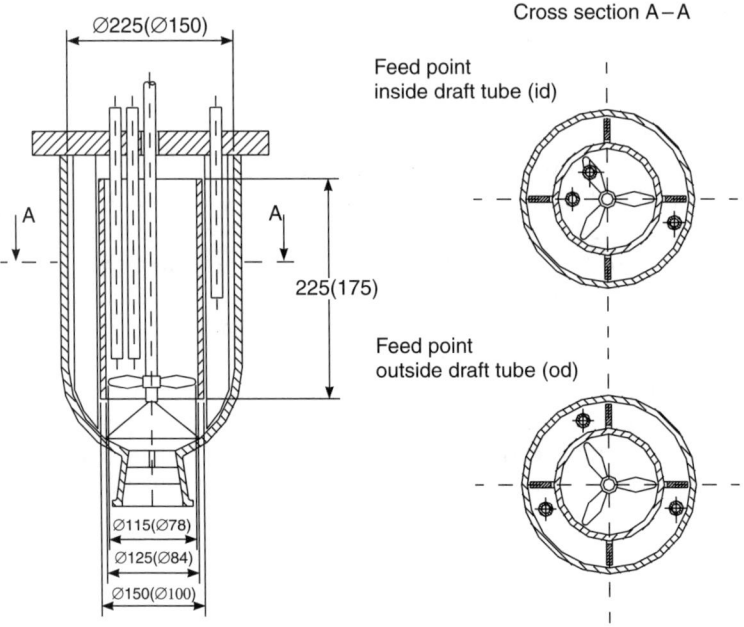

Figure 8.2 *Dimensions of the 4.3 l and 12 l large-scale reactors (after Zauner and Jones, 2000b)*

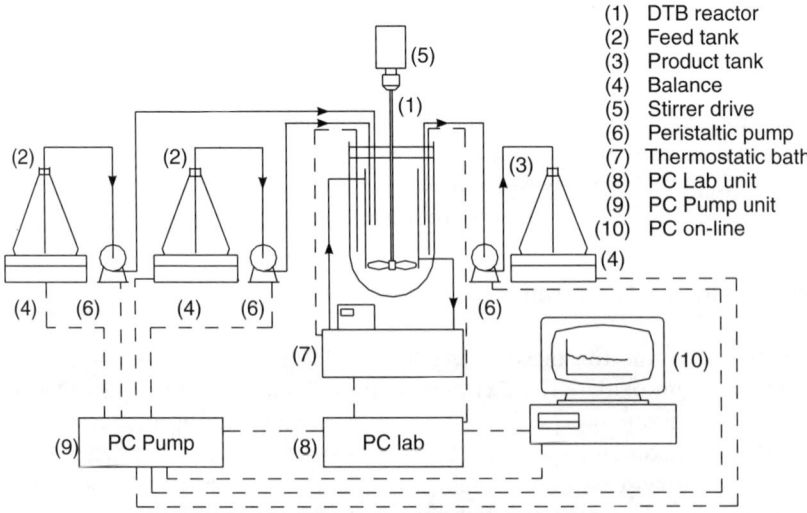

Figure 8.3 *Experimental set-up for scale-up experiments (after Zauner and Jones, 2000b)*

Volume mean size L_{43}

In Figure 8.4, the volume mean size is plotted versus the specific power input for the 300 ml, 4.3 l and 12 l reactors. Both the experimental results and model predictions show a maximum of L_{43} for a power input between 0.4 and 0.5 W/kg. Such a maximum was observed earlier for the precipitation of barium sulphate (Kim and Tarbell, 1996). The steep decrease of the volume mean size for specific power inputs greater than 1 W/kg is probably due to very high disruption rates under these vigorously agitated conditions. The model predicts the experimental data satisfactorily, but no clear trend with scale-up can be determined. Under this particular set of conditions, scale-up with constant mean specific power input could be a reasonably good scale-up criterion. In this case, the curves on different scales would thus merge into one curve. The ideal MSMPR model, in which it has been assumed that the nucleation rate and the growth rate are constant and that secondary nucleation does not occur, is not capable of predicting any of the experimentally observed effects of power input on mean size, as one of the assumptions for an MSMPR crystallizer is well-mixedness.

Semibatch mode of operation

The experimental set-up used for the semibatch experiments is similar to that used for the continuous experiments but without use of the offtake. The particle

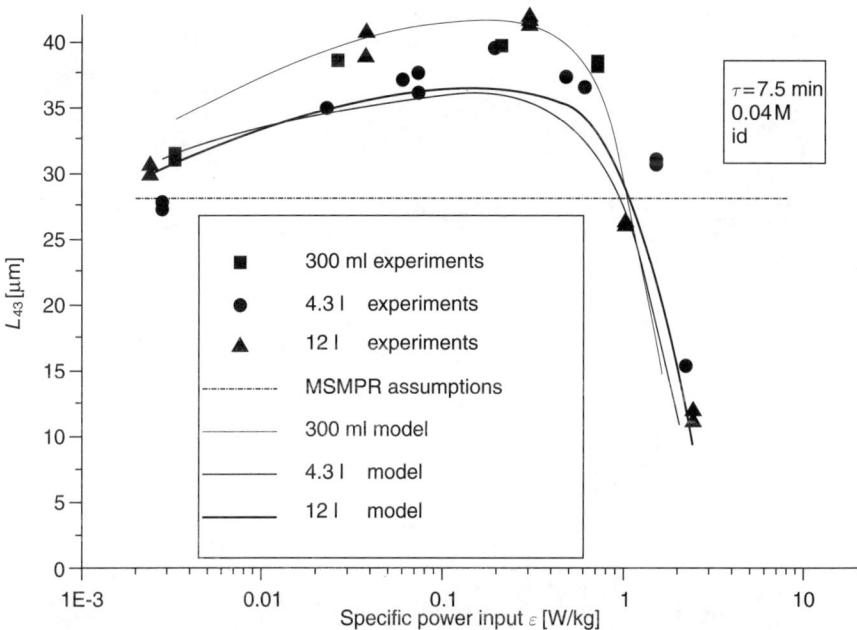

Figure 8.4 *Scale-up of continuous calcium oxalate precipitation: volume mean size L_{43} (300 ml, 4.3 l and 12 l; 0.04 M, 7.5 min, id). (Zauner and Jones, 2000b)*

Figure 8.5 *SEM micrographs of calcium carbonate agglomerates (a, vaterite and calcite; b, calcite). (After Zauner and Jones, 2000c)*

size distribution is measured using a Coulter Counter Multisizer II and/or a Sympatec Helos laser scattering analyser. The vessel is equipped with four baffles, but in contrast to the reactor used for the continuous experiments does not contain a draft tube.

In the continuous precipitation of calcium carbonate, only spherical vaterite was observed (Zauner and Jones, 2000b). In the semibatch experiments, however, a mixture of both calcite and vaterite was obtained (Figure 8.5a). Both polymorphs tend to form agglomerates consisting of a mixture of calcite and vaterite. When left in solution for several hours (e.g. overnight), the metastable vaterite crystals change to stable calcite to form agglomerates consisting only of calcite (Figure 8.5b).

In comparison with the continuous mode of operation, the mean size was found to depend to a greater degree on the mixing conditions on all scales in the semibatch mode.

In Figure 8.6, the results for the reference conditions (Rushton turbine, 40-min feed time, feed point position close to the impeller, total concentration 0.008 M) for calcium oxalate confirm this observation.

By changing the energy input, the volume mean size was varied over a wide range from 7 to 26 μm. For the two reactor scales of 1 l and 5 l, scale-up with constant specific power input seems appropriate, while for the 25 l scale smaller particle sizes are obtained in the industrially important range from 0.1 to 1 W/kg. It is interesting to note that such reductions in mean particle size with increasing scale of operation due to local supersaturation variation are the opposite to that observed from the effect of scale on secondary nucleation

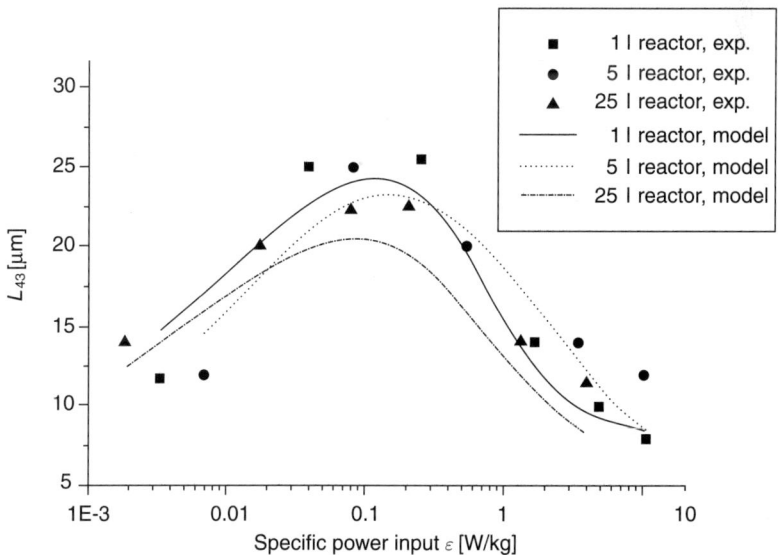

Figure 8.6 *Mean particle for calcium oxalate precipitation (after Zauner and Jones, 2000b)*

(Chapter 5), and may have deleterious effects on subsequent solid–liquid separation and customer acceptance.

The kinetic parameters, which were determined from laboratory-scale continuous experiments as a function of the energy input and/or supersaturation, were applied to the semibatch mode of operation without any adjustments or parameter fitting. The SFM slightly underestimates the mean particle size in the range between 0.01 and 1 W/kg, but correctly predicts the smaller particle size obtained experimentally for the 25 l reactor. On the same scale, the model also predicts a lesser degree of dependence of the particle size on the specific power input due to the interactions of mixing and the precipitation kinetics. This behaviour has also been observed experimentally in this research.

Small particle sizes obtained at low energy inputs are probably a result of local zones with very high levels of supersaturation and therefore high nucleation rates. At high values of energy input, in contrast, breakage might act as a size-reducing process, leading to smaller particles.

Effect of feed point position

In order to investigate the dependence of the particle size on the feed point position, one set of experiments was carried out with a feed point close to the impeller and another set with a feed point close to the surface. At low power inputs, the mean particle size obtained for the feed point in a highly turbulent zone near the Rushton turbine resulted in larger particles, giving mean particle sizes up to twice as large as those produced with a feed point close to the surface (Figure 8.7). Åslund and Rasmuson (1992) investigated feed points near the impeller, in the bulk and close to the surface and observed the same behaviour: at low stirrer speeds substantially larger particles were produced with the feed

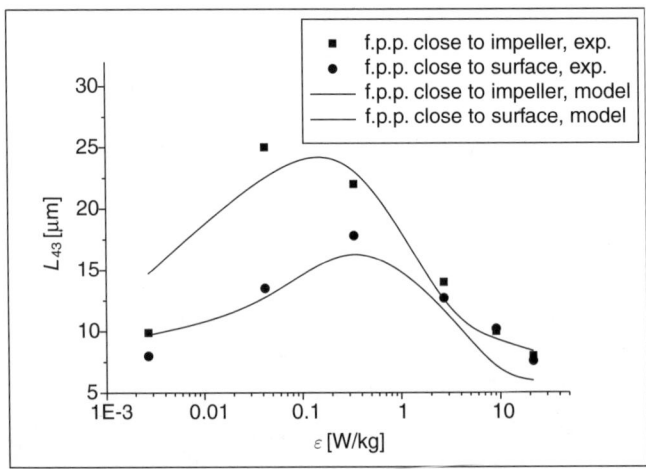

Figure 8.7 *Mean particle size versus specific energy input for different feed point positions (f.p.p.) (CaOx, Rushton turbine, 40 min feed time, total concentration 0.008 M.) (After Zauner and Jones, 2000b)*

point close to the impeller, while at high stirrer speeds almost no influence of the feed point position on the weight mean size was found.

Scale-up criteria

Various chemical engineering scale-up criteria are available in the literature and have been met with greater of lesser success in the literature. Each is examined in detail below.

Constant stirrer speed

This scale-up criterion is based on achieving a constant pumping rate per unit volume with scale-up and therefore leads to similar macromixing on different scales, as the circulation time in the reactor remains constant.

$$N_1 = N_2 \tag{8.26}$$

Constant tip speed

If the tip speed of the impeller blades is kept constant with scale-up, i.e.

$$N_1 \pi d_{s1} = N_2 \pi d_{s2} = \text{constant} \tag{8.27}$$

The criterion for scale-up becomes

$$N_2 = N_1 \left(\frac{V_2}{V_1}\right)^{-\frac{1}{3}} \tag{8.28}$$

Scale-up with constant tip speed, which implies constant shear in the impeller region, can be considered an approximation to scale-up with constant mesomixing, as the blending of the incoming reactant with the bulk or second reactant is closely linked to the shear field in the mixing zone. As Oldshue (1985) pointed out, however, with larger scales, the maximum shear rate in the impeller zone increases while the average shear rate in this zone decreases. Thus, the shear rate distribution changes with scale-up and this criterion is therefore only an approximation, based on mean shear rates in the impeller zone.

Constant power input per unit volume

The scale-up criterion that is probably most widely used for mixing-limited unit operations is based on constant power input per unit volume according to (Harnby et al., 1992).

$$\varepsilon = \frac{k_{power} d_s^5 N^3}{V} = \text{constant} \tag{8.29}$$

This leads to

$$N_2 = N_1 \left(\frac{V_2}{V_1}\right)^{-\frac{2}{9}} \tag{8.30}$$

The importance of the local energy dissipation and thus the specific power input for micromixing has already been referred to, where the micromixing time was related to the Kolmogoroff length scale of mixing. Even though it was possible to predict properties on different scales using this criterion, it fails with respect to other conditions.

The failure of 'conventional' criteria may be due to the fact that it is not only one mixing process which can be limiting, rather for example an interplay of micromixing and mesomixing can influence the kinetic rates. Thus, by scaling up with constant micromixing times on different scales, the mesomixing times cannot be kept constant but will differ, and consequently the precipitation rates (e.g. nucleation rates) will tend to deviate with scale-up.

The conventional scale-up criteria 'scale-up with constant stirrer speed', 'scale-up with constant tip speed' and 'scale-up with constant specific energy input' are all based on the assumption that only one mixing process is limiting. If, for example, the specific energy input is kept constant with scale-up, the same micromixing behaviour could be expected on different scales. The mesomixing time, however, will change with scale-up; as a result, the kinetic rates and particle properties will be different and scale-up will fail.

SFM methodology

In order to account for both micromixing and mesomixing effects, a mixing model for precipitation based on the SFM has been developed and applied to continuous and semibatch precipitation. Establishing a network of ideally macromixed reactors if macromixing plays a dominant role can extend the model. The methodology of how to scale up a precipitation process is depicted in Figure 8.8.

The model is able to predict the influence of mixing on particle properties and kinetic rates on different scales for a continuously operated reactor and a semibatch reactor with different types of impellers and under a wide range of operational conditions. From laboratory-scale experiments, the precipitation kinetics for nucleation, growth, agglomeration and disruption have to be determined (Zauner and Jones, 2000a). The fluid dynamic parameters, i.e. the local specific energy dissipation around the feed point, can be obtained either from CFD or from LDA measurements. In the compartmental SFM, the population balance is solved and the particle properties of the final product are predicted. As the model contains only physical and no phenomenological parameters, it can be used for scale-up.

For large-scale predictions, kinetic information can be obtained from laboratory-scale experiments. Only CFD simulations (or LDA measurements) need to be carried out to obtain the local energy dissipation in the feed zone(s) on the large scale. Using this information, the SFM predicts the particle properties on the large scale. Scale-up thus becomes possible without time-consuming and costly large-scale precipitation experiments. This methodology is very efficient as it combines the advantages of both a CFD and a population balance approach without having to solve the coupled equations with consequentially high computational demand and simulation time required for large vessels. The

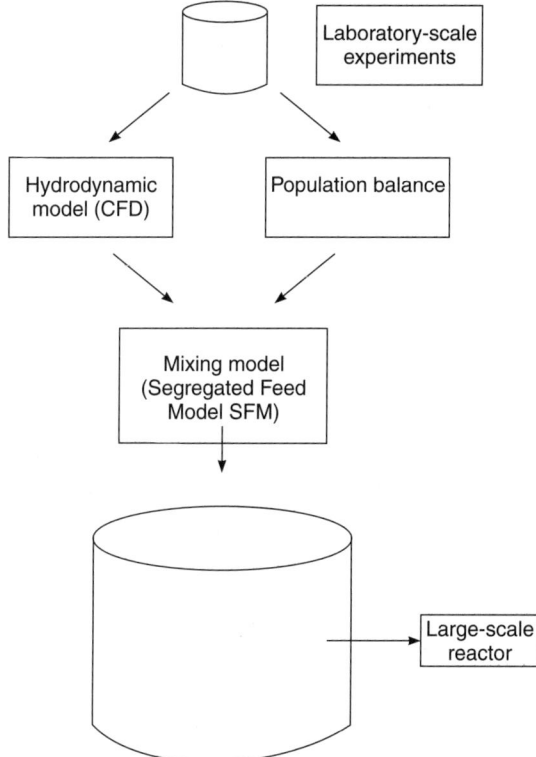

Figure 8.8 *Precipitation process scale-up methodology (after Zauner and Jones, 2000b)*

results of the model have been successfully validated with experiments under a wide range of conditions for the precipitation of calcium oxalate for both continuous and semibatch precipitation.

In an alternative approach, Jézéquel (2001) proposes the concept of the 'scaleable reactor' based on the observation that in highly non-linear systems such as precipitation, scale-up is better assured when all the critical parts of the process are exactly at the same scales. Thus, the critical parts are scaled by replication, whilst the main holding vessel in which slow processes take place is scaled in volume (Figure 8.9).

Jézéquel (2001) illustrated the procedure for precipitation of several sizes and morphologies of silver halide crystals.

Torbacke and Rasmuson (2001) report the empirical influence of different scales of mixing in reaction crystallization of benzoic acid in a loop reactor. The authors infer that the process is mainly governed by mesomixing in terms of liquid circulation rate but find anomalous behaviour in respect of feed pipe diameter.

Probability Density Function (PDF) methods

Baldyga and Orciuch (2001) studied the precipitation of barium sulphate from unpremixed feed in a 2D tubular precipitator (pipe 2 m × 0.0320 m i.d.). Crystal

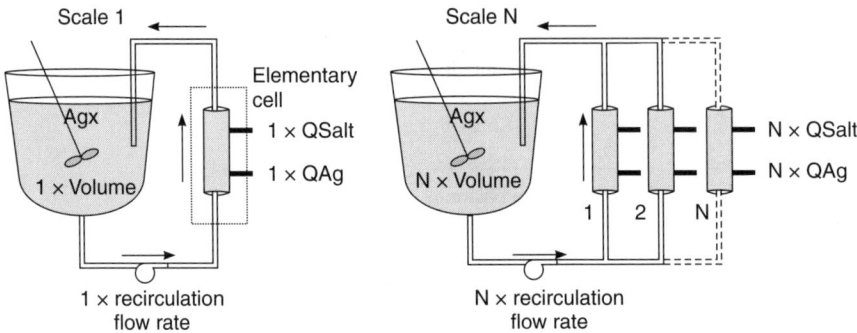

Figure 8.9 *The scaleable reactor concept (Jézéquel, 2001)*

size distribution is predicted using a Beta PDF (Probability Density Function) adapted from Baldyga and Orciuch (1997) incorporating a turbulent population balance, and the turbulent mixing model implemented in a k-ε CFD code (Fidap 8.5). The authors conclude that their method enables the particle size distribution to be predicted effectively when compared with experimental data in the absence of aggregation. Figure 8.10 gives an example of the crystal size distribution; both measured and calculated using both the full model and that which neglects concentration fluctuations. The difference is clearly observed; predicted and measured distributions agree better for the full model. The effect of mixing should be better observed at still larger concentrations, but then the

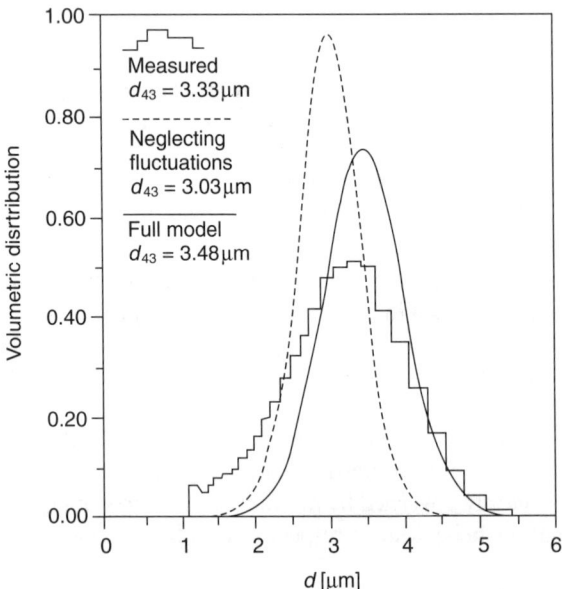

Figure 8.10 *Predicted and measured (averaged) crystal size distributions for barium sulphate* (Re = 30,000, $C_{AO} = 0.015\,\text{kmol m}^{-3}$, $C_{BO} = 1.500\,\text{kmol m}^{-3}$, Ru = 1). *(Baldyga and Orciuch, 2001)*

authors observed significant particle aggregation. The observation that micromixing effects are present only at high concentrations (so for low values of the time constants for precipitation compared to time constants for mixing) agree well with considerations presented by Baldyga and Bourne (1999) (see their Chapter 14) in relation to choosing an adequate model of mixing for modelling of precipitation processes.

Falk and Schraer (2001) similarly present a PDF model of precipitation reactors. The micromixing model considered is the IEM. Precipitation reactions are nucleation, growth and aggregation. Based on Monte Carlo simulations, the method is able to produce, at any point in the flow, reactants concentration and supersaturation fields, and the particle size distribution field, by the simple moments method. The main advantage claimed for the method is its capability to treat multi-dimensional (viz. size, morphology, chemical properties, etc.) population balances as efficiently as it treats high dimensional PDF. The PDF code is again coupled with a k-ε CFD code (FLUENT) to calculate the flow in the absence of two-phase effects. Typical integration times are reported to be several hours on a 300 mHz PC. The procedure was illustrated by the aggregation in absence of supersaturation of silica crystals in a 2D tubular vessel (pipe 0.50 m × 0.050 m id.).

Gas–liquid crystal precipitation

Precipitation of crystals promoted by gas–liquid reactions, such as oxidation or carbonation, finds application in fields as diverse as the production of fine chemicals, biotechnology and gas cleaning. This class of process, however, has not been so extensively studied until recently, despite its undoubted industrial importance – the oxidation of p-xylene to precipitate terephthalic acid crystals used for the manufacture of Dacron^{TM} and Terylene^{TM} being a significant example. Gas–liquid precipitation of methane hydrates can cause problems of pipeline plugging, as in the formation of, in the oil and gas industry, though they may also be a potential resource. The physical characteristics of the product particles of such reactions may be of significance for their effects on downstream processing (Jones, 1991) and/or technical application and are determined by the processing conditions employed and their formation kinetics (Jones, 1994).

A review article by Wachi and Jones (1995) aims to draw a coherent picture of the state-of-the-art of analysing and predicting the overall process of gas–liquid precipitation on which this section is largely based. In the early stages, i.e. during the induction period, mass transfer and chemical reaction proceed and generate supersaturation. Nucleation and subsequent crystal growth then commerce at rate depending on the level of supersaturation obtained. When such crystal formation is relatively rapid compared to the dispersion of the reaction product, significant non-uniformities may occur throughout the solution and mixing effects thus become important. Secondary nucleation may also occur due to particle breakage or attrition. The size of solid particles can also increase by agglomeration as well as by molecular or ionic growth at the crystal surface. During prolonged contact with the mother liquor, further ageing processes can affect the final characteristics of the product; small crystals begin to dissolve and the resulting solute is transported to the larger crystals, which continue to grow, i.e. Ostwald ripening occurs, or solution enclosed inside agglomerates oozes out, i.e. sweating takes place.

Each stage of particle formation is controlled variously by the type of reactor, i.e. gas–liquid contacting apparatus. Gas–liquid mass transfer phenomena determine the level of solute supersaturation and its spatial distribution in the liquid phase; the counterpart role in liquid–liquid reaction systems may be played by micromixing phenomena. The agglomeration and subsequent ageing processes are likely to be affected by the flow dynamics such as motion of the suspension of solids and the fluid shear stress distribution. Thus, the choice of reactor is of substantial importance for the tailoring of product quality as well as for production efficiency.

Several reported chemical systems of gas–liquid precipitation are first reviewed from the viewpoints of both experimental study and industrial application. The characteristic feature of gas–liquid mass transfer in terms of its effects on the crystallization process is then discussed theoretically together with a summary of experimental results. The secondary processes of particle agglomeration and disruption are then modelled and discussed in respect of the effect of reactor fluid dynamics. Finally, different types of gas–liquid contacting reactor and their respective design considerations are overviewed for application to controlled precipitate particle formation.

Industrial gas–liquid precipitation reactions

There is an increasing number of industrially important gas–liquid reaction precipitation systems (see Kirk-Othmer, 1993), including the following:

Ammonium phosphate is produced by reaction of ammonia with phosphoric acid resulting in the formation of the mono or di-basic salts:

$$NH_3(gas) + H_3PO_4(liquid) \rightarrow NH_4H_2PO_4(solid)$$

and

$$2NH_3(gas) + H_3PO_4(liquid) \rightarrow (NH_4)_2HPO_4(solid)$$

The mono salt tends to produce needles while the di-basic salt results in crystals that are more granular. Ammonium phosphate finds application as a fertilizer.

Ammonium sulphate is traditionally produced by reaction of by-product ammonia from coke ovens with sulphuric acid:

$$2NH_3(gas) + H_2SO_4(liquid) \rightarrow (NH_4)_2SO_4(solid)$$

Ammonium sulphate is increasingly produced by reaction of synthetic ammonia with sulphuric acid; from the process for production of caprolactam; and via a method using gypsum. Ammonium sulphate is also used as a fertilizer.

Barium carbonate can be prepared by carbonation of either barium hydroxide (Yagi et al., 1988) or barium sulphate (Kubota et al., 1990):

$$Ba^{2+} + CO_3^{2-} \rightarrow BaCO_3$$

The product crystals find industrial application as a component raw material for optical glass, fibreglass, Braun tubes, electric condensers, barium ferrite, etc. Needles shaped crystals are obtained at high pH, while pillar-shaped crystals are formed at neutral pH. The formation of carboxyl ions is via hydroxy ions at high pH, but at neutral pH it may accompany the production of hydrogen sulphide, as

$$CO_2 + 2SH^- + H_2O \rightarrow CO_3^{2-} + 2H_2S$$

Calcium carbonate is manufactured by bubbling carbon dioxide through milk of lime suspension in a 'carbonator', according to

$$Ca(OH)_2 \text{ (aq. sol.)} + CO_2 \text{ (gas)} \rightarrow CaCO_3 \text{(solid)} + H_2O$$

Precipitated calcium carbonate (PCC) has varieties of markets including the chemical industries, building, refractory, agricultural and highway construction. Although the individual processes of crystallization (Nancollas and Reddy, 1971; Plummer and Busenberg, 1982; Hostomský and Jones, 1991) and gas–liquid reaction (Juvekar and Sharma, 1973; Sada et al., 1985) have been studied extensively, the combined effects of these two processes remained unexplored until recently. Hexahedral crystals of calcite are produced either in a bubbling stirred tank reactor (Kotaki and Tsuge, 1990) or in flat interface reactors (Yagi et al., 1984; Yagi et al., 1988; Wachi and Jones, 1991b). While the size of individual crystals is in the range of 1–10 μm, agglomeration leads to growth in particle size up to a few tens of microns.

Calcium fluoride was precipitated from calcium nitrate and hydrogen fluoride (Naumova et al., 1990), as

$$CaNO_3 \text{ (liquid)} + HF \text{ (gas or liquid)} \rightarrow CaF_2 \text{ (solid)}$$

The size of crystals produced in the gas–liquid system varied from 10 to 100 μm by controlling the level of supersaturation, while the liquid–liquid system produced crystals of 5–30 μm. The wide variation of crystal size is due to the marked sensitivity of the nucleation rate on the level of supersaturation, while the impurity content is another variable that can affect the crystal formation.

Gypsum (calcium sulphate): The gas–liquid reaction between sulphur oxide and calcium hydroxide has been analysed based on the film model (Sada et al., 1977), but the formation of gypsum crystals was not closely investigated. The liquid phase reaction between sodium sulphate and calcium dichloride produces elongated crystals of gypsum. Gypsum is of commercial use for plaster, cement or fertilizer. For the purpose of pollution control in the burning of high sulphur coal, sulphur dioxide present in stack gases is removed by absorption into limewater with the formation of calcium sulphate–sulphite mixtures.

Geothite is produced by air oxidation of alkaline suspension of ferrous hydroxide (Sada et al., 1988). As a starting material for ferrous oxide, the preparation of fine particles with prescribed size, size distribution and shape is required in its application to magnetic materials for recording tapes and disks. With increasing oxidation rate, the crystal size decreases and the size distribution becomes sharper (Sada et al., 1988). The first step of the reaction, i.e. oxidation of $HFeO^{2-}$ to produce ferric hydroxo-complex, takes place only in the liquid film around the gas bubble:

$$Fe(OH)_2 + OH^- \rightarrow HFeO^{2-} + H_2O$$
$$HFeO^{2-} + O_2 \rightarrow Fe_2(OH)_3^{3+}$$

The second step of the reaction proceeds in the bulk solution, i.e. ferric hydroxo-complex is hydrolyzed to form geothite as:

$$Fe_2(OH)_3^{3+} + H_2O \rightarrow \alpha\text{-}FeOOH$$

The kinetic rate is first order with respect to the concentration of oxygen and independent of the ferric ion concentration.

Potassium chloride is crystallized from sea bitterns containing chlorides of potassium, sodium and calcium by ammoniation (Jagadesh et al., 1992). This process is less energy intensive and more efficient than by fractional crystallization by evaporation, as the ammonia used is recovered by distillation. Crystallization produces a better quality product in terms of both size and purity than by other methods.

Sodium bicarbonate is made by the carbonation of salt and ammonia in carbonation tower (Solvay process):

$$NaCl + NH_4OH + CO_2 \rightarrow NaHCO_3 + NH_4Cl$$

The sodium bicarbonate produced is heated to $175\,°C$ in rotary dryers to give light soda ash (Na_2CO_3). Holes are left in the crystals obtained, as the carbon dioxide is liberated. Dense soda ash used by the glass industry is manufactured from the light ash by adding water and drying.

Strontium carbonate was produced by the carbonation of strontium hydroxide (Yagi et al., 1988), as

$$Sr(OH)_2 + CO_2 \rightarrow SrCO_3 + H_2O$$

The product crystals were agglomerates of needles or dendrites. Loose flocs of dendroid strontium carbonate are compacted by agitation, which is an important factor in controlling the habit of product particles. Semi-batch operation produces larger particles compared to batch or continuous operation.

Terephthalic acid is made by air oxidation of *p*-xylene in acetic acid with cobalt and manganese salts of metal bromide at $200\,°C$ and $400\,\psi$.

$$CH_3(C_6H_4)CH_3 + 3O_2 \rightarrow COOH(C_6H_4)COOH$$

The crude terephthalic acid is cooled and crystallized. For polyester grade, it must be purified to 99.9 per cent by subsequent recrystallization.

Types of reactor

Gas–liquid contactors may be operated either by way of gas bubble dispersion into liquid or droplet dispersion in gas phase, while thin film reactors, i.e. packed columns and trickle beds are not suitable for solid formation due to the deposition of scale onto the vessel surface. Shah *et al.* (1982), Deckwer (1985) and Ueyama (1993) have reviewed flow dynamics of bubble column reactors whilst gas–liquid–solid reactors have been reviewed by Moroyama and Fan (1985) and Fan (1989). Typical reactors applied for gas–liquid precipitation are illustrated in Figure 8.11.

The type of reactor is selected in the light of numerous factors such as volume ratio of gas to liquid, pressure drop, heat transfer and mixing intensity.

The stirred tank reactor is often chosen since it can achieve an excellent near-uniform suspension of crystal particles. The mass transfer coefficient in flat interface stirred tank is dependent on the stirring rate as (Hikita and Ishikawa, 1969)

$$Sh = 0.322 Re^{0.7} Sc^{\frac{1}{3}} \qquad (8.31)$$

where $Sh = k_L/D_M$, $Re = d^2 N \rho / \mu$ and $Sc = \mu/(\rho_L D_M)$.

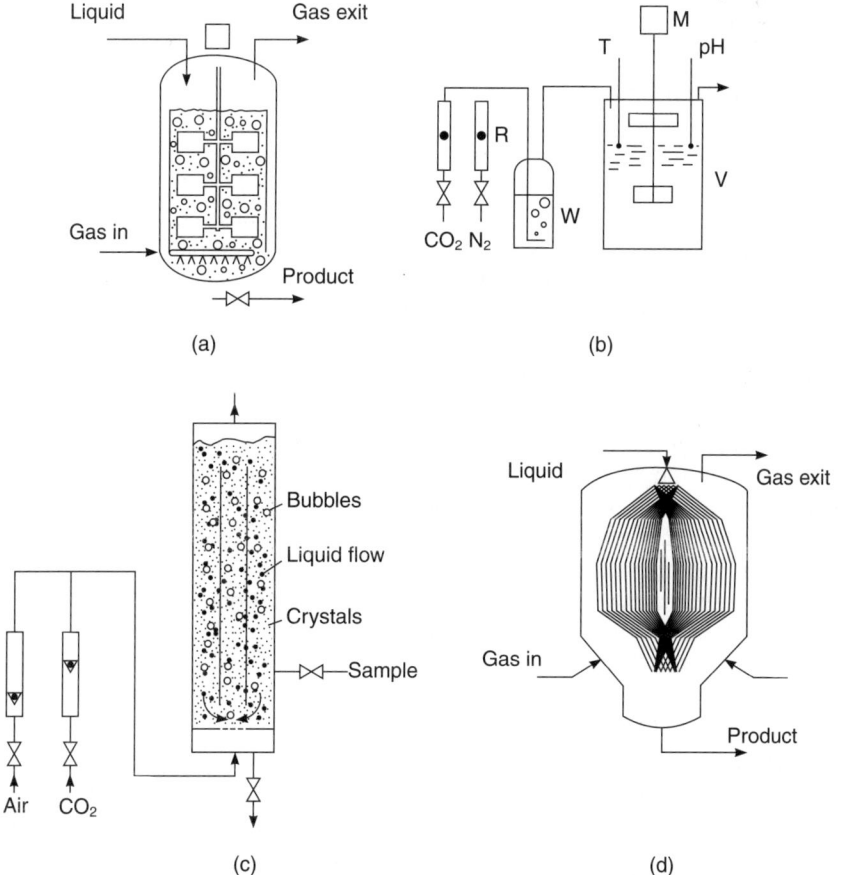

Figure 8.11 *Types of reactors for gas–liquid precipitation:* (a) *bubbling stirred tank;* (b) *flat interface stirred tank, draft-tube bubble column;* (d) *spray column (after Wachi and Jones, 1994)*

The dynamics of reactor flow is also important for its effect on the crystal agglomeration, since the intensity of turbulent shear dominates the orthokinetic mechanism for both processes of aggregation and disruption. The mean shear rate is estimated as (see Harnby *et al.*, 1992)

$$\dot{\gamma} = \sqrt{\frac{P_I}{\mu V}} \tag{8.32}$$

where the power input, P_I, in a stirred tank fitted with for example a two blade paddle is given as

$$P_N = \frac{P_I}{N^3 d^5} = 2.0 \quad (\text{for } Re = d^2 N \rho \frac{L}{\mu} > 10^4) \tag{8.33}$$

The power input in bubble column, on the other hand, is evaluated from isothermal gas expansion as (Jones, 1985)

$$P_I = Q_M RT \ln \frac{p_b}{p_t} \tag{8.34}$$

Role of gas–liquid mass transfer

It is now well established that mass transfer can determine the overall performance of gas–liquid chemical reactors (Astarita, 1967; Danckwerts, 1970). Similarly, during mass transfer of substrates with chemical reaction to create solute supersaturation and subsequent precipitation, the mass transfer process often controls the overall behaviour, as noted by Danckwerts (1958), but its effect on particle characteristics has only been studied in depth relatively recently. Under such situations, the spatial distribution of concentration at the micro-level can influence the size of product crystals. This is due to the highly sensitive kinetics of nucleation and crystal growth to the level of supersaturation. For liquid phase precipitation systems, Pohorecki and Baldyga (1983) and Garside and Tavare (1985), respectively, have reported studies of this phenomenon where the extent of micromixing is variously modelled. Different considerations apply, however, to precipitation in gas–liquid reaction systems, viz. the mass transfer resistance appears at the gas–liquid interfacial region rather than within the bulk. Rapid chemical reaction between substrates of gas and liquid leads to highly concentrated sparingly soluble products in the region of liquid film and consequent high spatial variations in supersaturation.

A non-ideal MSMPR model was developed to account for the gas–liquid mass transfer resistance (Yagi, 1986). The reactor is divided into two regions; the level of supersaturation in the gas–liquid interfacial region (region I) is higher than that in the main body of bulk liquid (region II), as shown in Figure 8.12.

A steady-state population density of crystals in the region II is

$$V(1 - \varepsilon_I) \frac{dG_{II} n_{II}}{dL} + F n_{II} = f n_I \tag{8.35}$$

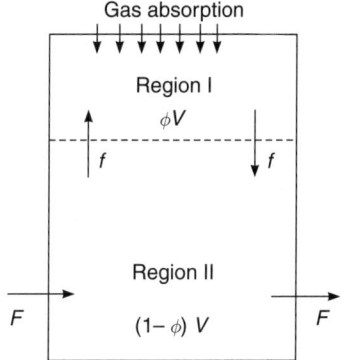

Figure 8.12 *Gas–liquid reactor model (Yagi, 1986)*

where

$$V\varepsilon_I B_I + V(1-\varepsilon_I)B_{II} = F \int_0^\infty n_{II} dL \qquad (8.36)$$

The model predicts that due to negligibly small nucleation rates in the main body of liquid, the number of small crystals is reduced compared with the ideal MSMPR model.

A more rigorous scheme of gas–liquid chemical reaction and absorption followed by precipitation is described based on the film model (Englezos *et al.*, 1987a,b; Wachi and Jones, 1990; 1991a; Skovborg and Rasmussen, 1994). The concept is illustrated in Figure 8.13.

Material balances of reactants and product are

$$\frac{dA}{dt} = D_M \frac{d^2 A}{dx^2} - kAB \qquad (8.37)$$

$$\frac{dB}{dt} = D_M \frac{d^2 B}{dx^2} - kAB \qquad (8.38)$$

$$\frac{dC}{dt} = D_M \frac{d^2 C}{dx^2} + kAB - G' - B' \qquad (8.39)$$

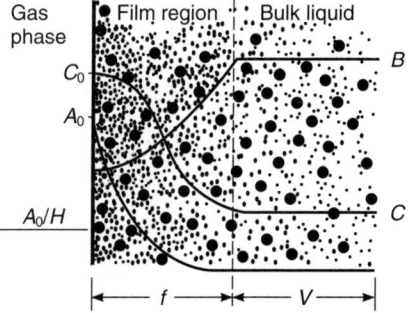

Figure 8.13 *Conceptual concentration profiles and particle distribution in the liquid film region (Wachi and Jones, 1991a)*

The population balance of the precipitated particles is

$$\frac{dn}{dt} + G\frac{dn}{dL} = Dp\frac{d^2n}{dx^2} \qquad (8.40)$$

Rates of nucleation and crystal growth may be given, respectively, as

$$J_n = k_n \Delta c^n \qquad (8.41)$$

and

$$G = k_g \Delta c^g \qquad (8.42)$$

where the level of solute supersaturation $\Delta c = c - c^*$ where c^* is the equilibrium saturation concentration. (Alternatively, precipitation kinetics employing solubility products may be adopted, Jones *et al.*, 1992). The diffusivity of crystal particles of size L is given by the Stokes–Einstein equation:

$$D_p = \frac{k_B T}{3\rho\mu L} \qquad (8.43)$$

The set of equations was solved numerically for the carbonation of limewater with the conditions of batch operation. In an alternative approach, the penetration model has also been applied (Hostomský and Jones, 1993).

Film-crystal model concentration profiles of *A*, *B* and *C* and particle number density distributions are shown in Figure 8.14(a).

Due to the diffusion resistance, the level of supersaturation is high in the liquid film, and falls toward the bulk liquid. This tendency is more pronounced by an increase of specific bulk liquid volume, i.e. by the decrease of gas–liquid contacting area. The smallest particles, i.e. crystal nuclei, exhibit significant spatial distribution in the liquid film since the nucleation rate is very sensitive to the level of supersaturation. The larger particles, however, show quite uniform distribution, dispersing during the period of crystal growth. Crystal size distributions are shown in Figure 8.14(b).

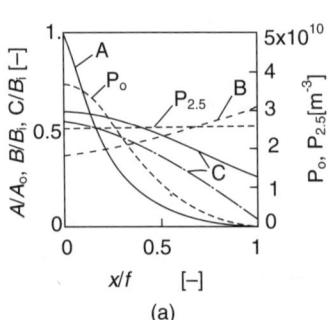
(a)

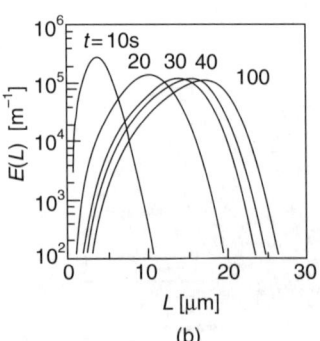
(b)

Figure 8.14 Predicted (a) concentration profiles in the film region; and (b) mean particle sizes during gas–liquid precipitation of $CaCO_3$ (Wachi and Jones, 1991a). Gas–liquid precipitation cell

Figure 8.15 *Experimental gas–liquid flat interface reactor (Wachi and Jones, 1991b)*

Wachi and Jones (1991b) used a gas–liquid flat interface reactor as a semi-batch precipitation cell for the experimental measurement of calcium carbonate precipitation, as shown in Figure 8.15.

Carbon dioxide gas diluted with nitrogen is passed continuously across the surface of an agitated aqueous lime solution. Clouds of crystals first appear just beneath the gas–liquid interface, although soon disperse into the bulk liquid phase. This indicates that crystallization occurs predominantly at the gas–liquid interface due to the localized high supersaturation produced by the mass transfer limited chemical reaction. The transient mean size of crystals obtained as a function of agitation rate is shown in Figure 8.16.

The size of crystal increases with time gradually approaching an asymptotic value. The higher the stirring rate, the larger the primary crystal sizes.

The data plotted in the figure clearly support the predicted positive dependence of crystal size on agitation rate. Precipitation in the crystal film both enhances mass transfer and depletes bulk solute concentration. Thus, in the 'clear film' model plotted by broken lines, bulk crystal sizes are initially slightly smaller than those predicted by the crystal film model but quickly become much larger due to increased yield. Taken together, these data imply that while the initial mean crystal growth rate and mixing rate dependence of size are

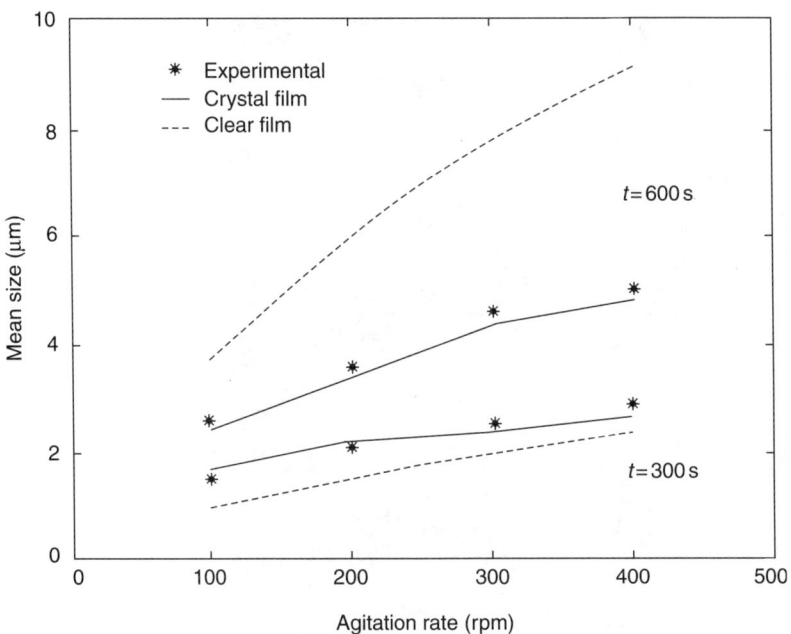

Figure 8.16 *Measured dependence of* $CaCO_3$ *crystal size on agitation rate (Jones et al., 1992b)*

successfully predicted, the terminal crystal size is overestimated implying an underestimation of the nucleation rates in this system by the kinetics adopted.

To fully grasp the main feature of the gas–liquid precipitation phenomena, however, it is not enough just to account for the formation of crystal particles, evaluation of the Hatta number is also essential.

$$\text{Ha} = \frac{\sqrt{kBD_M}}{k_L} \qquad (8.44)$$

The mass transfer effect is relevant when the chemical reaction is far faster than the molecular diffusion, i.e. $\text{Ha} > 1$. The rapid formation of precipitate particles should then occur spatially distributed. The relative rate of particle formation to chemical reaction and/or diffusion can as yet be evaluated only via lengthy calculations.

The maximum particle size obtained in an experimental study using a stirred tank (Figure 8.15) and bubble column (Figure 8.11c) respectively is plotted against the estimated average shear rate in Figure 8.17.

In the stirred tank, the final mean size of particles was reduced by the increase of stirring rate, being consistent with increased fluid shear induced particle disruption relative to aggregation. Use of three different gas velocities in the bubble column, however, results in no significant difference in agglomerate size but since the size is relatively small, it may simply reflect an asymptotic value.

Another possibility is that the effect of shear may be localized, e.g. at the wake of bubbles which is independent of the gas flow rate. Tsutsumi *et al.*

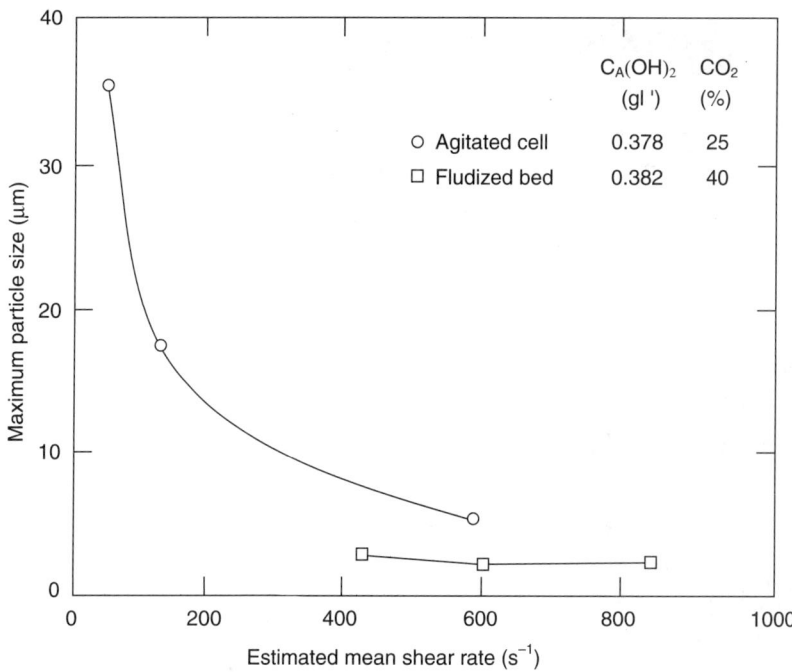

Figure 8.17 *Effect of estimated shear rate on the size of crystal agglomerates (Jones et al., 1992a)*

(1991) investigated the agglomeration of calcium carbonate crystals near bubble wakes. Both the conditions of turbulent shear and ionic concentration can contribute to crystal agglomeration.

In the stirred tank, on the other hand, the shear force will be highest near the impeller and directly determined by the stirring rate. It is also interesting to note that for a given estimated mean shear rate, the final particle size in the bubble column appears to be smaller than from the stirred tank. As mentioned above, however, it is likely that the local maximum, rather than the average, shear rates determine the final agglomerate size. The CFD modelling predictions (Brown and Boyson, 1987 and Al-Rashed et al., 1996) indicate that such velocity gradients can vary considerably throughout an agitated vessel.

In precipitation processes for the production of sparingly soluble products of rapid chemical reactions, crystal agglomeration competes with molecular/ionic crystal growth at the solid–liquid interface as the dominant growth mechanism. The resulting physical characteristics viz. particle size, shape, degree of agglomeration, internal surface area, structure and voidage, are all important for their effects on both product quality and downstream processing interactions. Electric repulsion forces due to particle surface charges are one of the key factors to avoid crystal aggregation. The level of electric charge measured as the 'zeta potential' varies with the concentration of ions, another topic in the field of colloid science (see Shaw, 1980). While small particles might collide during Brownian motion, i.e. according to the perikinetic mechanism, larger particles

Figure 8.18 *Precipitated calcium carbonate: (a), (b) crystals; and (c) agglomerates (Wachi and Jones, 1991b)*

are subject to imposed forces due to turbulence or gravity and are of more significance for their agglomeration, i.e. the orthokinetic mechanism.

In a study of crystal precipitation of calcium carbonate during the batch carbonation of lime water, individual crystals and agglomerated particles were observed as shown in Figure 8.18(a), (b) and (c), respectively (Wachi and Jones, 1991b).

Crystal agglomeration occurred only in the final stage. Time courses of pH, calcium ion concentration and modulus of zeta potential values are shown in Figure 8.19. The chemical reaction conversion is indicated by the decrease of pH and $[Ca^{2+}]$ values. In the early stage of the precipitation process, crystals exhibit a relatively high surface charge. The corresponding repulsion force may thus resist crystal aggregation and thereby inhibit the formation of agglomerates. The ζ-potential subsequently decreases virtually in step with the decrease in calcium ion concentration and pH. Although still quite high, the electric charge may then be sufficiently reduced to allow aggregation to commence and rapid particle growth occur (Figure 8.20). The concentration of calcium ion exhibited a minimum during the agglomeration period. The decrease of calcium concentration in the early stage clearly corresponds to the consumption of calcium by crystallization, while the increase in the later stage may suggest the subsequent dissolution of some part of the calcium carbonate crystals. (Note that the solubility of calcium carbonate changes according to the equilibrium of the solution chemistry.)

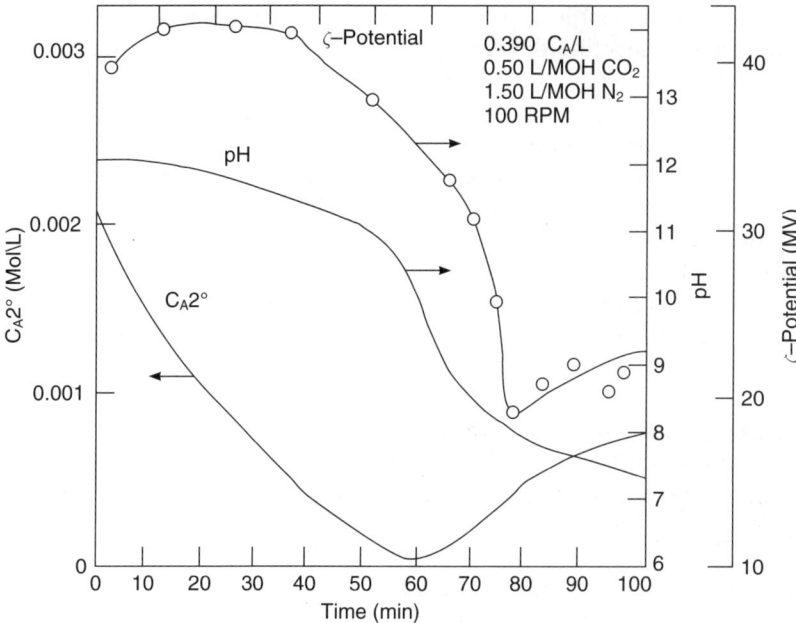

Figure 8.19 *Transient pH, ζ-potential and calcium ion concentration in an agitated reactor (Jones et al., 1992a)*

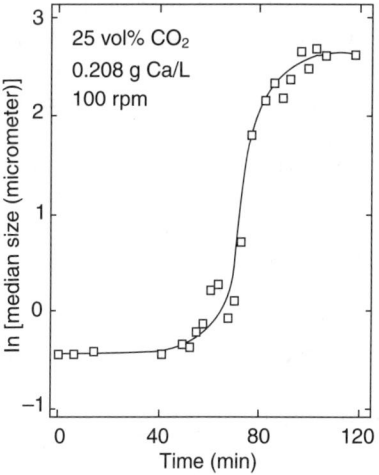

Figure 8.20 *Transient particle size in an agitated vessel reactor (Jones et al., 1992a)*

A dramatic increase of particle size occurs at around pH = 8. At the higher pH(>8), all precipitated particles were within the size range below 1.9 μm, whereas at lower pH(<8), 60–80%wt of particles were larger than 10 μm. The critical pH value is common to all other stirring conditions. The formation of larger particles was avoided by stopping the chemical reaction before the pH value declined to 8.

During the agglomeration process, crystal inter-growth may play an important role in cementing the aggregated crystals. Thus, the agglomeration process may then be inhibited by the insufficient or even negative supersaturation i.e. the aggregated particles may not become fused. The subsequent disruption of large agglomerates due to fluid drag or shear forces (Synowiec *et al.*, 1993 – see Chapter 5) also inferred in MSMPR studies (Wójcik and Jones, 1997 – see Chapter 6) is a further factor which may limit the maximum agglomerate size obtained.

It was shown in Chapter 6 that if the crystals are sufficiently small, for which the aggregation rate is determined by the perikinetic mechanism, then the mean particle volume increases linearly with time:

$$L^3 = a + bt \tag{8.45}$$

where the coefficient a depends on the particle diffusivity. For large crystals, on the other hand, fluid shear forces may induce orthokinetic aggregation giving rise to a corresponding log-linear transient mean particle size:

$$\log L = a' + b't \tag{8.46}$$

where the coefficient b' depends on the liquid shear rate (velocity gradient).

The data for the experimental transient log particle size of calcium carbonate agglomerates shown in Figure 8.20 exhibit a degree of linearity during the middle period of the operation and their growth is thus again consistent with orthokinetic aggregation. Of course, the agglomeration process cannot proceed indefinitely. The final size during the precipitation will be constrained by the release of supersaturation and the consequent end of crystal formation needed to bind the aggregated particles. The maximum ultimate size of agglomerates is also limited by the disruption of large particles due to applied shear stresses, as already noted.

A size distribution of agglomerate particles is shown in Figure 8.21. The peak population within the size range 10–20 μm represents the agglomerated particles, while another peak in the smallest range represents primary crystals. The absence of intermediate sized particles suggests that the disruption of agglomerates occurs by the mode of attrition rather than equal volume breakage. Thus, weakly adhered primary crystals may become detached under the shear field while others, of course, may not even have attached at all, possibly due to growth rate dispersion.

Crystal agglomeration has been investigated experimentally not only from analysis of particle size distributions but also from microscopic observation by SEM and by surface area measurements such as dye adsorption or the BET method. Primary crystal size and the degree of agglomeration often determine important properties of the product particles such as specific surface area and settling velocity, respectively. To predict particle size distribution, kinetics of the multi-step phenomena of nucleation, crystal growth and agglomeration have been incorporated to the population balance model. Normally, only the overall particle size is considered and the primary particles 'coalesce' thereby loosing their identity within the agglomerate, however, without distinction between agglomerated particles and single crystals of equivalent size. Thus,

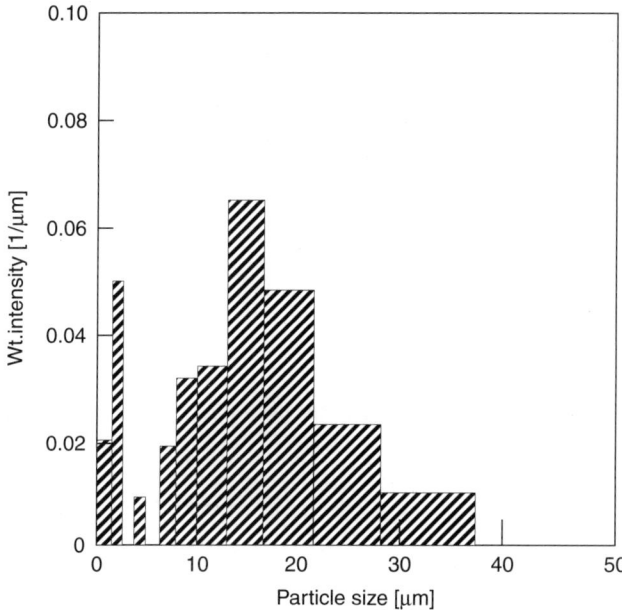

Figure 8.21 *Particle size distribution of precipitated calcium carbonate (Wachi and Jones, 1991b)*

important secondary characteristics such as total (i.e. internal plus external) specific surface area cannot then be predicted.

An alternative theory has been developed to model precipitation with agglomeration where, beside the overall particle size, an additional co-ordinate of crystal number within an agglomerate is introduced (Wachi and Jones, 1992). Figure 8.22 shows the concept of agglomeration and disruption respectively.

The significance of this novel attempt lies in the inclusion of both the additional particle co-ordinate and in a mechanism of particle disruption by primary particle attrition in the population balance. This formulation permits prediction of secondary particle characteristics, e.g. specific surface area expressed as surface area per unit volume or mass of crystal solid (i.e. m^2/m^3 or m^2/kg). It can also account for the formation of bimodal particle size distributions, as are observed in many precipitation processes, for which special forms of size-dependent aggregation kernels have been proposed previously.

Employing two co-ordinates of overall particle size, L, and degree of agglomeration, S (which is, of course, proportional to the mean primary particle size) to define the population density, $n(S, L, t)$, the population balance during precipitation with agglomeration is described as:

$$\frac{dn}{dt} + G\frac{dn}{dL} = B_n + B_a + B_d + D_a + D_d \tag{8.47}$$

Rates of nucleation, B_n, and crystal growth, G, are respectively evaluated as the functions of supersaturation. The level of supersaturation, ΔC, is determined

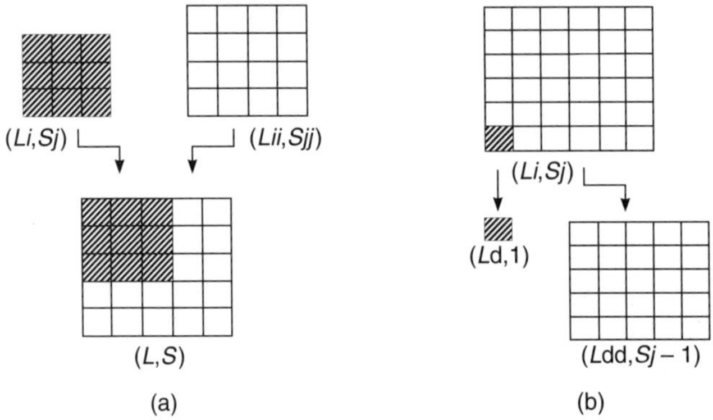

(Li,Sj)   (Lii,Sjj)

(L,S)

(a)

(Li,Sj)

(Ld,1)

(Ldd, Sj – 1)

(b)

Figure 8.22 *Schematic concept of (a) aggregation; and (b) disruption (Wachi and Jones, 1995)*

by a balance between the supply due to chemical reaction, r, and its consumption by the crystallization as:

$$\frac{\mathrm{d}\Delta C}{\mathrm{d}t} = r - \rho\alpha_v L_0^3 B_n - \frac{1}{2}\rho\alpha_s \int_{L=0}^{\infty}\int_{S=1}^{\infty} L^2 Gn\mathrm{d}L\mathrm{d}S \tag{8.48}$$

Aggregation between two particles of size and crystal number (L_i, S_j) and (L, S), respectively, produces a large particle (L_{ii}, S_{jj}).

Conservation of mass and crystal number lead to

$$L_i^3 + L^3 = L_{ii}^3 \tag{8.49}$$
$$S_j + S = S_{jj} \tag{8.50}$$

Thus, particle death and birth rates, respectively, due to aggregation are given by

$$D_\mathrm{a}(L_i, S_j) = n(L_i, S_j)\int_{L=0}^{\infty}\int_{S=1}^{\infty}\beta_\mathrm{a} n(L, S)\mathrm{d}S\mathrm{d}L \tag{8.51}$$

$$B_\mathrm{a}(L_{ii}, S_{jj}) = \frac{1}{2}\int_{L=0}^{L_{ii}}\int_{S=1}^{S_{jj-1}}\beta_\mathrm{a} n(L, S)n(L_i, S_j)\mathrm{d}S\mathrm{d}L \tag{8.52}$$

A semi-empirical form of orthokinetic aggregation kernel due to Thompson (1968) is adopted for illustrative purposes only, viz.

$$\beta_\mathrm{a} = \alpha_v K_\mathrm{a}\frac{(L_i^3 - L_{ii}^3)^2}{L_i^3 + L_{ii}^3} \tag{8.53}$$

Disruption of an agglomerated particle (L, S), produces a single primary crystal $(L_\mathrm{d}, 1)$ and the residual particle $(L_\mathrm{dd}, S-1)$. Conservation of crystal mass balance requires the following equations to be satisfied:

$$L_\mathrm{d}^3 \times S = L^3 \tag{8.54}$$
$$L_\mathrm{d}^3 + L_\mathrm{dd}^3 = L^3 \tag{8.55}$$

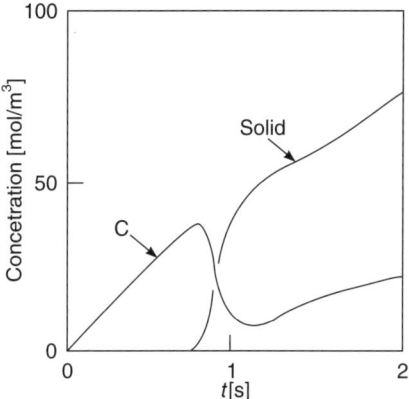

Figure 8.23 *Predicted time courses of supersaturation and magma density (Wachi and Jones, 1992)*

Particle death rate due to disruption and the resulting birth rates of the residual agglomerate and removed primary single crystal, respectively, are given by

$$D_a(L, S) = \beta_d n(L, S) \tag{8.56}$$

$$B_d(L_d, 1) = \int_{S=1}^{\infty} \beta_d n(L, S) dS \tag{8.57}$$

$$B_d(L_{dd}, S - 1) = \beta_d n(L, S) \tag{8.58}$$

where the disruption kernel is assumed to be proportional to the particle volume according to

$$\beta_d = K_d \alpha_v L^3 \tag{8.59}$$

Predicted time courses of supersaturation and magma density during the precipitation are shown in Figure 8.23. After a linear increase of supersaturation due to chemical reaction, the level of supersaturation falls coincident with the rapid appearance of solid phase. Then, as further reaction and crystallization proceed, the supersaturation gradually increases again until the end of chemical reaction. This gradual increase of supersaturation may be due to the decrease of the surface area available for crystal growth consequent on agglomeration, but may be expected to fall again towards equilibrium subsequently.

Figure 8.24 shows predicted particle size distributions at various times. During the initial stage of the precipitation i.e. induction period, only small particles, largely comprising primary crystals of size around 1 μm are produced. Then, following the induction period, the particle size exhibits a bimodal distribution. The second peak at the large size, within the size range 10–40 μm, indicates the presence of agglomerated particles, similar in form to that observed experimentally in Figure 8.21.

As observed earlier, the size of agglomerates may continue to increase until limited by aggregation-disruption equilibrium. Thus, while molecular or ionic growth of primary crystals increases their size slightly during the very first stage, crystal aggregation and disruption (secondary agglomerative growth) is

248   *Crystallization Process Systems*

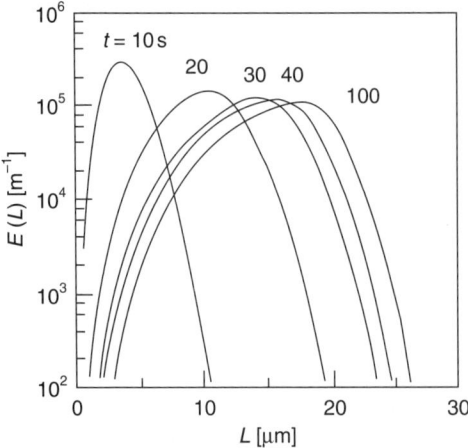

Figure 8.24 *Predicted transient particle size distribution during the batch precipitation of calcium carbonate crystals (Wachi and Jones, 1992)*

most significant in determining the overall particle size. According to this mechanism, even though the size of agglomerates becomes relatively large, the predicted surface area of the primary crystals within them (assuming point contact, which represents a theoretical maximum) remains high.

Having predicted both agglomerate size distribution and primary particle number per agglomerate (related to the mean primary particle size and specific surface area), the next step in modelling the internal structure of agglomerating precipitates is to predict the primary crystal size *distribution* within the agglomerate particle size distribution. This was achieved by Hostomský and Jones (1993a) using the statistical technique known as Monte Carlo simulation.

Monte Carlo simulation

The term Monte Carlo is used to describe any approach to a problem where a probabilistic analogue to a given mathematical problem is set up and solved by stochastic sampling. This invariably involves the generation of many random numbers, so giving rise to the name. A general simulation for particle processes has been developed and the precise mathematical connection between differential population balances and the Monte Carlo approach has been established (Shah *et al.*, 1977; Ramkrishna, 1981). Sengupta and Dutta (1990) investigated growth rate dispersion in an MSMPR, with crystal shape and growth rate as the internal co-ordinates. In their work, they simply simulated one crystal at a time. This approach allowed them to build up a picture of the PSD of an exit stream from an MSMPR by giving each simulated crystal a residence time, randomly selected from the residence time distribution. Wright and Ramkrishna (1992) examined the self-preserving particle size distribution (PSD) in aggregating batch systems.

Hostomský and Jones (1991) described a numerical procedure for a noniterative solution of the steady-state MSMPR crystallization, where both the growth rate and the collision coefficient are size-independent. A Monte Carlo stochastic approach was then used to model sizes of individual primary particles within the agglomerates. Gooch and Hounslow (1996) used the Monte Carlo approach to simulate batch and continuous crystallization processes in which simultaneous nucleation, growth rate dispersion, and aggregation occur without particle disruption. The Monte Carlo predictions were shown to agree well with existing numerical and analytical solutions to the population balance and with experimental data for the potassium sulphate system.

Monte Carlo methodology

In this model, the size of the primary particles is proportional to their individual residence time within the crystallizer and this, together with the identification number of the agglomerate to which the particle is attached, forms a state matrix containing detailed information about each particle. In the calculation, within a unit volume of suspension during any time interval Δt:

- The number of collisions leading to the formation of new agglomerates is proportional to $1/2\beta m_0 \Delta t$.
- Similarly, the number of agglomerates (or free primary particles) leaving the MSMPR crystallizer is proportional to $m_0/\tau \Delta t$.

and the number of primary particles formed by nucleation is proportional to $B\Delta t$.

In the Monte Carlo technique, a random selection is made (as in the Principality!) of those particles that will either take part in collision events or be withdrawn from the crystallizer in the time interval Δt. Thus, the evolution of size, size distribution and surface area (knowing the surface shape factor) of both primary and secondary particles together with their mean characteristics can be evaluated numerically. The results of some sample calculations are illustrated in Figures 8.25 and 8.26 respectively.

In Figure 8.25 the distribution of residence times (∞ sizes) of primary particles within and agglomerate is plotted for agglomerates composed of 5 and 20 particles respectively. It is apparent that the number of particles of residence time greater than the mean is relatively small.

Figure 8.26 shows that with increasing number of constituent primary particles, the specific surface area of agglomerates decreases. This occurs since the average size (age) of the primary particles increases with average number of primary particles within the agglomerate.

Falope *et al.* (2001) extended the MSMPR model of agglomerative crystal precipitation based on the Monte Carlo simulation technique to account for particle disruption by considering two alternative particle size reduction mechanisms – one representative of particle splitting into two parts of equal volume, the other representative of micro attrition.

A comparison of the predicted crystallizer PSD performance using the Monte Carlo method with data for both calcium oxalate and calcium carbonate precipitation (Figure 8.27).

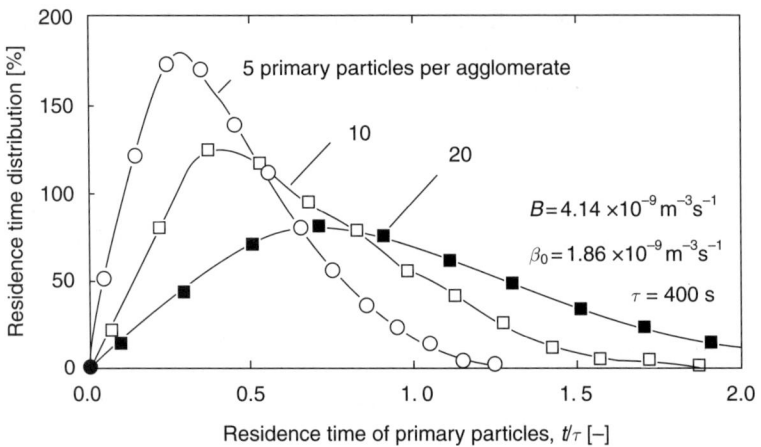

Figure 8.25 *Monte Carlo simulation of distribution of primary particle residence times (∞ size) within MSMPR precipitated agglomerates of 5 and 20 crystals (Hostomský and Jones, 1993a)*

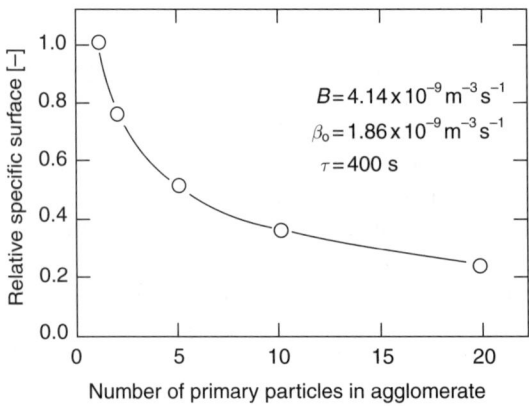

Figure 8.26 *Monte Carlo simulation of MSMPR precipitated crystal agglomerate specific surface area (Hostomský and Jones, 1993a)*

Reasonable agreement is demonstrated consistent with crystal agglomeration occurring via aggregation with particle attrition.

Gas–liquid reactor modelling

Al-Rashed et al. (1996) and Al-Rashed and Jones (1999a,b) presented a CFD-based model to predict the effects of mixing during batch-wise gas–liquid reactive precipitation in the flat interface cell used by Wachi and Jones, 1991b. A 2D flow simulation was developed for the chemical reaction with

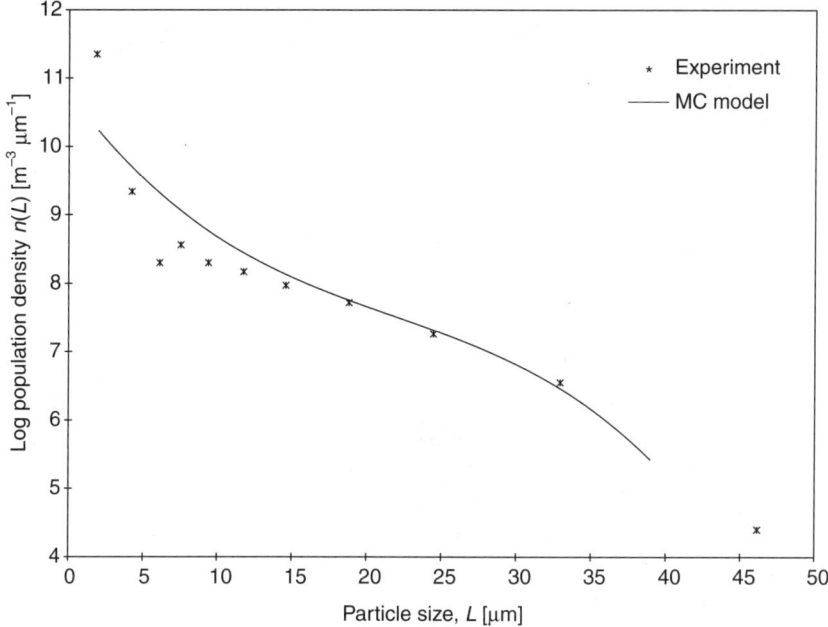

Figure 8.27 *Comparing Monte Carlo model predictions with MSMPR experimental data for calcium carbonate due to Hostomský and Jones, 1991 (Falope et al., 2001)*

precipitation process using the moment transformation of the population balance as an auxiliary set to the CFX code (Harwell, UK).

Computational Fluid Dynamics modelling predictions (Al-Rashed et al., 1996) indicate that such velocity gradients can vary considerably throughout a vessel, as illustrated in Figure 8.28.

Although the experimental and simulation time scales differ, the CFD simulation (Figure 8.29(a),(c),(e)) for the zeroth moment (M_0) indicates that once the particles reach the observable size, they will appear approximately in the experimentally observed regions (Figure 8.29 (b),(d),(f)). Predicted velocity vectors are superimposed on supersaturation profiles in Figure 8.30.

As implied in Figures 8.28–30, the precipitated particles of $CaCO_3$ are assumed to follow the flow pattern. At the tip of the impeller the flow speed vectors proceed in the radial direction until they hit the wall of the vessel where they split into two circulations, above and below the mid-plane of the impeller. In the upper circulation, the velocity vectors hit the interface at the right corner therefore the precipitated particles have less chance to disperse into the bulk from that side. However, the flow continues its journey from the right corner to be parallel to the interface for a short distance and then starts to come down to complete the circulation. The flow, on its way to the interface, brings fresh reactants and carries the precipitated particles into the bulk of the liquid. Computation time using DNS was excessive, however, leading subsequently to the adoption of a hybrid CFD-mixing model to predict bubble column performance.

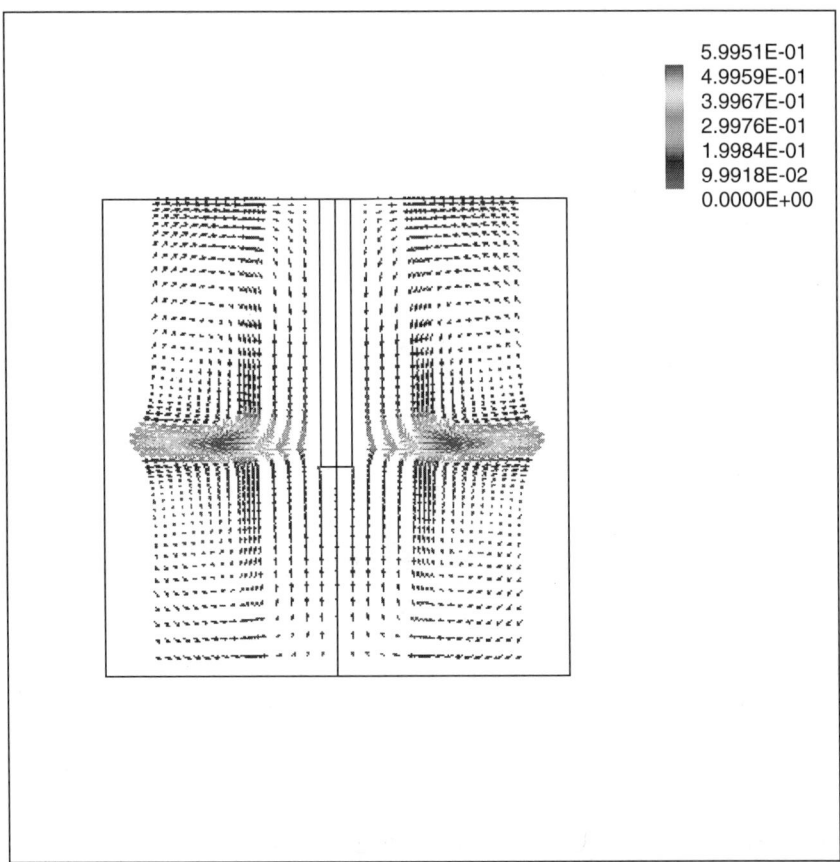

Figure 8.28 *Predicted variation in fluid velocities in an agitated vessel* (*Velocities in* m/s). (*Al-Rashed et al., 1996*)

Bubble column reactor modelling

Rigopoulos and Jones (2001) presented a dynamic model of a bubble column reactor (Figure 8.31) with particle formation, accomplished by adopting a hybrid CFD-reaction engineering approach. The CFD is employed for estimating the hydrodynamics and is based on the two-phase Eulerian–Eulerian viewpoint (stationary observers). The model is developed to account for the different reaction environments emerging in the interface and bulk together with the effect of their interaction on particle size. To accomplish this, the principles of three chemical engineering concepts are combined: the tanks-in-series concept (see Meklenburgh and Hartland, 1975), the penetration theory of gas–liquid mass transfer (Higbie, 1935), and the particle population balance (Hulburt and Katz, 1964).

According to the penetration theory, liquid elements or 'parcels' remain in contact with the gas for a limited time and subsequently are mixed with the bulk. This concept is particularly suited to bubble flows because it reflects their

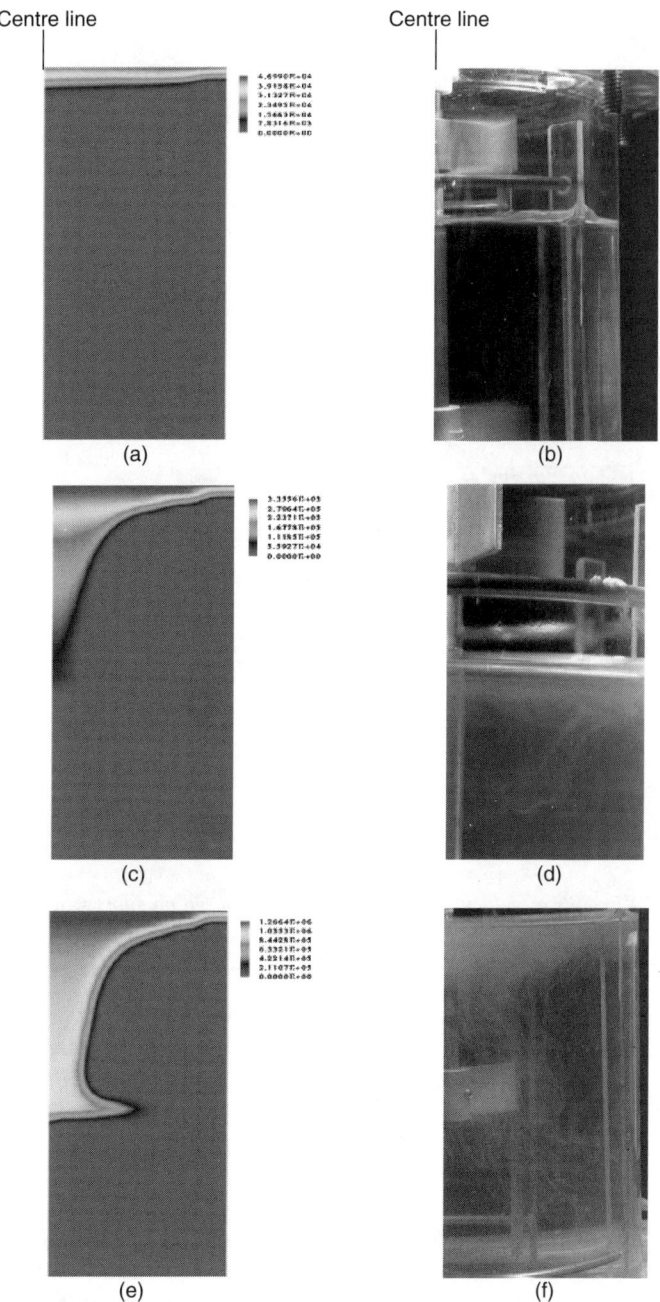

Figure 8.29 *The gas–liquid interface in the early stages of the precipitation process at 450 rpm. LHS: CFD model M_0 (a: $t = 0.1$ s; c: $t = 1$ s; e: 10 s), RHS: Experimental (b: $t = 0$; d: $t = 1$ min; f: $t = 3$ mins). (After Wachi and Jones, 1991b; Al-Rashed and Jones, 1999)*

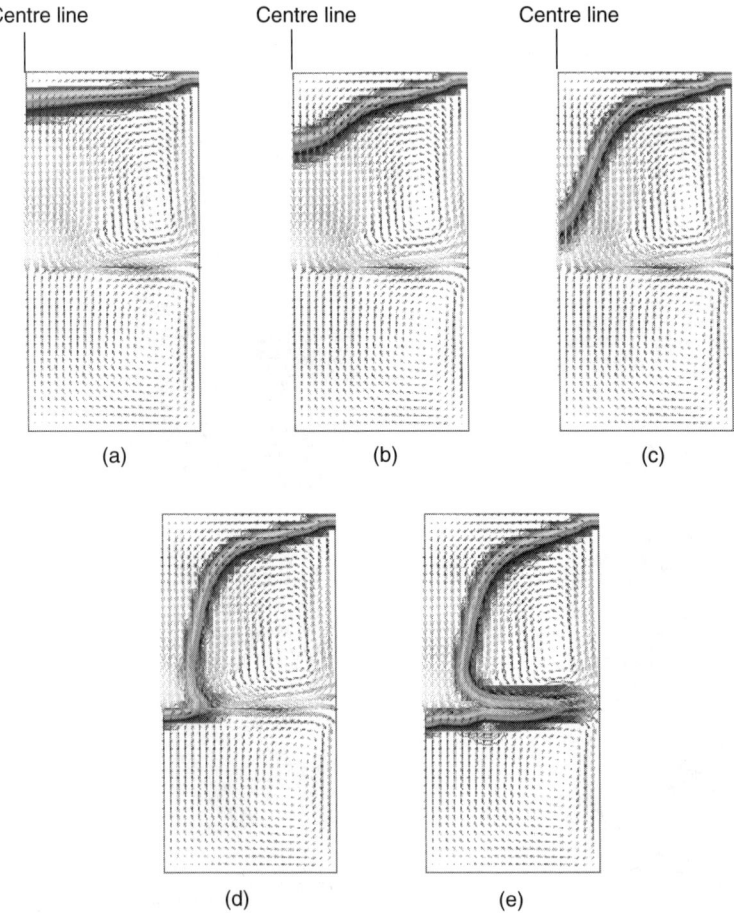

Figure 8.30 *CFD images of velocity vector* (m/s) *superimposed on supersaturation* (mol/m³) *contours* ($Re = 2.1 \times 10^4$), *for* $t = 0.04, 0.4, 0.6, 0.9$ *and* 10.0 s *for a , b, c, d and e, respectively* (*Al-Rashed and Jones, 1999a*)

apparent mechanism of bulk-interface interaction: a rising bubble, having a positive slip velocity, is in contact with its surrounding liquid for a short time and subsequently overcomes it. Then, the only parameter, the contact time, can be theoretically estimated from the bubble Sauter-mean diameter and terminal velocity:

$$\tau = \frac{d}{u_b} \qquad (8.60)$$

These concepts were implemented according to the following scheme: the liquid element surrounding the bubble and the bulk are considered as two separate dynamic reactors that operate independent of each other and interact at discrete time intervals. In the beginning of the contact time, the interface is being detached from the bulk. When overcome by the bubble, it returns to the bulk and is mixed with it. Hostomský and Jones (1995) first used such a framework for crystal precipitation in a flat interface stirred cell. To formulate it for a

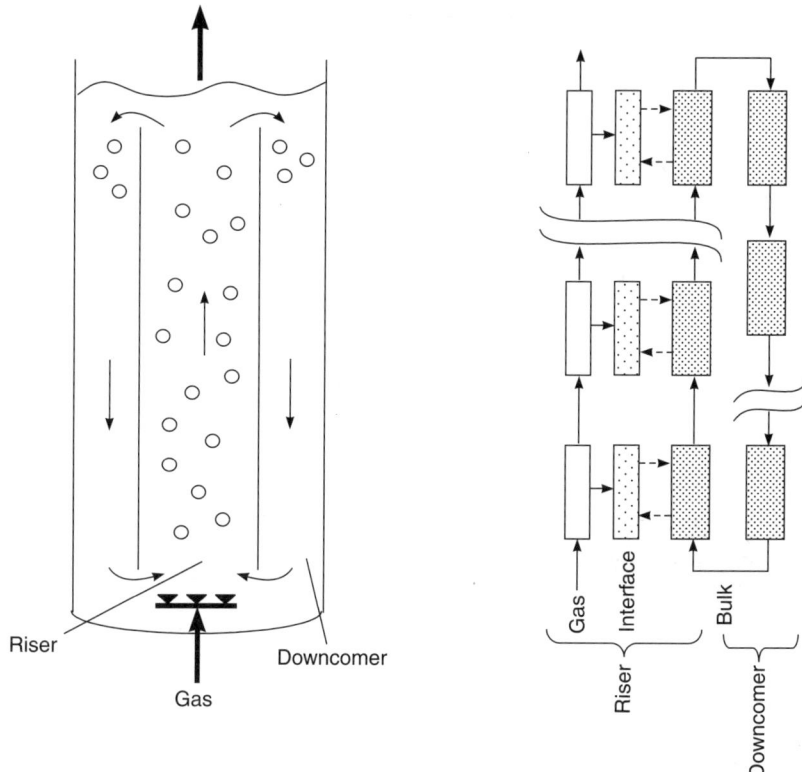

Figure 8.31 *Network of compartments in the bubble column reaction engineering model (after Rigopoulos and Jones, 2001)*

bubble column, Rigopoulos and Jones (2001) employed a series of bulk cells in contact with the interfacial reactors, which absorb the volatile components from the gas phase. This network is illustrated in Figure 8.31.

The reaction engineering model links the penetration theory to a population balance that includes particle formation and growth with the aim of predicting the average particle size. The model was then applied to the precipitation of $CaCO_3$ via CO_2 absorption into $Ca(OH)_2$aq in a draft tube bubble column and draws insight into the phenomena underlying the crystal size evolution.

Comparison of the simulations with experimental results (Figure 8.32) showed reasonable agreement prior to the onset of agglomeration. The ultimate aim of the model is to enable particle product design by the ability to relate particle size distribution to equipment design and operating conditions.

Reactor design

A strategic structure for reactor development is illustrated in Figure 8.33. To design a commercial reactor, knowledge of the fluid dynamics should be combined with the kinetics of microscopic phenomena, viz. chemical reaction,

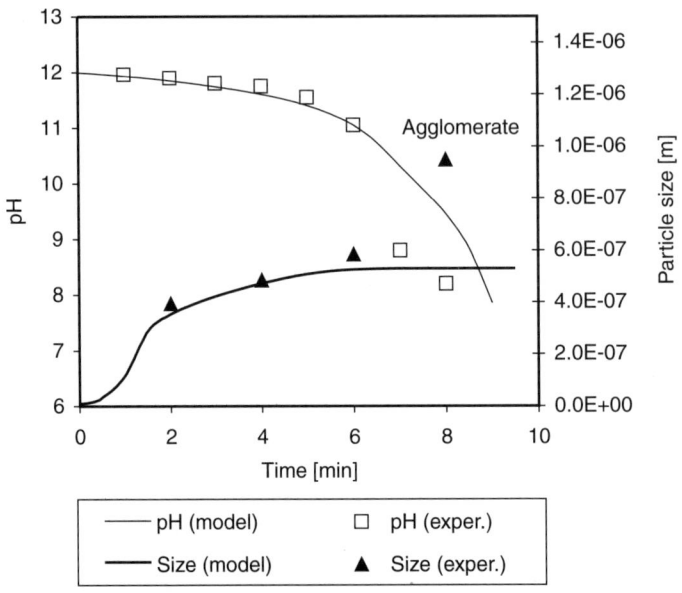

Figure 8.32 *Evolution of pH and mean crystal size, model versus experimental data of Jones et al. (1992) (Rigopoulos and Jones, 2001)*

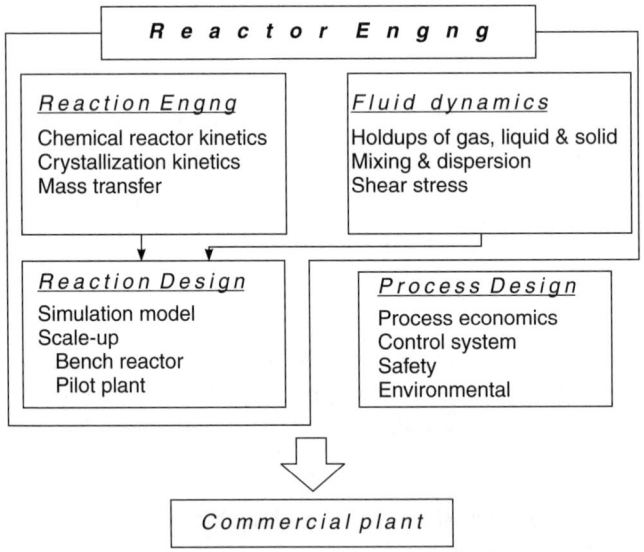

Figure 8.33 *Schematic development of an industrial gas–liquid reactor (Wachi and Jones, 1994)*

crystallization and mass transfer. The global balances of mass, heat, energy and pressure are described by a set of differential equations. For convenience of numerical computation, a stage-wise model can be employed instead of a differential one, e.g. Wachi et al. (1987). Since the rigorous formulation of all these factors requires enormous computational capacity, however, model analysis should presently be divided into several steps.

Microcrystallizers

The synthesis of powders with controlled shape and narrow particle size distribution is increasingly a major challenge for the chemical industry. It has been shown that whilst the conventional mixed suspension crystallizer is versatile and flexible in operation, the size distribution of the product is often far from ideal from either the downstream units or customer viewpoints. It can also give rise to difficulty on scale-up since the required power input to maintain perfect mixedness is difficult to attain. Plant scale product is often of differing characteristics in crystal shape, size distribution and degree of agglomeration to that obtained from the laboratory scale.

Lemaître et al. (1996, 1997) conceived a new tubular reactor, known as the Segmented Flow Tubular Reactor (SFTR), which operates in the plug flow regime (Figure 8.34). The SFTR avoids the mixing problems encountered in batch reactors leading to inhomogeneous reaction conditions and resulting in broad particle size distributions. Narrower particle size distributions, enhanced control of particle morphology, polymorph selectivity and better stoichiometry control are the claimed advantages of the SFTR process.

The SFTR reactor uses two immiscible fluid phases (gas or liquid) to create individual micro volumes of previously well-mixed reactants. A static mixer is placed upstream, which ensures efficient and reproducible mixing of the reactants before they enter the reactor. The reactor is believed to be potentially suitable for the forced precipitation of crystals where the mixing step is of major importance in determining their chemical and physical characteristics. Instead of scaling 'up' by increasing vessel size, the SFTR is to be scaled 'out' by replication of similarly sized configurations, as production demands.

The SFTR consists of three parts: (1) the mixer; (2) the segmenter; and (3) the tubular reactor. For the case of forced precipitation, the co-reactants are

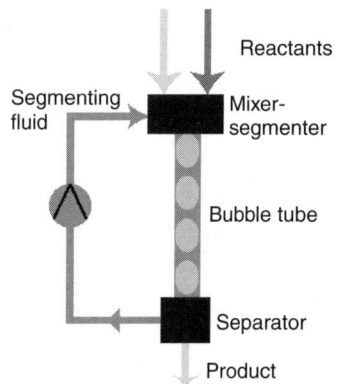

Figure 8.34 *Segmented flow tubular reactor (SFTR). (After Lemaître et al., 1996, 1997)*

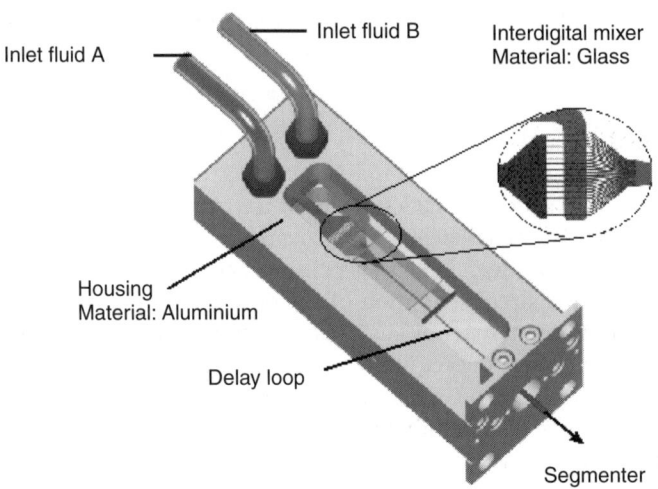

Figure 8.35 *Interdigital mixer (Schenk et al., 2001)*

introduced through the mixer, where the initial supersaturation is created. Air or an immiscible solvent then segments the reacting mixture. These small suspension volumes move as a procession of micro batches through the tubular reactor, the length of which defines the residence time. Each reactor volume is well mixed and equi-sized thereby avoiding the mixing problems of scaled up conventional batch crystallizers. Although each reactor volume behaves batch wise, a steady-state flow is obtained in which each step, mixing and precipitation, is separated.

The reactor has been successfully used in the case of forced precipitation of copper and calcium oxalates (Jongen et al., 1996; Vacassy et al., 1998; Donnet et al., 1999), calcium carbonate (Vacassy et al., 1998) and mixed yttrium-barium oxalates (Jongen et al., 1999). This process is also well adapted for studying the effects of the mixing conditions on the chemical selectivity in precipitation (Donnet et al., 2000). When using forced precipitation, the mixing step is of key importance (Schenk et al., 2001), since it affects the initial supersaturation level and hence the nucleation kinetics. A typical micromixer is shown in Figure 8.35.

Experiments with the SFTR demonstrate that crystals can be produced in a wide variety of sizes, ranging from tens of nanometres to several micrometers (Figure 8.36).

Schreiner et al. (2001) modelled the precipitation process of $CaCO_3$ in the SFTR via direct solution of the coupled mass and population balances and CFD in order to predict flow regimes, induction times and powder quality. The fluid dynamic conditions in the mixer-segmenter were predicted using CFX 4.3 (Harwell, UK).

Calculation of the induction time is crucial, since gaining a stable and continuous process requires residence times in the mixer < precipitation induction times in order to prevent incrustation in the mixing device. The induction

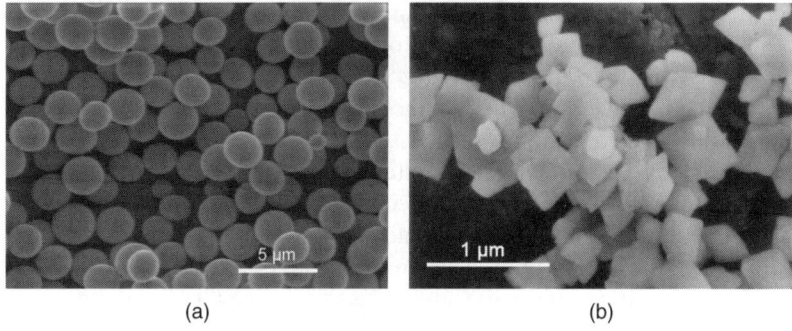

Figure 8.36 *Crystals form the SFTR:* (*a*) *Vaterite*; (*b*) *Y-Ba oxalate* (*courtesy www.bubbletube.com*)

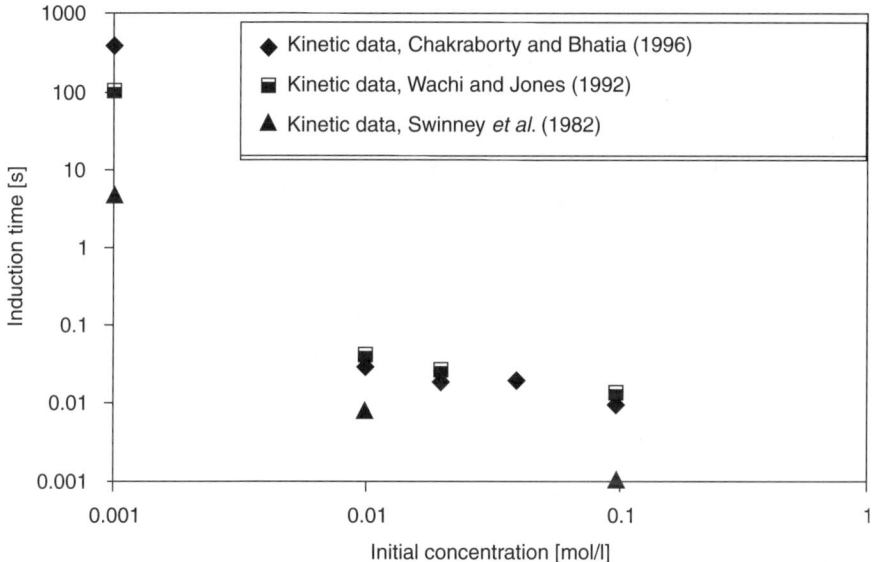

Figure 8.37 $CaCO_3$ *induction times* (*Schreiner et al., 2001*)

time is closely connected with the kinetics of the chemical system. It was seen in Chapter 5 that a chemical system with 'fast' crystallization kinetics would have 'short' induction times. Mersmann *et al.* (1990) derived a correlation for describing the induction time as a function of the supersaturation, the diffusion coefficient and the particle size. Computed results for calcium carbonate using published kinetic data are shown in Figure 8.37.

Summary

For liquid–liquid crystal precipitation systems where the particle formation processes are fast, mixing becomes an important determiner of performance with a subtle interplay of micro- and mesomixing, which changes as scale of operation increases. Kinetic information can be obtained from laboratory-scale experiments. The CFD simulations need to be carried out to obtain the local energy dissipation in the bulk and feed zone(s). Using this information, the SFM predicts the particle properties on the large scale. Scale-up thus becomes possible without time-consuming and costly large-scale precipitation experiments. Secondary processes including crystal agglomeration and breakage can also be taken into account. Ultimately, direct solution of the combined population and momentum balance equations may provide the best solution, especially for very fast reacting systems. The alternative hybrid methodology is very efficient, however, as it combines the advantages of both a CFD and a population balance approach without having to solve the equations together, which is currently expensive in terms of computational demand and simulation time required. The results of the model have been successfully validated with experiments under a wide range of conditions for precipitation in both continuous and semibatch operation.

Gas–liquid precipitation via chemical reaction has been widely adopted for the industrial manufacture of fine particles. Due to mass transfer resistance, the formation kinetics and crystal size distribution are dominated by the spatial distribution of supersaturation at the gas–liquid interfacial region. Product particles can again enlarge by agglomeration as well as molecular growth, often providing very rapid increases in particle size. Overall particle morphology i.e. void fraction, internal surface area and particle size are determined by the primary crystal size within agglomerates, whose overall particle size is affected by the turbulent shear field.

Alternative microcrystallizer configurations are being developed that seek to avoid the mixing problems asociated with conventional agitated vessels and offer the potential of consistent precipitation of high quality crystal products.

9 Design of crystallization process systems

Industrial crystallization facilities often comprise crystal formation and subsequent separation steps. The systems approach, that is consideration of a processing system as a complex whole as well as of individual units, is now well-established in many areas of chemical process engineering, such as vapour/liquid separation processes including distillation/heat exchange networks (see Rudd and Watson, 1968; Rudd et al., 1973; Douglas, 1988), for which the growth of computer power has made integrated process design a reality. Substantial progress has also been made since the first steps were made towards a holistic approach to systematize the design of particulate solids processes and the effects of crystallization conditions on filtrability (Neville and Seider, 1980; Yates, 1981; Söhnel, O. and E. Matejcková, 1981; Giorgio and Kern, 1983; Jones, 1984; Rossiter and Douglas, 1986; Rossiter, 1986; Barton and Perkins, 1988; Rajagopal et al., 1988, 1992; Evans, 1989; Jones, 1991; Gruhn et al., 1997; Hill and Ng, 1997). Due its industrial importance, increasing numbers of studies have been made of industrial crystallization/separation systems. The design of crystallization process systems is seen as an important latter part of a wider product design process involving four steps viz. needs, ideas, selection and manufacture (Cussler and Moggridge, 2001).

Thus, methods are now becoming available such that process systems can be designed to manufacture crystal products of desired chemical and physical properties and characteristics under optimal conditions. In this chapter, the essential features of methods for the analysis of particulate crystal formation and subsequent solid–liquid separation operations discussed in Chapters 3 and 4 will be recapitulated. The interaction between crystallization and downstream processing will be illustrated by practical examples and problems highlighted. Procedures for industrial crystallization process analysis, synthesis and optimization will then be considered and aspects of process simulation, control and sustainable manufacture reviewed.

Crystallization process systems design

A common starting point is that the process engineer is given a brief from which to determine a crystallization plant design viz. some specification of the product and process (e.g. mean particle size, production rate) and characteristics of the feed solution (e.g. composition, temperature etc.), Figure 9.1.

The first step in carrying out the design is to select amongst alternative types of equipment available (crystallizers, filters etc.) and to determine data

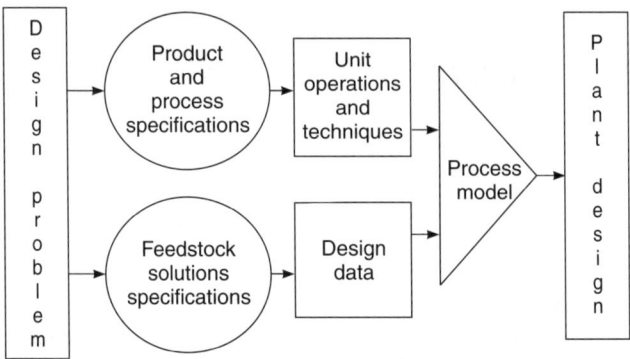

Figure 9.1 *Typical crystallization process systems design procedure*

(solubility, kinetics etc.). These two sets of information are brought together in a model of the process from which the required plant capacity is calculated following which control systems are added, the process costed, and HSE and sustainability issues considered. If necessary, this process can be repeated to consider alternative designs, even for different starting specifications of both feedstock and product until an acceptable scheme is devised.

Crystallization process equipment

As mentioned above, the unit operation of crystallization rarely exists in isolation but is normally part of a wider particulate processing system as illustrated schematically, in a very simplified form, in Figure 9.2. A particular feature of such processes is the variety of unit operations and the range of equipment types that may be employed at any stage.

Solids processes clearly differ from vapour liquid systems in two main ways. Firstly and most obviously, solid systems have a particle size or size distribution added to their specification. Secondly, solids processing systems use different

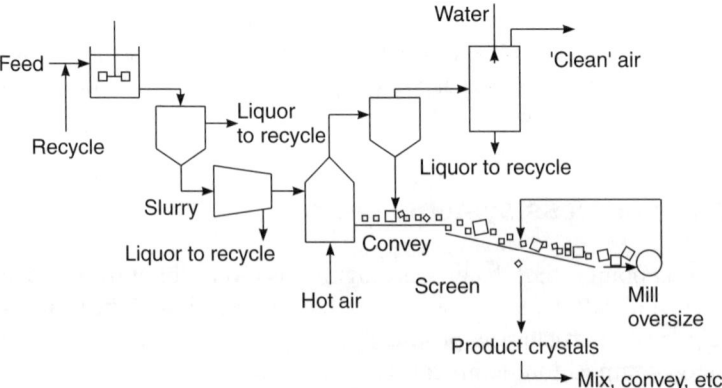

Figure 9.2 *A typical industrial particulate crystallization process system (schematic)*

types of unit operation for separation duties e.g. centrifuges, hydrocyclones and filters.

A complete crystallization process system can comprise several distinct steps:

- feed preparation
- feed purification
- crystallization
- solid–liquid separation
- drying
- solids handling.

The feed preparation stage e.g. by mineral extraction and evaporation, chemical reaction etc. can give rise to both dissolved and suspended solid impurities, either of which may affect the crystallization step. Removal of suspended solids, e.g. by filtration, is usually the easier process. Dissolved impurities can have by far the more pronounced effect, however, and may have to be removed e.g. by chemical means or by adsorption. Such impurities may, of course, actually be beneficial to the process by inducing nucleation, habit modification etc.

Crystallization itself can be achieved by either of several means e.g. cooling, evaporation, drowning out etc. (see Chapter 3). Subsequent solid–liquid separation normally entails a slurry thickening stage followed by either filtration or centrifugation followed by drying e.g. in a fluidized bed (Chapter 4). Dry solids handling stages follow, such as screening, milling, mixing, granulating, tabletting, bagging, packaging etc.

Crystallization design data

Crystallization process systems design and operation is a complex matter requiring extensive data for systematic evaluation. Whilst simplified design methods and heuristics are available, the simple fact remains that the more and better the data input, the better the final design and reliability of the plant.

Ideally, amongst the data required are the following:

1. *Crystal characteristics*: composition, form, habit, shape factors, and solid density (Chapter 1).
2. *Solution characteristics*: composition, equilibrium relationships (solubility), metastable zone width, purity, partition coefficient, liquid density, viscosity, and their temperature dependence (Chapter 3).
3. *Slurry characteristics*: settling rates, filterability, drying rates and final moisture contents (Chapter 4).
4. *Crystallization kinetics*: crystal nucleation, growth, aggregation and disruption kinetics (Chapters 5 and 6).

This list is not exhaustive, particular systems often require other specific data to be collected. Such data can take considerable time to collect experimentally unless efficient experimentation is employed. Crystallization data are published in divers sources including textbooks and journals, many of which have been referred to throughout the text and are listed at the end of this book. Published sources are often helpful in providing at least part of the information. Sometimes data from a similar substance may be a helpful starting point for use in simulation whilst more data are being collected.

Crystallization process models

At the crystallization stage, the rates of generation and growth of particles together with their residence times are all important for the formal accounting of particle numbers in each size range. Use of the mass and population balances facilitates calculation of the particle size distribution and its statistics i.e. mean particle size, etc.

Crystals suspended in liquors emerging from crystallizers are normally passed to solid–liquid separation devices such as gravity settlers or thickeners that may subsequently feed filters to remove yet more liquid prior to drying. Here the transport processes of particle motion and the flow of fluids through porous media are important in determining equipment size, the operation of which may be intensified by application of a centrifugal force.

Because the physical characteristics of the particulate solids both affect, and are affected by each unit operation, there exists an interaction between them. The study of particulate systems thus provides methods for synthesizing and evaluating economically optimum process configurations.

To recapitulate, the population balance (Chapter 2) is

$$\frac{\partial n}{\partial t} + \frac{\partial (nG)}{\partial L} + \frac{n - n_0}{\tau} = B_a - D_a + D_d - D_d + \int_0^\infty \delta(L - L_u) dL \qquad (9.1)$$

In principle, given expressions for the crystallization kinetics and solubility of the system, equation 9.1 can be solved (along with its auxiliary equations – Chapter 3) to predict the performance of continuous crystallizers, at either steady- or unsteady-state (Chapter 7). As is evident, however, the general population balance equations are complex and thus numerical methods are required for their general solution. Nevertheless, some useful analytic solutions for design purposes are available for particular cases.

Crystal size distribution

If crystal agglomeration and breakage can be neglected and crystal growth is invariant then at steady state the familiar analytic form is

$$n(L) = n^0 \exp\left(\frac{-L}{G\tau}\right) \qquad (9.2)$$

Median crystal size

By combining expressions from MSMPR theory above, it can be shown that the median crystal size in the absence of agglomeration and disruption is given by

$$L_{50} = 3.67 \left[\frac{M_T^{1-j} t^{i-1}}{6 f_v \rho_c k_n} \right]^{\frac{1}{(i+3)}} \qquad (9.3)$$

where k_n is the nucleation rate coefficient and $i(=b/g)$ is the relative kinetic index (a useful result thereby eliminating the absolute need for knowledge of the working level of supersaturation). Analytic expressions are also available for other defined mean crystal sizes.

Crystallizer volume

Crystallizer volume is proportional to residence time, thus assuming constant crystallization kinetics and residence time

$$V = K_p L_{50}^n \qquad (9.4)$$

where

$$n = \frac{i+3}{i-1} \qquad (9.5)$$

Thus for a given constant of proportionality, K_p, the crystallizer volume V and hence capital cost is related to the required median crystal size (see later).

Solids hold-up

The solids hold-up, M_T, is related to the CSD by the third moment equation

$$M_T = f_v \rho_c \int_0^\infty n(L) L^3 dL \qquad (9.6)$$

where f_v is the crystal volume shape factor and ρ_c is the crystal density. This value of M_T integrated over the total crystal size distribution can thus be compared with that determined by mass balance from the change in solution concentration between inlet and outlet, viz.

$$Q_o M_T = Q_i C_i - Q_o C_o \qquad (9.7)$$

Solid–liquid separation

It was shown in Chapter 2 that the simplest models of solid–liquid separation are those based of the Carman–Kozeny equation for filtration in which the bed permeability (filtrability), F, may be expressed by

$$F = \frac{\left(\frac{1}{K}\right) \varepsilon^3}{(1-\varepsilon^2)} S_p^2 \qquad (9.8)$$

where S_p is the specific surface area of particles (surface area of single particle per unit crystal volume) and is thus related to the crystal size distribution. Average bed voidage, ε, in a filter bed of area A and depth H is defined by

$$\varepsilon = 1 - \frac{M_c}{(HA_f \rho_c)} \tag{9.9}$$

where M_c is the mass of crystals.

Thus, in principle, it is possible to link crystallization and filtration equations via the crystal size distribution to the crystallization kinetics and mean residence time.

For example, the permeability of a precipitate from a continuous MSMPR crystallizer may be given by an expression of the form (Söhnel et al., 1991)

$$\log B = i \log \tau \tag{9.10}$$

where the coefficient i depends on the growth rate mechanism and bed porosity.

In general bed voidage is dependent on the packing characteristics of the material (a function of the orientation of the particles in the bed), which is both particle size and shape dependent. Unfortunately, at present it is only possible to predict bed porosity for certain distributions of regular particles. Porosity can be related to crystallizer solids hold-up empirically, however, by an expression of the form

$$(1 - \varepsilon)\rho_c = bM_T + d \tag{9.11}$$

where b and d are empirical coefficients. Thus, in practice both bed voidage and permeability are determined empirically. Given these data, the required area for various types of filter can be determined. For example, the required area of a rotary vacuum filter to separate a continuously formed precipitate at a production rate P is given by

$$A = Pf(c, \tau) \tag{9.12}$$

where $f(c, \tau)$ is related empirically to the conditions of precipitation and residence time (for more details, see Söhnel et al., 1991).

Particle characteristics

In Chapter 3 it was shown that two separate cases can be distinguished for practical purposes namely crystallization and precipitation respectively. They are used here to distinguish between the methods of supersaturation generation viz. cooling and/or evaporation, or via addition of a precipitant (e.g. salting- or drowning-out agent, or reactant); precipitation normally inducing much the higher level of supersaturation. In both cases, however, new crystals are generated by nucleation and subsequent growth processes and may agglomerate to a greater or lesser extent. If conditions permit, crystal inter-growth may continue within agglomerates further changing their particle characteristics. A potential drawback of precipitates is that, being normally of small size, they often exhibit much lower filtrabilities. This deficiency may be overcome, however, by appropriate control of operating conditions.

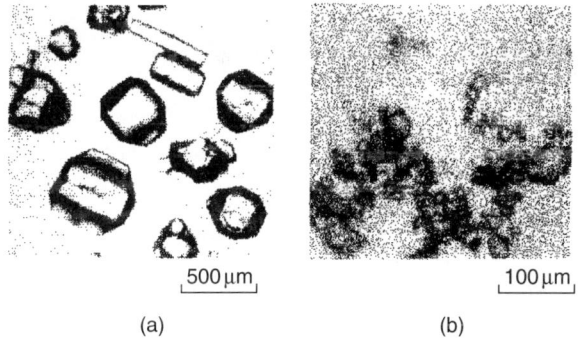

Figure 9.3 *Potassium sulphate crystals formed by:* (a) *cooling;* (b) *drowning-out with acetone (Jones et al., 1987)*

These aspects are illustrated by studies of crystal shape and degree of agglomeration of potassium sulphate (Budz *et al.*, 1987a,b; Jones *et al.*, 1987) and the change in crystallization kinetics and subsequent filtrability on moving from simple cooling crystallization to precipitation by drowning-out (Jones and Mydlarz, 1989, 1990; Mydlarz and Jones, 1989). Cooling produces relatively large, mostly discrete, crystals while, as a consequence of the change in crystallization kinetics referred to above during drowning-out, precipitation produces much smaller crystals with a higher degree of agglomeration (Figure 9.3b) and consequent lower permeability (see later).

Crystallization kinetics

Experimental MSMPR population density distributions for cooling crystallization are illustrated in Figure 9.4.

The overall linear growth rate determined from these data is given by the following empirical expression (Jones and Mydlarz, 1989)

$$G = 0.744 \exp\left(\frac{-38.4}{RT}\right)\left(1 + 2L^{\frac{2}{3}}\right)\sigma^2 \tag{9.13}$$

The overall growth kinetics is thus second order and exhibits diffusion dependence, with intrinsic surface integration kinetics (not shown) being greater than second order implying a polynuclear growth mechanism.

Nucleation rates during cooling crystallization are given empirically by (Jones and Mydlarz, 1989)

$$B = 1.14 \times 10^{24} \exp\left(\frac{-62.5}{RT}\right) M_T \sigma^{2.25} \tag{9.14}$$

The observed dependence of nucleation rate on solids hold-up and the low supersaturation order both imply the secondary mode occurring.

The presence of the second solvent during drowning-out crystallization has a marked effect on the crystallization kinetics and consequent particle

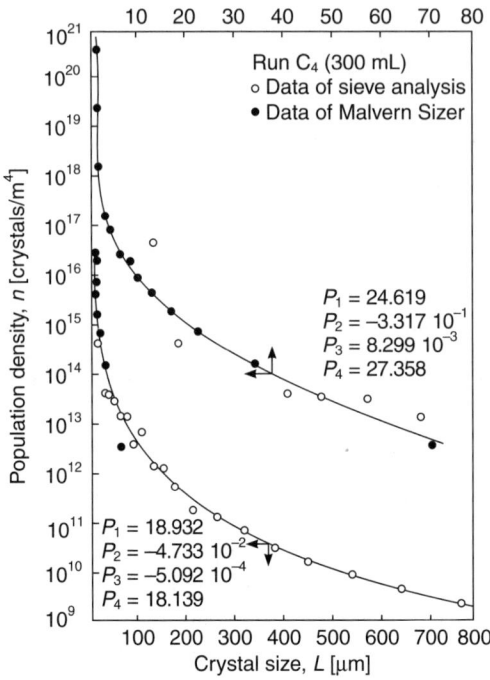

Figure 9.4 *Population density distribution of potassium sulphate crystals from continuous cooling crystallization (Jones and Mydlarz, 1989)*

characteristics (Figure 9.3). Thus, the overall crystal growth rate is reduced viz. (Jones and Mydlarz, 1989)

$$G = 1.34 \times 10^{-6} \exp(-13.96x^{0.5} + 8.72x) \times (1 + 2L^{\frac{2}{3}})\sigma^{2.75} \tag{9.15}$$

where x is the mass fraction of acetone (kg acetone/kg water).

Similarly, the nucleation rate dependence on solids hold-up is decreased while the order of nucleation is increased implying the additional occurrence of primary nucleation (e.g. Jones and Mydlarz, 1989)

$$B = 8.62 \times 10^{14} M_T^{0.2} \sigma^{2.74} \tag{9.16}$$

Crystal slurry filtrability

As referred to above, cooling crystallization produces relatively large crystals. As a consequence of the change in crystallization kinetics, however, precipitation produces much smaller crystals with a higher degree of agglomeration (Figure 9.3), and consequent lower permeability; filter cake permeability (Figure 9.5), being found to decrease with increasing mean particle size and increasing coefficient of variation (Jones and Mydlarz, 1989).

Both operating conditions (viz. precipitant dosage rate) and mixing conditions (viz. precipitant dilution and vessel size) affect the crystal size distribution

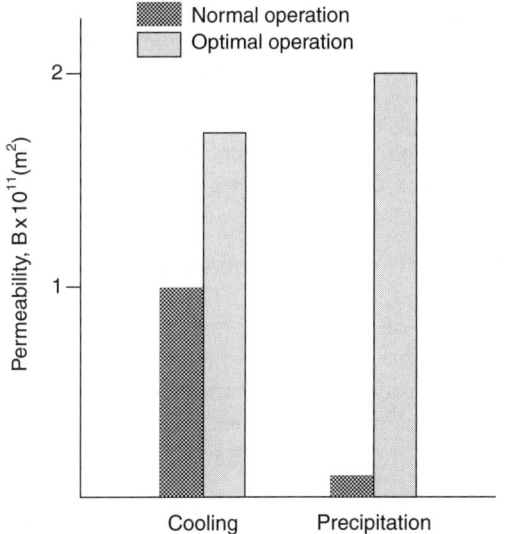

Figure 9.5 *Permeabilities of potassium sulphate crystal beds (Jones et al., 1987)*

of the precipitates (not shown). Suitable control of these factors considerably increases the permeability of the precipitates, thought to be due to both increasing particle size and the presence of agglomerates. Size distribution of the agglomerates alone is then an inadequate indicator, however, and currently measurement of permeability itself provides a better test to assess separability with cake porosity being found in the range 0.4–0.7 and Kozeny coefficients being in the range 2–8.

Mathews and Rawlings (1998) successfully applied model-based control using solids hold-up and liquid density measurements to control the filtrability of a photochemical product. Togkalidou et al. (2001) report results of a factorial design approach to investigate relative effects of operating conditions on the filtration resistance of slurry produced in a semi-continuous batch crystallizer using various empirical chemometric methods. This method is proposed as an alternative approach to the development of first principle mathematical models of crystallization for application to non-ideal crystals shapes such as needles found in many pharmaceutical crystals.

Improving solid/liquid separability

To improve the filterability of crystals, the factors that are generally considered are:

- Particle size (the larger the particle, the lower the surface area to volume ratio and the better the filterability),
- Particle shape (for a given volume, the best shape for minimizing surface area and improving filterability is a sphere), and
- The size distribution of the particles (with particles of a uniform size, there is increased voidage between the particles and a lower tortuosity, that is deviation from the linear path for the fluid flowing between the particles, making for improved filterability), and

- Needle and plate shaped crystals, that is, crystals with a high aspect ratio, have a large surface area to volume ratio and generally poor filterability. Larger crystals are known to be generally easier to separate from solution.

Various techniques are used to obtain larger crystals to improve solid/liquid separability. In general, the aim is to avoid large numbers of particles being formed from the fixed amount of solute in the crystallizing solution. In other words, to avoid high nucleation. To do this the following are generally used:

- Slow cooling rates (controlled by the rate of cooling and the temperature of the coolant),
- Adequate stirring to even out regions of high local supersaturation whilst avoiding excessive stirring giving rise to unwanted primary or secondary nucleation. That is operating so that growth rather than nucleation occurs (and maintaining the solution within the metastable zone),
- Seed crystals to provide a surface area for crystal growth whilst avoiding new nucleation sites, and
- If it is economically feasible, relatively low concentration solutions.

Effect of downstream processing on product characteristics

Not only do particle characteristics determine the performance of downstream units, but the downstream units themselves may affect the particle characteristics. As illustrated by Bennett (1962), substantial changes to the

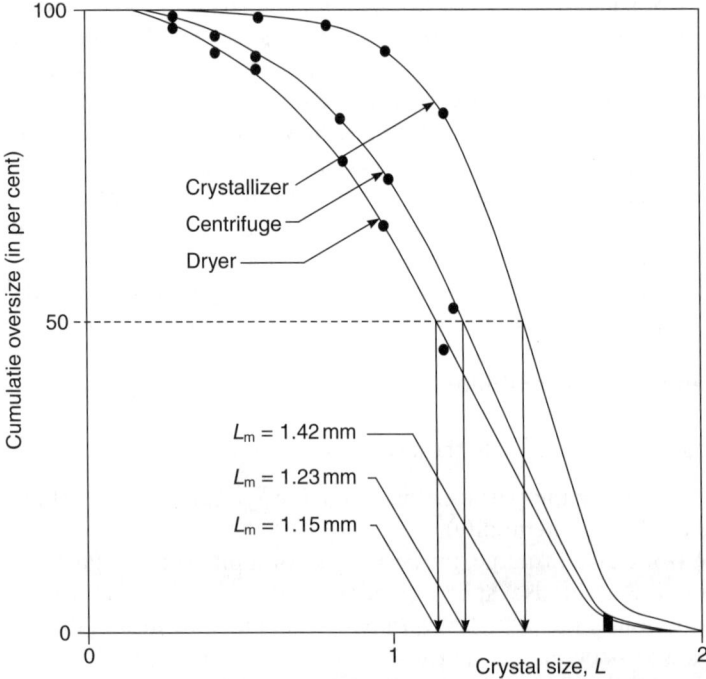

Figure 9.6 *Effect of process system on crystal product characteristics (after Bennett, 1962)*

crystal characteristics can occur as the crystal magma passes from the crystallizer to the thickener, filter/centrifuge and drier (Figure 9.6).

In the particular case cited, a 20 per cent reduction in median crystal size is obtained. The process of attrition and methods for its determination were considered in detail in Chapters 4–6. Similar changes may also occur to crystal shape and form.

Crystallization process synthesis

Rossiter and Douglas (1986) state that the first step in process design is to generate a basic structure for the flowsheet i.e. the choice of unit operations and interconnections which can be analysed, refined and costed, and then compared to alternatives. Thus, the generation of an industrial crystallization flowsheet gives rise to a number of optimization problems for which a systematic hierarchical decision process for particulate systems was proposed:

Level 1: Batch versus continuous
Level 2: Flowsheet input–output structure
Level 3: Crystallizer and recycle considerations
Level 4: Separation systems specification
Level 5: Product drying
Level 6: Energy systems

The initial aim of the procedure is to generate a reasonable 'base case' design that can be used for preliminary economic evaluation of the process. This can subsequently be optimized and/or compared with any process alternatives that are identified. The complete process is always considered at each decision level, but additional fine detail is added to the structure of the flowsheet at any stage. Established heuristics and equipment selection procedures are used together with new process synthesis insights to guide each flowsheet decision.

An analysis of this type is appropriate during the preliminary design stage when process alternatives may have to be screened economically, or when it is necessary to determine a feasible design as the basis for equipment trials and/or final designs.

As mentioned above, among many possible process variables, industrial crystallization frequently focuses on the optimization of particle size. In many cases this may be fixed by market demands, in others it may be a variable e.g. during the processing of intermediates.

In practice, industrial crystallization processes are subject to a number of constraints, which tend to limit equipment selection. For example, since particle size and purity tend to be such important variables, equipment and operating conditions that induce minimum particle breakdown or achieve maximum crystal purity are normally desirable.

Costs

The total annual cost (TAC) of a process plant comprises two factors viz. the capital costs of installed plant items and the operating cost of running them

$$\text{TAC} = \text{Capital cost} \times \text{annualization factor} + \text{Operating cost} \quad (9.17)$$

where the annualization factor includes an element for depreciation.

Capital costs from various sources are summarized by Rossiter (1986)

$$\text{Crystallizer capital costs} \propto V^{0.55} \quad (9.18)$$

where V is the crystallizer volume.

$$\text{Centrifuge capital costs} \propto D^{1.11} \quad (9.19)$$

where D is the diameter of the centrifuge.

$$\text{Dryer capital costs} \propto A_d^{0.5} \quad (9.20)$$

where A_d is the dryer bed area.

Similarly, various operating and utility cost estimates are quoted in the literature, although accurate data remain proprietary information.

Crystallization process optimization

Design level optimization

Rossiter and Douglas (1986) attempted to optimize a given flowsheet by determining best values of design variables. The scope of such studies can span from preliminary design; where process alternatives may have to be compared and screened economically, to determining the best conditions for equipment trails or for detailed final design. In both contributions, the focus was on the optimization of median size. Simplified cost correlations derived for various units were used to evaluate the impact of average size on system economics. Flowsheet decomposition analysis was carried out to identify the independent subsystems. From individual models for crystallizers, primary product separation and drying units, a cost model for the whole process was developed. The effects of perturbations in a given design parameter while keeping others constant were used to determine an optimal set of values for all the design parameters.

Rossiter (1986) demonstrated the procedure for the production process of crystalline common salt from brine. It was found that the optimal median size is determined by the entrainment limit in the crystallizer. The crystallizer had to be operated at maximum allowable temperature and the slurry density measured for quality constraints. It was also suggested that cost discontinuities should be imposed based on temperatures of the available heat sources, possible materials of construction and other intrinsic properties of the system.

The limitations of the procedure were identified as: (a) fixation of most design variables; and (b) limited number of variables due to the scarcity of the available design and cost relationships.

Operational level optimization

As mentioned above, the solids process synthesis approach (Rossiter and Douglas, 1986) has been applied to the optimization of a continuous salt crystallization plant similar to that depicted in Figure 9.2 (Rossiter, 1986). In this process, the required crystallizer volume increases strongly with required particle size. Moreover, volume is generally minimized by maximizing slurry density, subject to crystal quality constraints. In addition, the cross-sectional area of an evaporative crystallizer is often constrained by the need to avoid excessive liquor entrainment in the vapour leaving the boiling zone thus giving rise to a minimum allowable cross-sectional area for a given duty. Since the height to diameter ratio of many commercial crystallizers is fixed, however, the 'entrainment limit' effectively imposes a minimum on crystallizer volume.

The results of applying this technique is summarized in Figure 9.7 in which the TAC of the process and the entrainment temperature are plotted against median particle size (L).

Taken on its own, the TAC versus L curve passes through a shallow minimum at a value of L of about 250 μm. This figure is substantially below the lower bound imposed by the feasible zone determined by the entrainment limit, however, which ranges from about 420 to 650 μm. Thus the notional optimum value of L at 250 μm cannot be achieved in practice at the temperatures within the range considered, although there is clearly an economic incentive to do so.

Similarly, Rajagopal *et al.* (1988) found a stronger optimal dominant crystal size for a potash plant, which depended on a trade-off between the crystallizer cost and the filter cost (Figure 9.8).

In a further attempt to systematize the crystallization processing system, Chang and Ng (1998) considered units both upstream and downstream of a crystallizer. The proposed procedure guides the designer in a systematic manner to generate alternative flowsheets for a given crystallizer output. Firstly, the required unit operations are determined by comparing the product specification (production rate, product purity, etc.) with the crystallizer product (occlusions, inclusions, crystal size etc.). Secondly, the destinations of the reaction solvent, mother liquor, wash liquid, recrystallizing solvent and drowning-out solvent are assigned. Thirdly, the solvent recovery system is considered to recover the solvents and unconverted reactants and purge impurities from the system. Fourthly, short-cut models are used to screen process alternatives. Rules and heuristics are provided for each of the following steps.

Step 1: Selection of the required unit operations
Step 2: Assignment of solvent destination
Step 3: Solvent recovery system
Step 4: Process enhancement
Step 5: Evaluation of flow sheet alternatives

The methodology was illustrated with reference, amongst others, to processes for the crystallization of adipic acid, including reaction, crystallization, filtration, washing, recrystallization, recycle and drying steps (Figure 9.9).

Woinaroschy *et al.* (1994) presented an operational level optimization based on artificial neural networks (ANN), where the objective was to determine best operating conditions for a crystallizer. Steady-state experimental data obtained under a range of operating conditions for reactor temperature, feed concentration and mean residence time was used to develop an ANN model capable of optimizing operation of a $CaCO_3$ precipitation process in a 1 litre MSMPR

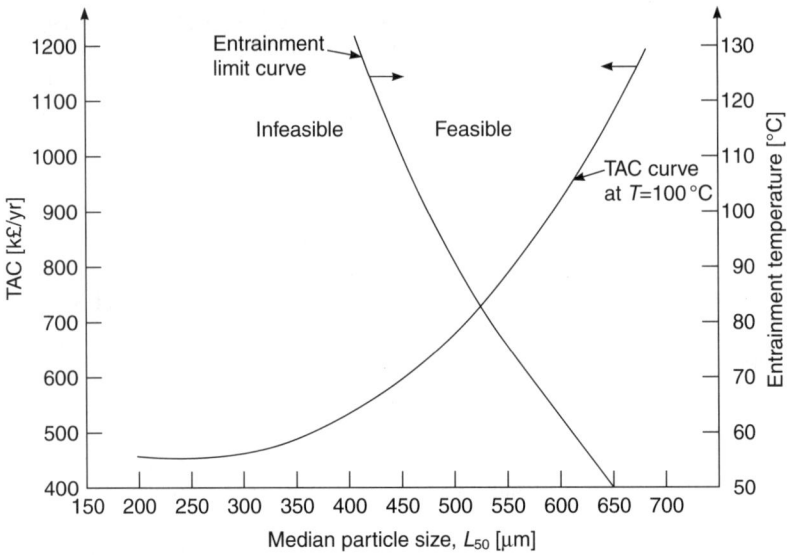

Figure 9.7 *Crystallizer-centrifuge-dryer system optimization (Rossiter, 1986)*

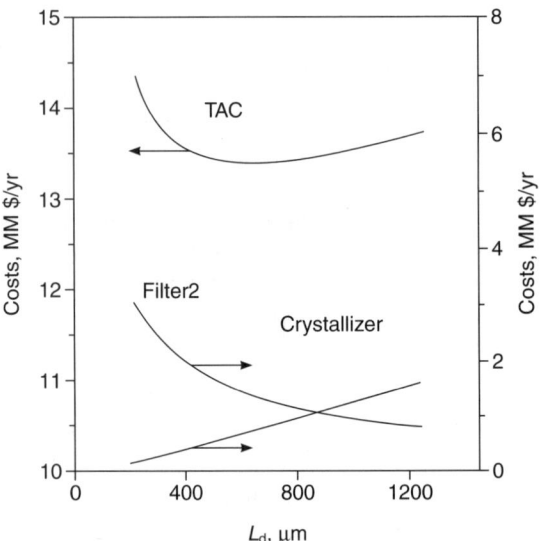

Figure 9.8 *Crystallizer-filter system optimization (Rajagopal et al., 1988)*

crystallizer. The choice of ANN was based on the limitations of formulating an adequate simplified representation of the process from theoretical considerations to account for all the variables of interest.

The ANN model had four neurones in the input layer; one for each operating variable and one for the bias. The output was selected to be cumulative mass distribution; thirteen neurones were used to represent it. A sigmoid functional

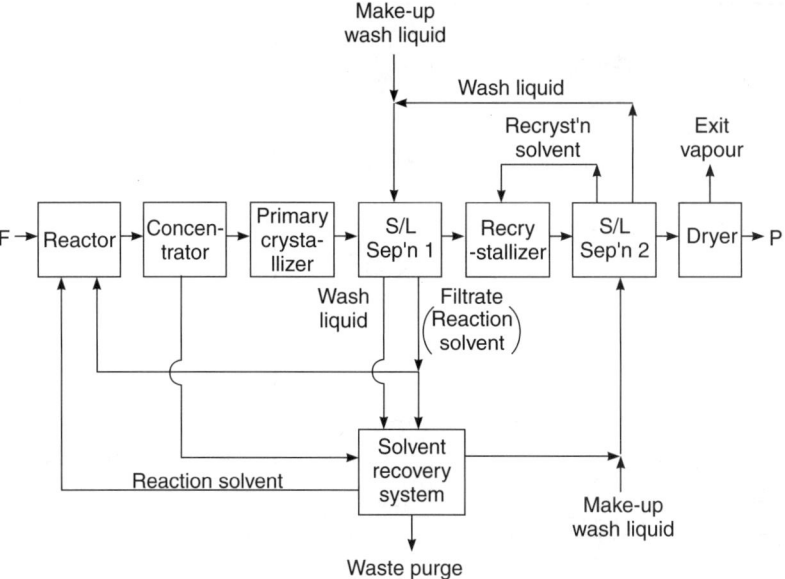

Figure 9.9 *Adipic acid reactor-crystallizer-separation system optimization (Chang and Ng, 1998)*

form was chosen for the transfer function between inputs and outputs in the hidden and output layers. The back propagation rule was used in the learning process for training the network from 28 sets of experimental data. A compound objective function was formulated with the aim of finding vectors of operating conditions leading to a certain fixed value of mean size with minimum CSD dispersion. An adaptive search method that involves a number of steps (details can be found in Luss and Jakoola, 1973) was used to find the optima. The steps included: (1) preliminary search of the random variable range in which the minimum value of objective function might be found; (2) selection of these ranges to search for global minimum; and (3) further refinements in the vector found. The results were found to be in good agreement with the qualitative suggestions from theory and practice and showed that the optimum occurred at low temperatures. Concentration had to be increased and residence time reduced if low median sizes were desired. As expected, longer residence times resulted in larger crystals.

Multi-component systems

Crystallization-based separation of multi-component mixtures has widespread application. The technique consists of sequences of heating, cooling, evaporation, dilution, diluent addition and solid–liquid separation. Berry and Ng (1996, 1997), Cisternas and Rudd (1993), Dye and Ng (1995), Ng (1991) and Oyander et al. (1997) proposed various schemes based on the phase diagram. Cisternas (1999) presented an alternate network flow model for synthesizing crystallization-based separations for multi-component systems. The construction

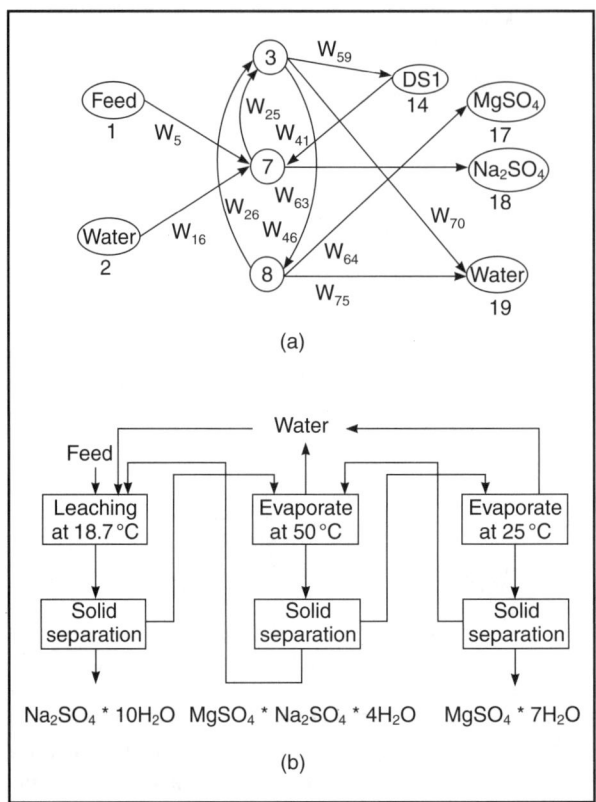

Figure 9.10 (a) Solution flow; and (b) corresponding flow sheet for the separation by crystallization of astrakanite (Cisternas, 1999)

of the network is based on the identification of feasible thermodynamic states. The technique is illustrated with potash production from sylvinite (47.7 per cent KCL, 52.3 per cent NaCl), magnesium sulphate and sodium sulphate from astrakanite (Figure 9.10) and separation of a potassium chloride, nitrate and sulphate mixture.

Biotechnological systems

Crystallization is used as both a particle formation and separation process in many biotechnological applications. Penicillin production is the model for a large group of antibiotic products, including cephalosporins, which are based on β-lactams. Penicillin can be made either synthetically or microbiologically. In the biotechnological route, mutants of *Penicillium chrysogenum* are grown in 100 000-L aerated fermenters that are primarily charged with lactose, corn steep liquor, and calcium carbonate. After about seven days, the broth contains typically 80 mg of penicillin per litre of broth.

Cussler and Moggridge (2001) describe the key to the purification of penicillin (Figure 9.11) as the recognition that when the pH is above 5.5 the

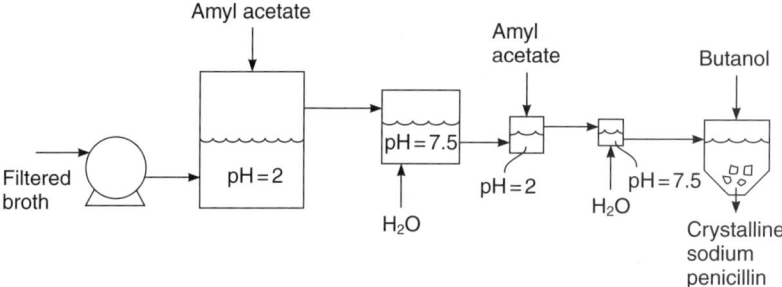

Figure 9.11 *A schematic process for penicillin purification (after Cussler and Moggridge, 2001)*

carboxylic acid groups ionize and the penicillin becomes water-soluble. When the pH is below 5.5, the carboxylic group remains protonated, and the penicillin is more soluble in organic extraction solvents. Then the penicillin is re-extracted with water and finally precipitated with butanol to form sodium or potassium penicillin crystals.

Crystallization process synthesis procedure

Wibowo and Ng (2000) presented a unified procedure for synthesizing crystallization-based processes. This is achieved by recognizing the fact that four basic crystallization-related movements in composition space viz. (1) heating or cooling; (2) stream combination or splitting; (3) solvent addition and removal; and (4) addition or removal of a mass separating agent (MSA, e.g. a diluent), are sufficient to represent a variety of crystallization processes. Suitable movements are selected based on the relevant features of the phase diagram to construct feasible flowsheets.

The procedure (Figure 9.12) is based on six steps:

Step 1: Definition of process objectives
Step 2: Generation of separation core structure
Step 3: Selection of separation sequence and unit operations
Step 4: Addition of further units to the process structure
Step 5: Selection of crystallizer type
Step 6: Economic evaluation and feasibility check of generated alternatives.

Rules are provided to aid decision making at each step. The procedure is illustrated with examples including separation of amino acids, *p*- and *m*-cresols, chorobenzoic acids, calcium carbonate and magnesium oxide from dolomite, and the production of salt.

Crystallization process systems simulation

Since crystallization is a particulate process, the model representing size distribution is written in terms of functions. Evans (1989) and Hounslow and

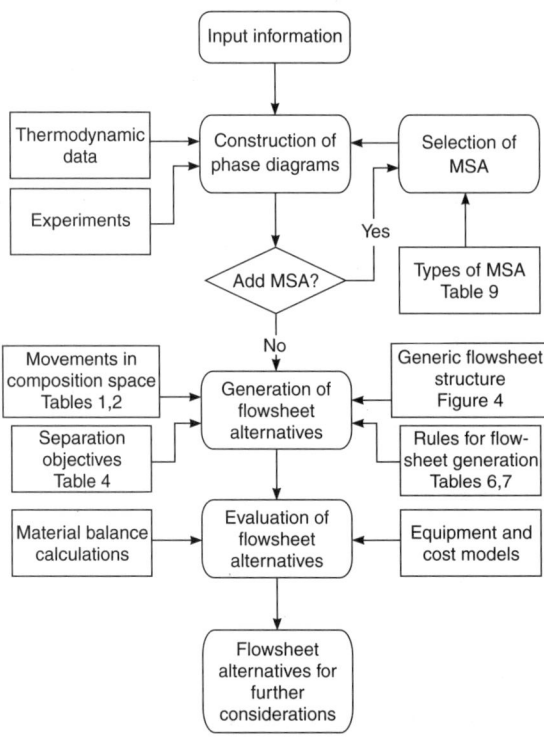

Figure 9.12 *Summary of the unified procedure for synthesizing crystallization-based separation processes* (*Wibowo and Ng, 2000*)

Wynn (1992, 1993) have presented short-cut studies of solids process flowsheeting. Custom-written software to solve the model equations has been used by a number of authors. It has also been proposed that symbolic manipulation packages, like Mathematica^{TM}, can be used to solve the population balance equations; as a stand-alone modelling environment or as a set of modules to be linked to a generic process simulation package such as SPEEDUP^{TM} (Hounslow, 1989; Sheikh and Jones, 1996).

The moment-based method has been tested on complicated particulate process flowsheets and compared with sophisticated discretization methods to reveal that moment based method gives acceptably accurate results with considerably less computer time.

Machine learning methodology

Process operational quality, which has emerged as an essential pre-condition to increase profitability by fundamentally improving the design and operation of the process, involves two complementary steps: (1) control within pre-specified limits; and (2) continuous improvement of operational performance (Saraiva and Stephanopoulos, 1992a). The first step deals with the rectification of abnormal process behaviour (as a result of special causes) through efficient control. The final level of performance thereby achieved, however, is a result of common and sustained causes within the process itself that are not avoidable. It has been estimated that only 6–30 per cent of production problems are due to special causes while the remaining 80–94 per cent are due to common and sustained causes (Hart and Hart, 1989). The magnitude of common cause contributions can only be reduced through the introduction of appropriate changes in operating conditions and strategies by searching for better levels and ranges of decision variables.

Crystallization processes have seldom been subjected to process improvement techniques at operational level. Rather, process improvement studies have been restricted to the design level through simple 'design and cost' relationships (Rossiter and Douglas, 1986; Jones, 1991), for they are characterized by mathematical complexities associated with the adequate representation of CSD and functional discontinuities in kinetic processes, which are beyond current optimization algorithms (Cuthrell and Biegler, 1987; Biegler *et al.*, 1995). These calculations require a thorough understanding of the process so that it can be effectively represented in simplified mathematical form to find the best solution to a process within the imposed constraints.

Machine learning, the study and computer modelling of learning processes in their multiple manifestations, has also been used for the task of developing and analysing systems to improve performance from existing data, often from a less model driven standpoint (Saraiva and Stephanopoulos, 1992b; Saraiva 1995). Sheikh and Jones (1997) presented a revised machine learning methodology applied to a simulated crystallization process flowsheet for continual improvement of its performance by generating and analysing process data. The aim is to identify bands of crucial decision variables leading to zones of best average process performance. The methodology comprises two components viz., symbolic induction and case based reasoning. It uses an incremental algorithm to update performance classification rules (Figure 9.13).

These concepts and procedures were illustrated by application to a potassium nitrate crystallization process comprising a MSMPR crystallizer, a hydrocyclone and a fines dissolver. By identifying and establishing ranges for the most crucial decision variables viz., feed concentration, flowrate and cooling stream temperature, three zones leading to an improvement of nearly 12 per cent on nominal average performance is detected within two generations of classification rules.

Stage-wise crystallization process synthesis

Continuous operation of crystallizers, though beneficial in reducing capital and running costs through smaller, less expensive, equipment and reduced maintenance, is marked by some undesirable crystal properties. None more so than the exponential form of CSD, particularly when nearly all applications require a uniform crystal size. A common approach for tackling this problem is to design for poor mixing so that the resulting CSD gradients could be exploited by fines dissolution and product classification loops. Such equipment, however, is difficult to design, requires abnormally high in-process inventories and its

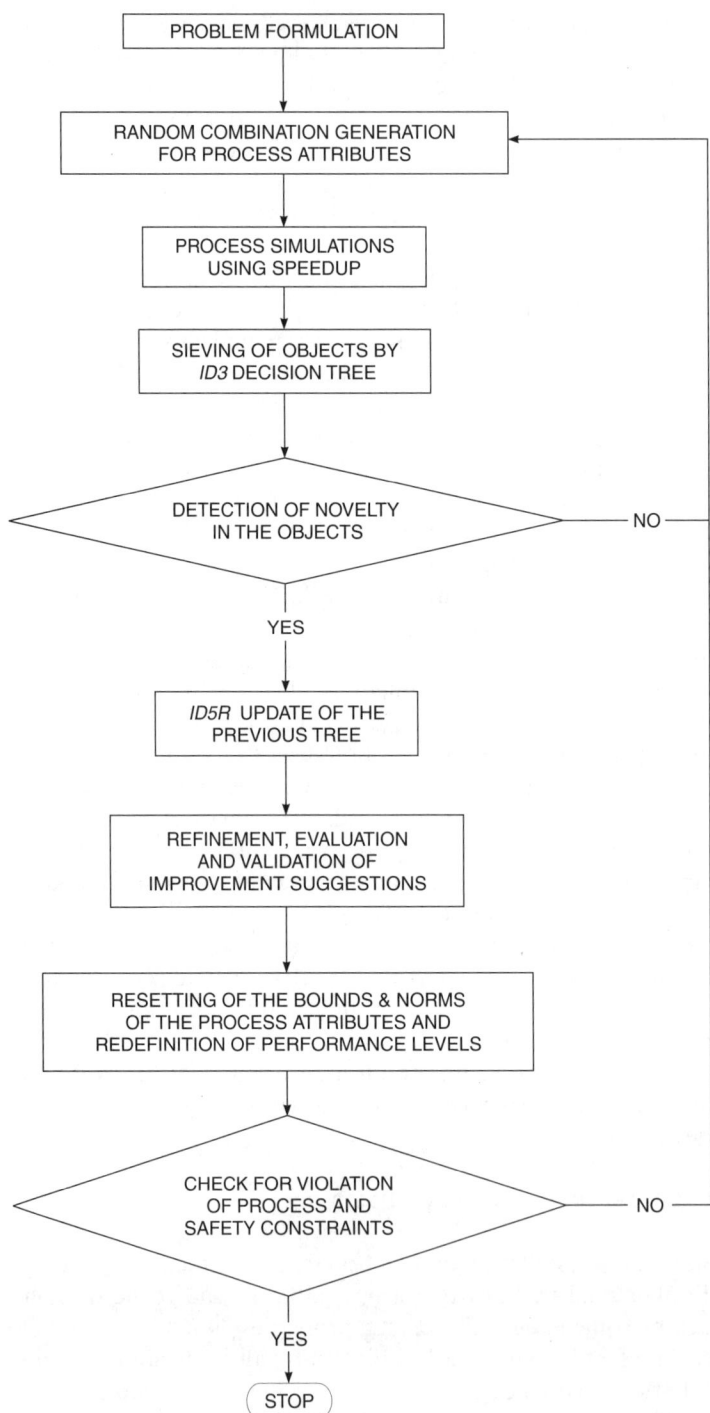

Figure 9.13 *Revised machine learning scheme as applied to a crystallization problem* (*Sheikh and Jones, 1997*)

operation is subject to cyclic behaviour (Randolph and Tan, 1978). A series of MSMPR crystallizers in cascade offers a viable alternative not only capable of narrowing the CSD, but also providing with improvements such as flexible operation of temperature regimes, the possibility of using larger cooling surface and economies of energy consumption. This type of design has been used from early this century for the crystallization of gypsum from phosphoric acid (McCabe, 1929a,b) in order to improve filterability through uniform size and occlusion free crystals.

The performance of a cascaded configuration is determined by the supersaturation level not only dependent on the operating temperature but also on the feed concentration and mean residence time which in turn is specified by volume and throughput for each crystallizer in the network. All these variables have to be determined through population balance based simulations of the network to optimize performance. Whilst design techniques for such networks have been devised (Hounslow and Wynn, 1993; Nývlt, 1992; Randolph and Tan, 1978) with varying levels of simplifications, there has only been one notable attempt (Larson and Wolff, 1971) for optimizing a pre-defined network. In their work, Larson and Wolff (1971) simulated cascades of up to three crystallizers for different combinations of crystallizer volumes and fraction of solute crystallized in a specified configuration to improve certain characteristics of the CSD. The adequacy of their procedure, however, is strongly dependent on the assumptions made during the derivation of the models and the exhaustiveness of the combinations simulated. Furthermore, due to the *a priori* specification of the network structure and the fact that crystallization system is designed in isolation with the energy and separation system, the solution thus obtained would be suboptimal even if the above concerns were addressed.

Network synthesis methodology

Significant recent approaches to chemical reactor network synthesis can be classified into two categories, viz. superstructure optimization and network targeting. In the former, a superstructure is postulated and then an optimal sub-network within it is identified to maximize performance index (Kokossis and Floudas, 1990).

The superstructure approach is limited by both the question of completeness of the network and the possibility of overlooking a better configuration due to limitations of the superstructure. It also leads to very large mixed integer non-linear (MINLP) formulations because global optimization techniques have to be applied to obtain an acceptable solution. The targeting approach originated from the work of Linhoff and Hindmarsh (1983) for the integration of heat exchangers networks. It is commonly known as 'pinch technology'. For reactor networks, the targeting methodology due to Balakrishna and Biegler (1992a) is based on mixing between different reacting environments. It proceeds by determining the maximum possible performance from a network that cannot be extended with further mixing and or reaction without explicit specification of the network structure. A network capable of achieving this target is then devised by extending the concept of attainable regions, through simple optimization formulations. Though the simple formulations do not allow for parallel reactor structures, they have been coupled with energy integration and separation sequences with relative ease (Balakrishna and Biegler, 1992b; Balakrishna and Biegler, 1996) to provide a framework for integrated process design. An extensive review of reactor network synthesis strategies can be found in Hildbrandt and Biegler (1994).

Problem formulation for crystallizer networks

Particulate products, such as those from comminution, crystallization, precipitation etc., are distinguished by distributions of the state characteristics of the system, which are not only function of time and space but also some properties of states themselves known as internal variables. Internal variables could include size and shape if particles are formed or diameter for liquid droplets. The mathematical description encompassing internal co-ordinate inevitably results in an integro-partial differential equation called the population balance which has to be solved along with mass and energy balances to describe such processes.

In many situations of engineering interest, for instance design and control, knowledge of the complete distribution is unnecessary; rather some average or total quantities with regards to internal variable such as average size and mass concentration are sufficient. Moment transformation is frequently used to obtain a lower dimension description of the population balance by converting it into a set of ordinary differential equations with the first four related to total number, length, area, and volume of crystals per unit volume of suspension respectively. The other major difference between crystallization and chemical reactions is that in the former two kinetic processes occur simultaneously, viz. nucleation of crystals and their subsequent growth. The level of supersaturation, which in turn depends on both the solute concentration and temperature, governs both these processes.

An optimal control problem analogous to a continuous flow reactor (CFR) formulation can be developed for crystallizer networks on the basis of moment transformation to the population balance and the assumption that effects of crystal agglomeration and disruption are negligible (Shiekh and Jones, 1997). Control profiles in temperature, $T(\alpha)$, fraction of crystal entering, $f(\alpha)$ and volumetric flow fraction, $q(\alpha)$, at each stage are obtained from orthogonal collocation on finite elements using the Lagrange basis function. This method has been shown to be the preferred technique (Cuthrell and Biegler, 1987) because it leads to networks easier to achieve practically and the integrals within the optimization problem are evaluated automatically at Gaussian quadrature points. Furthermore, state variable constraints can be imposed directly and there is no need to derive adjoint equations for them.

The two states viz. disappearing mass of solute and number of crystals formed not only show opposing trends in time but also differ largely in magnitude. These attributes of the system together with the exponential form of nucleation kinetics necessitate very fine discretization, which, coupled with the fact that the differential equation cannot be solved off-line, would result in a very large and highly ill conditioned non-linear programme. The model can be simplified by assuming the nucleation rate expression to be independent of varying solute concentration within a finite element while allowing supersaturation to vary with temperature through saturation concentration. This in effect uncouples number balance from mass balance and the former can be solved independently just as a function of temperature. This procedure simplifies the problem and reduces its size, since expressions for Lagrange basis functions and their derivatives along with the interpolating function are eliminated for number balance.

The consequences of this simplification will include an over prediction of the number of crystals and a less than proportionate increase in the mass of crystals formed, since growth rate is still a function of both the actual concentration and temperature. The assumption can, however, be justified on the basis that major variations in nucleation rate are likely to be resulting from changes in temperature effecting saturation concentration rather then the slowly diminishing solute concentration. With the above simplifications, the following non-linear programme can be formulated to provide optimal control variables at the collocation points using Lagrange interpolation basis function for approximating state variables over each finite element $\Delta \alpha_i$

$$\max_{q_i, f_{ij}, T_i} J(X_{\text{exit}}, \tau)$$

$$\sum_k X_{ik} L'_k(\alpha) - R(X_{ij}, T_i, n_i j) \Delta \alpha_i = 0 \tag{9.21}$$

$$n_i = B_i \Delta \alpha_i q_i + \left(\left(1 - \left(\frac{f_{i-1}}{q_{i-1}}\right)\right) \times n_{i-1}\right) \tag{9.22}$$

$$B_i = A \exp\left(\frac{-c}{\ln\left(\frac{X_{i,0}}{X_{\text{sat}}(T_i)}\right)}\right) \tag{9.23}$$

$$R(X_{ik}, T_i, n_i) = n_i [\tau_{ik} k_g (X_{ik} - X_{\text{sat}}(T_i))]^3 \tag{9.24}$$

$$X(0) = X_0 \tag{9.25}$$

$$X_{i,0} = \phi_i X_0 + (1 - \phi_i) X_{i-1,\text{end}} \tag{9.26}$$

$$X_{i,\text{end}} = \sum_k X_{ik} L_k(\alpha_{i+1,0}) \tag{9.27}$$

$$X_{\text{exit}} = \sum_i \sum_j X_{ij} f_{ij} \tag{9.28}$$

$$\tau = \sum_i \sum_j \alpha_{ij} (q_i - f_{ij}) \tag{9.29}$$

$$Q_{ij} = \sum_i \sum_j (q_i - f_{ij}) \tag{9.30}$$

$$\phi_i = \frac{Q_{i,0}}{Q_{ij}} \tag{9.31}$$

$$\sum_i q_i = 1.0 \tag{9.32}$$

$$\sum_i \sum_j f_{ij} = 1.0 \tag{9.33}$$

$L_k(\alpha)$ is the interpolation function and $L'_k(\alpha)$ its derivative. Their detailed description and evaluation procedure can be found in Rice and Do (1995). The value of state variable X, within an element is extrapolated to find its magnitudes at the end through equation 9.25. The balance between feed and crystallizing streams is represented by equation 9.26.

General performance specifications in terms of CSD parameters do not exist and therefore the objective functions for such problems often depend upon the end use envisaged for the product. For the present problem, a composite objective function was developed comprising total number of crystals, average size and coefficient of variance; all three with equal weighting. The first two features of size distribution manifest the effects of nucleation and growth rates and there is always an optimal level of supersaturation beyond which they do not increase simultaneously, rather the average size falls while the number increases. Coefficient of variance is crucial in quantifying on the effectiveness of network in narrowing crystal size distribution. All these parameters depend on super-saturation and residence time, which can be determined optimally by control profiles. The appropriate relationships are as follows

$$n_{\text{total}} = \sum_i \sum_j n_i f_{ij} \tag{9.34}$$

$$L_{\text{av}} = \frac{L_{\text{total}}}{n_{\text{total}}} \tag{9.35}$$

$$\text{CV} = \left(\frac{n_{\text{total}} A_{\text{total}}}{L_{\text{total}}^2} - 1 \right)^{\frac{1}{2}} \tag{9.36}$$

$$L_{\text{total}} = \sum_i \sum_j \alpha_{ij} (q_i - f_{ij}) n_i f_{ij} G_{ij} \tag{9.37}$$

$$A_{\text{total}} = \sum_i \sum_j (\alpha_{ij} (q_i - f_{ij}))^2 n_i f_{ij} (G_{ij})^2 \tag{9.38}$$

When crystallization is carried out in stages it is normal practice to have a descending profile for temperature to compensate for diminishing supersaturation by reducing saturation concentration (Nývlt, 1992). This relationship between temperature and concentration profiles would satisfy the modified sufficiency condition to interpret the above CFR type solution as an optimal in providing maximum achievable performance, consequently eliminating the need to check for recycle crystallizer extensions.

The optimal control profiles identified by the solution of the non-linear programme were used to simulate the network through rigorous distributed parameter models on SPEEDUP® to obtain a detailed description of its performance. The models are based on an analytical solution to simplified population balance that ignores effects of crystal agglomeration and disruption. It provides the complete CSD in each crystallizer and does not suffer from errors in mass balances, often observed in discretized solutions to the population balance (Lister et al., 1995). The equation takes the following form

$$\frac{\partial n_i}{\partial L} = \frac{n_{i-1} - n_i}{G_i \tau_i} \qquad (9.39)$$

subject to the boundary condition $n_i(0) = V_i B_i^0 / G_i \tau_i$

This simple steady-state form is amenable to analytical solution and the application of an integrating factor followed by subsequent integration yields

$$n_i(L) = \exp\left(-\frac{L}{G_i \tau_i}\right)\left(\frac{V_i B_i^0}{G_i \tau_i} + \int_0^L \frac{n_{i-1}}{G_i \tau_i} \exp\left(-\frac{L}{G_i \tau_i}\right) dL\right) \qquad (9.40)$$

Details of the complex kinetic models and other constituent equations used in conjunction with the above equations can be found in Sheikh and Jones (1997).

The stage-wise potassium nitrate crystallization process illustrates these concepts and procedures. The optimal solution determined is used to simulate a network of MSMPR crystallizers using rigorous models capable of predicting the complete crystal size distribution. Optimal flow distribution for the network in each of the finite elements, arising from optimal specification of control variables, $q(\alpha)$ and $f(\alpha)$ are depicted in Figure 9.14 while the temperature profile along the network is shown in Figure 9.15.

The novel provision of side feeds promotes mixing between feed and crystallizing streams and increases solute concentration. This not only eliminates the need for equal volume (or residence time) of each crystallizer in the network but may also reduce the energy requirements for cooling the suspension. The magnitude of such reductions will depend, however, on the exact mixing profiles between the crystallizers.

Figure 9.16 shows predicted CSD from the crystallizers and in the product stream. It can be seen that slopes for the first three crystallizers do not change significantly i.e. there is little reduction in CV, the number of crystals, however, increases by a few orders of magnitude. The number of crystals increases because of increasing levels of supersaturation achieved by reduction in temperature. These crystals, however, do not grow significantly because of rather small residence time within these crystallizers, which does not help reduce the coefficient of variance largely. For these reasons, significant improvements in CV are observed in vessels 4 and 5 as the crystals are allowed more time to grow.

The optimal network increases total residence time by 48 per cent when compared with an equivalent MSMPR of the same volume and throughput. This increase would translate into a similar increase in mean crystal size and a 78 per cent increase in yield. Exactly the same residence time as for the single crystallizer have been reported from simple cascade configurations previously designed for stage-wise crystallization processes for slight improvements in

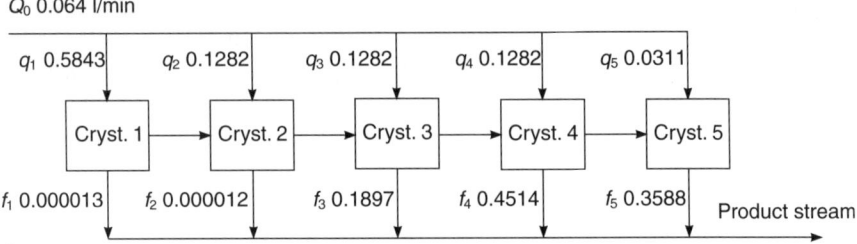

Figure 9.14 *Optimal flow profiles for a five stage KNO_3 crystallization process (Sheikh and Jones, 1997)*

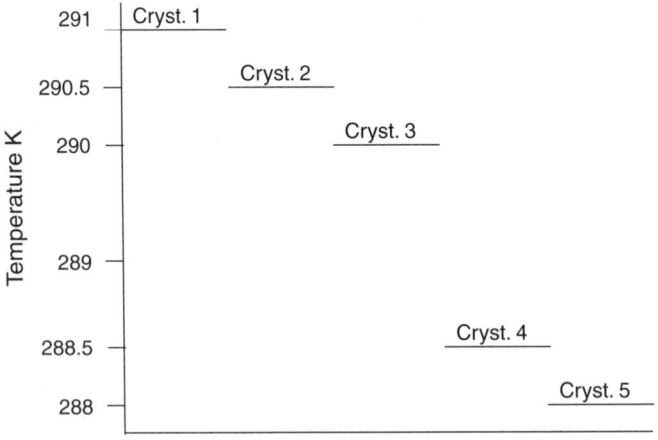

Figure 9.15 *Temperature profile within a crystallizer network (Sheikh and Jones, 1997)*

yield and average size at the expense of degradation in co-efficient of variation (Larson and Wolff, 1971).

The reductions in residence time is obtained due to the plug flow reactor (PFR) type structure of the network which reduces the residence time (or volume) of the crystallizer for similar yield (Levenspiel, 1972), while optimal allocation of feed and removal of product stream further help in improving the yield without the degradation of co-efficient of variation. Since average crystal size is directly proportional to residence time at constant growth rate, while the yield increases cubically, the network is expected to result in an increase of nearly 40 per cent and 78 per cent in average size and yield respectively when compared with an MSMPR of similar volume operated at similar supersaturation and throughput. Conversely, the single crystallizer would have to be 20 litres in volume. The benefits of operating at different temperatures have to be added to these advantages for the true value of the new design scheme, because varying temperatures in subsequent crystallizers have already been shown to improve both average size and yield.

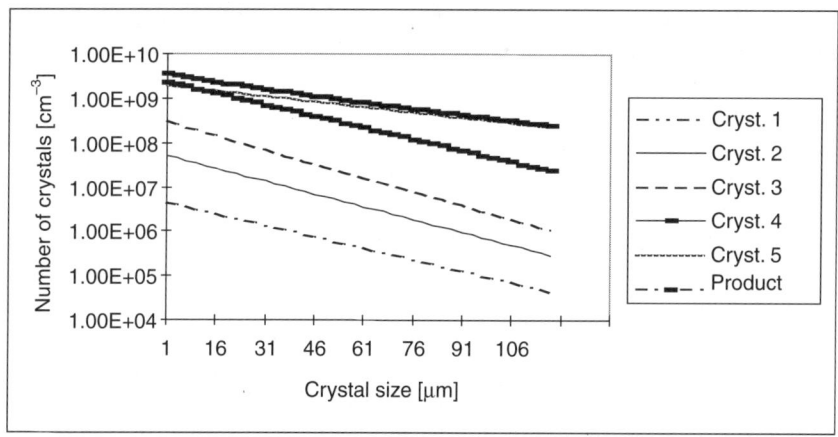

Figure 9.16 *Crystal size distributions from each crystallizer and product stream* (*Sheikh and Jones, 1997*)

Crystallization process instrumentation and control

Crystallization process control is desirable from a number of standpoints. The primary objective is often to meet customer requirements by achieving consistent product quality to a desired specification of crystal size, size distribution and purity. Secondly, process requirements often dictate maintenance of stable crystallizer operation, the avoidance of fines and encrustation, and the minimization of subsequent downstream processing.

Several measures for these objectives have been recommended in the literature. These include minimizing the level of supersaturation during batch operation (Griffiths, 1925), minimizing the extent of nucleated crystal mass (Mullin and Nývlt, 1971), maximizing the final size of seed crystals from batch crystallization (Jones, 1974), maximizing the mean size and minimizing the second moment of the CSD (Ajinkya and Ray, 1974), maximizing the number average size (Chang and Epstein, 1982), minimizing the ratio of nucleated-crystal mass to seed crystal mass (Eaton and Rawlings, 1990) and maximizing the weight mean size (Ma et al., 1999). Thus, crystallization process control has almost exclusively been concerned with the control of the crystallizer *per se* rather than the whole process.

The combination of non-ideal phase behaviour of solutions, the non-linearity of particle formation kinetics, the multi-dimensionality of crystals, their interactions and difficulties of modelling, instrumentation and measurement have conspired to make crystallizer control a formidable engineering challenge. Various aspects of achieving control of crystallizers have been reviewed by Rawlings et al. (1993) and Rohani (2001), respectively.

Batch crystallizers

Batch crystallizers are widely used for relatively small tonnage (say <10 kTe/yr) dyestuffs, fine chemicals, pharmaceuticals and other speciality chemicals. The vessel capacities are correspondingly small (typically 0.02–20 m^3). A typical forced-circulation batch crystallizer with external cooling is shown in Figure 9.17.

As seen in Chapter 7, the operation of batch crystallizers is inherently unsteady-state. Transient values occur of the major operating variables such as slurry density, supersaturation, temperature and mean particle size. Methods of operational control such as by use of seeding and temperature programming were also considered in detail.

Several attempts have been made to determine crystallizer properties and apply feed back control to batch crystallizers based on some measured chemical or physical quantity – either solution-based or from the magma. Thus, Doucet and Giddey (1966) and Virtanen (1984) describe methods implemented for the control of supersaturation via measurement of solution refractive index during the industrial crystallization of sucrose. Garside and Mullin (1966) highlight the important need for the ability to measure and control solution concentration accurately in the determination of supersaturation within crystallizers and describe a densitometric technique. Rohani *et al.* (1990) and Rohani and Bourne (1990a,b) applied crystal fines destruction in a closed loop feedback during batch cooling crystallization in which a slurry sample is heated while its temperature and light transmittance are measured thus enabling solids concentration to be determined. Farrell and Tsai (1994, 1995) and Vega *et al.* (1995) demonstrated the application of general model control (GMC) due to Lee and Sullivan (1998) for CSD trajectory regulation. They used an *in situ* laser particle counter with and without seeds, respectively, and concluded that rigorous mechanistic modelling of process and sensor with regard to the population balance is not necessarily a pre-requisite for implementation of model based control. Chang and Epstein (1982, 1987) also analysed feedback control strategies and predicted potential improvements in product CSD. Eaton and Rawlings (1990) developed tools for feedback control using non-linear models and applied them to simulate batch optimal crystallization control profiles. Bohlin and Rasmuson (1992) further analysed the application of controlled cooling and seeding in batch crystallization. Rawlings *et al.* (1993) reviewed model identification and control of solution crystallization processes. Difficul-

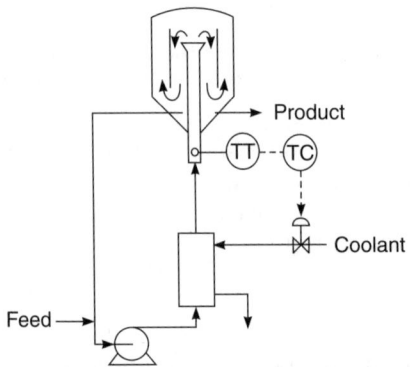

Figure 9.17 *Forced-circulation batch crystallizer (after Rohani, 2001)*

ties associated with laser light scattering measurements were considered in detail and the potential use of image analysis as an alternative was highlighted.

In an effort to reduce growth rate dispersion (Chapter 5) and improve product CSD, Heffels *et al.* (1994) proposed a control method for unseeded batch crystallization in which slow growing initially formed crystals are redissolved, controlled via laser light back scattering. Whilst noting reported restrictions on use of lasers for CSD determination, Monnier *et al.* (1997) demonstrated use of calorimetry for supersaturation measurement coupled with an on-line *in situ* laser sensor. Additionally, estimates of the final CSD were made via image analysis. Dunuwila and Berglund (1997) demonstrated the feasibility of using attenuated total reflection and fourier transform infra red (ATR FTIR) spectroscopy as a solution phase chemometric method for the determination of supersaturation *in situ*, and proposed it as a future control measure. Groen *et al.* (1999) report a vessel incorporating instruments for examining the batch crystallization of organic speciality chemicals. Mathews and Rawlings (1998) successfully applied model-based control using solids holdup and liquid density measurements. In open-loop control, inputs to a dynamic system are computed that optimize a specified performance criterion. Ma *et al.* (1999) proposed a method that quantifies the impact of parameter and control implementation inaccuracies on the performance of open-loop control policies. Such information can be used to decide whether more laboratory experiments are needed to generate parameter estimates of higher accuracy, or to define performance objectives for lower-level control loops that implement the optimal control policy. The methodology was illustrated on a batch crystallization process.

Continuous crystallizers

Continuous crystallizers find widespread application for the production of bulk chemicals such as fertilizers, fibre intermediates, salts and sugars. Production rates are typically 100 kTe/yr, or more with vessel capacities typically exceeding 20 m^3. Typical continuous crystallizers are depicted in Figure 9.18 and are described in more detail in Chapter 3.

It was shown in Chapter 7 that the performance of continuous crystallizers is determined by the characteristics of a feedback loop relating the output performance expressed as crystal size distribution and to the feed concentration and residence time. Thus, an increase in crystallizer residence time, or decrease in feed concentration, reduces the working level of supersaturation. This decrease in supersaturation results in a decrease in both nucleation and crystal growth. This in turn leads to a decrease in crystal surface area. By mass balance, this then causes an increase in the working solute concentration and hence an increase in the working level of supersaturation and so on. There is thus a complex feedback loop within a continuous crystallizer, as considered in Chapter 7 and illustrated in Figure 8.11.

The feedback loop within a crystallizer thus leads to changing conditions within the crystallizer and its product characteristics, both at start up and subsequently due to disturbances due to changes in feed stream concentrations,

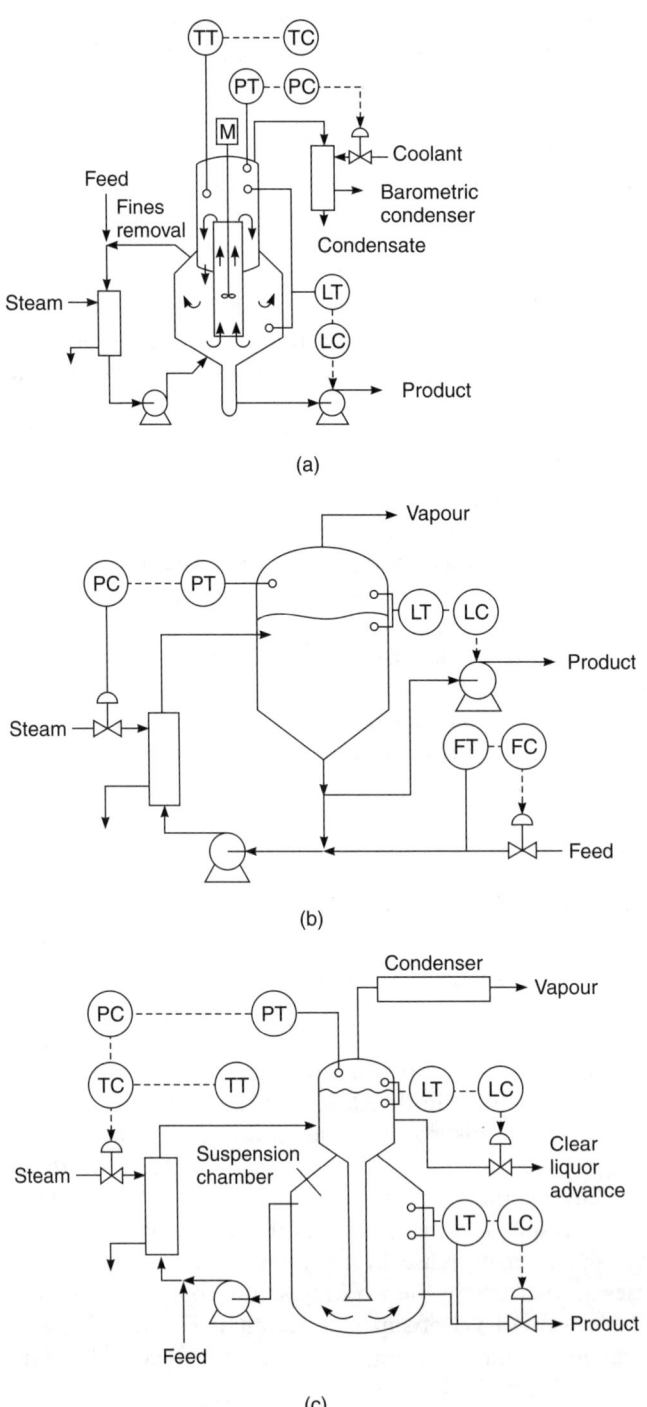

Figure 9.18 *Continuous crystallizers:* (a) *draft-tube and baffle* (DTB); (b) *single effect forced-circulation evaporative;* (c) *Oslo or Krystal type (after Rohani, 2001)*

flow rates, temperature etc. These crystallizer dynamics, which may become unstable under certain conditions, therefore require some form of control. Traditionally, this has been by manual intervention but, due to the 8–10 residence times taken to reach steady state after a disturbance (Chapter 7) and a typical residence time of up to one hour, it can be the following shift which experiences the effects of the changes. Thus considerable effort has gone into devising rational automatic control schemes for industrial crystallizers.

Three of the more important issues in achieving control of continuous crystallizers are:

1. identification of most suitable process inputs
2. use of CSD to determine a dynamic process model
3. on-line measurement techniques.

Most of the literature in control of continuous crystallizers is based on a single-input single output (SISO) control structure. Different controlled variables and manipulations have been suggested based on the relative ease and accuracy of on-line measurements and their efficiency in effectively addressing set-point tracking and disturbance rejections. Both linearized physical models and black-box models have been suggested for the controller design, as reviewed by Sheikh (1997) as follows.

Physical model based control

Several early theoretical studies demonstrated the need for CSD control in industrial crystallizers, later authors began to propose, and implement control schemes based on MSMPR population balance analyses (Saeman, 1956; Han, 1969; Gupta and Timm, 1971; Lei et al., 1971a,b; Beckmann and Randolph, 1977). Rovang and Randolph (1980) and Parks and Rousseau (1979), respectively, based control systems on CSD measurement aimed at determining nuclei density whilst Helt and Larson (1977) demonstrated the potential of using a differential refractometer for control via supersaturation determination. Attarian et al. (1976) applied feedback control of pH measured in a continuous crystallizer used for precipitation of calcium carbonate. Satisfactory feedback control of the pH was achieved using proportional plus integral action. Good control of particle size was demonstrated using a Coulter Counter^{TM}.

Rousseau and Howell (1982) considered the merits of using different measurements for stabilizing low order cycling of CSD in a continuous crystallizer with both fines destruction and product classification. The analysis was carried out on a simulated process using population and mass balances along with kinetic equations and employed finite difference techniques to solve the system. The main advantages of using a finite difference method in comparison with a linearized form of analytical solution were cited as: (a) no modifications to the models were necessary to accommodate different removal functions; and (b) any form of nucleation kinetics could be used.

To achieve stabilization objectives, Rousseau and Howell (1982) considered a control system designed to stabilize crystal growth rate rather than nuclei density. This was done using a proportional controller based on the deviations of supersaturation from steady state. A differential solution refractometer was suggested for measuring supersaturation, because its readings do not suffer from electrical noise or by the presence of crystals in the slurry. Though the measurements are easier to make than for nuclei density, it was found that a supersaturation based controller is more sensitive to measurement errors and required a controller constant nine times higher than for nuclei density based controller to dampen the oscillations.

Randolph et al. (1987) used light scattering measurements in a continuous fines stream to infer nuclei density from slurry density measurements. A theoretical equation was used to calculate nuclei density from density measurements. An 18-litre KCL crystallizer was used to implement the controller, which had the task of eliminating CSD transients (not oscillations) caused by outside disturbances such as flowrates, temperature or agitation. Using a proportional action control, manipulation of the fraction of the fines dissolver flow that is sent back to the crystallizer controlled the nuclei density inferred from secondary measurements. Only disturbances in vessel temperature were considered; these were suddenly introduced into the system at steady state to generate a nucleation pulse. Shutting off the cooling water disturbed the temperature until it elevated to the desired level. The cooling water was then reopened to initiate temperature control. Both open-loop and closed-loop experiments were performed and it was observed that the latter reduced root mean square of CSD fluctuations due to nucleation upset by a factor of 3.5. A recycle line was added to fines dissolver stream which enabled varying its flowrate without introducing upsets in the crystallizers due to changes in the cut size of the fines removed. It also helped establish relationships permitting linear changes in fines destruction through linear changes in the manipulated variables.

Rawlings et al. (1992) analysed the stability of a continuous crystallizer based on the linearization of population and solute balance. Their model did not depend on a lumped approximation of partial difference equations and successfully predicted the occurrence of sustained oscillations. They demonstrated that simple proportional feedback control using moments of CSD as measurements can stabilize the process. It was concluded that the relatively high levels of error in these measurements require robust design for effective control.

Jager et al. (1992) used a dilution unit in conjunction with laser diffraction measurement equipment. The combination could only determine, however, CSD by volume while the controller required absolute values of population density. For this purpose the CSD measurements were used along with mass flow meter. They were found to be very accurate when used to calculate higher moments of CSD. For the zeroth moment, however, the calculations resulted in standard deviations of up to 20 per cent. This was anticipated because small particles amounted for less then 1 per cent of volume distribution. Physical models for process dynamics were simplified by assuming isothermal operation and class II crystallizer behaviour. The latter implies a fast growing system in which solute concentration remains constant with time and approaches saturation concentration. An isothermal operation constraint enabled the simplification of mass and energy balances into a single constraint on product flowrate.

The solute balance reduced to a constraint on growth rate. The model did not result, however, in the required state-space representation essential for multi-variable model based control. A finite difference approximation of the crystal size axis was used to convert population balance into a set of ODEs, which was then linearized at an operating point to obtain the required state-space representation. The need for a product classification step, whereby coarse crystals are removed at a finite size in addition to fines dissolution, was highlighted. Hydrocyclones were suggested for reducing CSD dispersion by using variable underflow discharge diameter as an additional input for control.

Black box modelling for dynamics

Rohani and Paine (1991) developed a feedback controller, where process dynamics were obtained from step change responses of the outputs by fitting first-order plus dead time models. Again, the rate of fines dissolution/removal was the manipulated variable with fines suspension density being the output or control variable. Control variable was interpreted using a fines suspension density sensor (FSDS) which uses a sample cell containing a representative sample stirred and heated at an appropriate rate, while the transmittance of an infra-red light beam passing through the cell is being recorded (Rohani and Paine, 1987). Temperatures are measured both just outside the cell and within it until the transmittance measurement reaches a plateau, i.e. all the crystals have dissolved. These temperature measurements along with the solubility data are used to infer fines density in a simple manner. The fines suspension density sensor was used to design a PI controller for both the set point tracking and rejecting disturbances arising from temperature fluctuations in the crystallizer. The controller (Figure 9.19) gave acceptable performance in rejecting small temperature disturbances. It was adopted for the control of crystal purity (Redman and Rohani, 1992).

Redman et al. (1997) performed a detailed analysis of a crystallization system for control variables such as mean crystal size, weight per cent solids in product and fines streams, and supersaturation through manipulations in feed, product, recirculation and fines dissolution flowrates to determine the best pairings. Open-loop response to step changes in the manipulation variables were used for selecting and designing the controller. An in-line backward scattering laser light sensor determined mean size, while nuclei density was inferred by the FSDS. Supersaturation was determined through density measurements and solubility data. Experiments were performed in a $1\,m^3$ forced circulation, evaporative crystallizer. Bristol's relative gain array analysis (Bristol, 1966) was carried out on the findings to determine the pairings, which were found as:

1. fines dissolution flowrate with average size
2. product flowrate with weight per cent solids
3. feed flowrate with supersaturation.

A SISO controller was designed (for mean crystal size), even though relative gain array analysis showed possible interactions between all of the three control

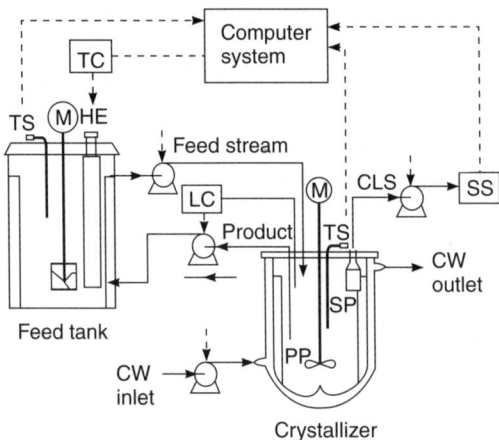

Figure 9.19 *Experimental apparatus for feedback control of KCl crystallization (after Redman and Rohani, 1992)*

variables. A cascade configuration was designed with the fines dissolution flowrate as control variable in the slave loop. The internal model control (IMC) controller design techniques were used to determine controller parameters. The controller was shown to be successful in tracking an increase in mean crystal size set point. A subsequent reduction in set point however, failed to bring the mean crystal size down to the new lower value, even though the fines flowrate was immediately reduced to its lower limit by the controller. This happened because the improved design had already minimized the nucleation process and thus rendered fines dissolution ineffective. The controller was also shown to be ineffective in rejecting product flowrate disturbances, where no response was observed until the disturbance had been removed.

Jager *et al.* (1992) developed a black-box model based controller for a 970-litre crystallizer again using a dilution unit and on-line measurement with light scattering devices together with mass flow meters. System identification techniques were used; these offer a lower order state space model for multi-variate analysis. A three-step identification procedure was employed to obtain the model. In the first step, the ARX model (Ljung, 1987) was used. The second step involved transformation of the ARX model into a state-space representation through approximate realization (Damen and Hajdasinski, 1982). Finally, the model was used as an initial parameter combinatory for fitting the actual data. Uncorrelated white noise signals were added to the data used, because they could be added to process inputs simultaneously without affecting the ability of the identification algorithm to distinguish the contribution of each individual input to the output signal, thus allowing effective use of data. A model derived through these steps was used to establish relationships between inputs including heat input, product flowrate and rate of fines dissolution, and the output measures, which were third moment and mass based average size.

It is evident that most studies reported to date have used number density, average size or weight per cent as control variables. Often these variables are inferred from other measurements, including density, solution supersaturation, refractive index etc. Inferential techniques have been shown to be particularly suitable for industrial scale applications where laser scattering devices for on-line size distribution measurement are not yet practical for industrial control purposes, although substantial progress is being made to that end. Even when usable, however, these measurement devices are often characterized by noise and require operation at very low solids concentration.

It has often been suggested (Jager et al., 1992; Sherwin et al., 1967; Redman et al., 1997) that an effective controller will have to control not a single state but multiple states through MIMO architectures. Based on their pioneering study into the dynamics of continuous crystallizers, Sherwin et al. (1967) suggested that it was necessary to control nuclei density along with any other property of CSD. Most other work in this area, however, is based on SISO controllers which at times are shown to be unable to address either set point tracking or disturbance rejection effectively because of their heavily restricted control objectives.

Fines dissolution flow rate has almost universally been used as the manipulated variable, even though its limitations are not only obvious but have been quantified by Jager et al. (1992) and Redman et al. (1997). Apex diameter has been suggested as an input variable to affect cut size according to the desired value for mean crystal size (Jager et al., 1992). Simple analysis of crystallization processes reveals that both nucleation and growth rates are strong functions of supersaturation and therefore ways of manipulating supersaturation to achieve desired CSD properties ought to be a key concern in control exercises. Changing feed concentration or crystallizer temperature or residence time can bring about such manipulations.

Crystallization in practice

Scale-up

Often, the performance of a large-scale crystallization plant does not match the performance obtained in the laboratory. Scale-up affects crystallizer performance in a number of ways, notably via mixing effects on secondary kinetic processes (such as secondary nucleation, agglomeration and breakage), and giving rise to regions of varying energy dissipation, supersaturation and temperature respectively. Clearly, these aspects are best taken into account at the design stage.

Models of secondary kinetic processes were considered in detail Chapters 5 and 6 whilst accounting for poor mixing effects by use of CFD-based methods were described and illustrated in Chapters 2 and 8. Such CFD codes can be linked via compartmental and mixing models, or coupled directly with the population balance if the precipitation processes are particularly fast in comparison to the mixing process.

As vessels increase in size the surface area:volume ratio decreases thereby increasing the need for high heat transfer rates with the possibility of enhanced encrustation on cooling surfaces or at evaporation zones as a consequence.

Crystal encrustation

Crystal encrustation (fouling, incrustation, salting, scaling) is frequently encountered during crystallization plant operation and this important problem is best addressed for at the design stage. Encrustation on surfaces can seriously reduce their heat transfer capacity thereby reducing production rate, increasing both costs and required retention time. Encrustation around the liquor/vapour interface ('wind and water line') can reduce vapour flowrates and enhance liquor entrainment. Encrustation on agitators or around the shaft increases their power requirements and causes damage to bearings through unbalanced loads. Problems can occur if large crystal lumps detach and block outlets, impact on pump impellors and add to line blockage. Encrustation on transfer lines leads to greatly increased pumping power requirements and can lead to complete blockage and shutdown of the plant followed by expensive cleaning operations. Sometimes, of course, a virtue is made of crystal encrustation, as in the operation of scraped surface chillers.

The most frequent site for crystal encrustation is on a compatible solid surface within a zone of high supersaturation and low agitation. Selection of a less compatible material having a smooth surface can avoid the major excesses of encrustation. Duncan and Phillips (1979) and Shock (1983), respectively, reveal a connection between the metastable zone width of crystallizing solutions and their propensity to encrust. It is well known that judicious crystal seeding can also help minimize encrustation. Simple laboratory tests are recommended to determine all these issues before the plant is built.

Crystal caking

A troublesome aspect of crystalline materials is their tendency to cake, or bind together, during storage. Most customers require crystal products to be free flowing, sugar and table salt being perhaps the most obvious examples. Sometimes to aid free flowing, additives are employed (such as magnesium carbonate and sodium hexacyanoferrate II added to table salt, or tricalcium phosphate added to icing sugar) whilst in other cases the material is granulated first, particularly if the product is naturally cohesive or dusty.

The most likely causes of caking are wide particle size distribution of non-uniform elongated crystals, dampness, and diurnal temperature cycles giving rise to crystal agglomeration each of which should be avoided if possible. Attaining conditions to achieve granular crystals of narrow size distribution, sieving out the fines, bagging in airtight containers and storage under constant temperature conditions are all ways that may help overcome this troublesome problem. Again, pre-production tests are desirable to determine these issues.

Design of crystallization process systems 297

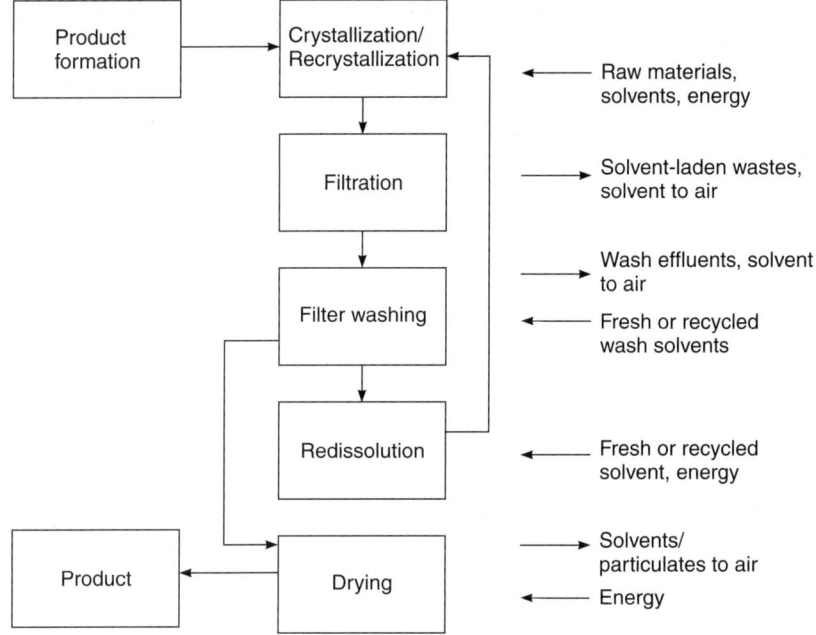

Figure 9.20 *Potential environmental impacts and resource usage associated with the formation of solid products by crystallization from solution (after Sharratt, 1996)*

Sustainability

Finally, Sharratt (1996) addresses the interaction between environmental concerns and the design and operation of a crystallization process system. The underlying environmental driving forces, and their influence on the need for materials substitution, reduction of waste generation during downstream processing and the requirement for recycling, are identified as key problem areas for the process engineer and industrial chemist to address (Figure 9.20).

Similarly, product and process design is part of a business process of yet wider sustainability implications (Allenby, 1999).

Summary

This chapter has been concerned with aspects of the design of industrial crystallization processes as systems. Methods for the prediction of crystal size distribution based on the population balance are now well established. Similarly, computational methods to predict crystal morphology and purity are being developed in which molecular modelling features strongly (Meyerson, 1999). Simple methods are also available to predict the required size of both crystallizer and subsequent solid–liquid separation units. Computer software packages are also becoming more available that can assist in calculating such quantities as crystal sizes, vessel volumes, energy dissipation and flow patterns, supersaturation and temperature profiles and power input. Considerable progress has also been made in the implementation of process control. Thus, provided sufficient data are available, process synthesis of the flowsheet permits the optimization of the 'base case' process design and the economic analysis of alternatives prior to detailed study of the sustainability of the final process.

References

Abegg, C.F., Stevens, J.D. and Larson, M.A., 1968. Crystal size distribution in continuous crystallizer when growth rate is size-dependent. *American Institute of Chemical Engineers Journal*, **41**, 188.
Adler, P.M., 1981. Heterocoagulation in shear flow. *Journal of Colloid and Interface Science*, **83**, 106–115.
Ajinkya, M.B., and Ray, W.H., 1974. On the optimal operation of crystallization processes. *Chemical Engineering Communications*, **1**, 181–186.
Allen, T., 1996. *Particle Size Measurement*, 5th edition. London: Chapman and Hall.
Allenby, B.R., 1999. *Industrial Ecology. Policy Framework and Implication.* New Jersey: Prentice Hall.
Al-Rashed, M.H. and Jones, A.G., 1999a. Validation of a CFD model of a gas–liquid precipitation system. In *Industrial Crystallization '99*. (Rugby: Institution of Chemical Engineers). Cambridge, 13–15 September 1999. Paper 157, p. 160.
Al-Rashed, M.H., Jones, A.G., 1999b. CFD modelling of gas–liquid reactive precipitation. *Chemical Engineering Science*, **54**, 4779–4784.
Al-Rashed, M.H. and Jones, A.G., Hannan, M. and Price, C., 1996. CFD Application on a simple geometry of batch precipitation of calcium carbonate. In *Industrial Crystallization '96*. Ed. B. Biscans, Progep, Toulouse, 16–19 September 1996, pp. 419–424.
Aoki, Y. and Nakamuto, Y., 1984. Penetration twins of potassium chloride. *Journal of Crystal Growth*, **67**, 579–586.
Aoun, M., Plasari, E., Davis, R. and Villermaux, J., 1999. A simultaneous determination of nucleation and growth rates from batch spontaneous precipitation. *Chemical Engineering Science*, **54**, 1161–1180.
Aquilano, D. and Franchini-Angela, M., 1985. Twin laws of calcium oxalate trihydrate (COT). *Journal of Crystal Growth*, **73**, 558–562.
Ash, S.G., Everett, D.H. and Radke, C., 1973. *Journal of the Chemical Society Faraday Transactions*, **69**(2), 1256.
Åslund, B.L. and Rasmuson, Å.C., 1992. Semi batch reaction crystallization of benzoic acid. *American Institution of Chemical Engineers Journal*, **38**, 328–342.
Astarita, G., 1967. *Mass Transfer With Chemical Reaction*. Amsterdam: Elsevier.
Attarian, C., Baker, C.G. and Svrcek, W.Y., 1976. pH control of a continuous precipitator. *Canadian Journal of Chemical Engineering*, **54**, 606–611.
Austin, L., Shoji, V., Jindal, V., Savage, K. and Kimpel, R., 1976. Some results on the description of size reduction as a rate process in various mills. *Industrial and Engineering Chemistry Process Design and Development*, **15**, 187–196.
Backhurst, J.R. and Harker, J.H., 1973. *Process Plant Design*. London: Heinemann.
Bakker, A., Laroche, R.D., Wang, M.H. and Calabrese, R.V., 1997. Sliding mesh simulation of laminar flow in stirred reactors. *Transactions of the Institution of Chemical Engineers*, **75**, 42–44.
Balakrishna, S. and Biegler, L.T., 1992a. Constructive targeting approaches for the synthesis of chemical reactors. *Industrial and Engineering Chemistry Research*, **31**(1), 300.
Balakrishna, S. and Biegler, L.T., 1992b. Targeting strategies for the synthesis and energy integration of non-isothermal reactor networks. *Industrial and Engineering Chemistry Research*, **31**(9), 2152.

Balakrishna, S. and Biegler, L.T., 1996. Chemical reactor network targeting and integration: An optimisation approach. In *Advances in Chemical Engineering*. Ed. J.L. Anderson, **23**, 243.
Baldyga, J. and Bourne, J.R., 1984a. A fluid mechanical approach to turbulent mixing and chemical reaction. Part I: Inadequacies of available methods. *Chemical Engineering Communications*, **28**, 231–241.
Baldyga, J. and Bourne, J.R., 1984b. A fluid mechanical approach to turbulent mixing and chemical reaction. Part II: Micromixing in the light of turbulence theory. *Chemical Engineering Communications*, **28**, 243–258.
Baldyga, J. and Bourne, J.R., 1984c. A fluid mechanical approach to turbulent mixing and chemical reaction. Part III: Computational and experimental results for the new micromixing model. *Chemical Engineering Communications*, **28**, 259–281.
Baldyga, J. and Bourne, J.R., 1989a. Simplification of micromixing calculations. I. Derivation and application of new model. *Chemical Engineering Journal*, **42**, 83–92.
Baldyga, J. and Bourne, J.R., 1989b. Simplification of micromixing calculations. II. New applications. *Chemical Engineering Journal*, **42**, 93–101.
Baldyga, J. and Bourne, J.R., 1992. Interactions between mixing on various scales in stirred tank reactors. *Chemical Engineering Science*, **47**, 1839–1848.
Baldyga, J. and Bourne, J.R., 1999. *Turbulent mixing and chemical reactions*. Chichester: John Wiley & Sons.
Baldyga, J. and Orciuch, W., 1997. Closure problem for precipitation. *Transactions of the Institution of Chemical Engineers*, **75A**, 160–170.
Baldyga, J. and Orciuch, W., 2001. Barium sulphate precipitation in a pipe – an experimental study and CFD modelling. *Chemical Engineering Science*, **56**(7), 2435–2444.
Baldyga, J. and Poherecki, R., 1995. Turbulent micromixing in chemical reactors – a review. *Chemical Engineering Journal*, **58**, 183–195.
Baldyga, J., Bourne, J.R. and Hearn, S.J., 1997. Interaction between chemical reactions and mixing on different scales. *Chemical Engineering Science*, **52**, 457–466.
Baldyga, J., Podgorska, W. and Pohorecki, R., 1995. Mixing-precipitation model with application to double feed semibatch precipitation. *Chemical Engineering Science*, **50**, 1281–1300.
Bamforth, A.W., 1965. *Industrial crystallization*. London: Leonard Hill.
Barton, G.W. and Perkins, J.D., 1988. Experiences with *SPEEDUP* in the mineral processing industries. *Chemical Engineering Research and Design*, **66**, 408.
Beckmann, J.R. and Randolph, A.D., 1977. Crystal size distribution and dynamics in a classified crystallizer. Part II. Simulated control of crystal size distribution. *American Institution of Chemical Engineers Journal*, **23**, 510–520.
Beddow, J.K., 1980. *Particulate Science and Technology*. New York: Chemical Publishing Co., Inc.
Bemrose, C.R. and Bridgwater, J., 1987. A Review of Attrition and Attrition Test Methods. *Powder Technology*, **49**, 97–126.
Bennett, R.C., 1962. Product crystal size distribution in commercial crystallizers. *Chemical Engineering Progress*, **58**(9), 76–80.
Bennett, R.C., Fiedelman, H. and Randolph, A.D., 1973. Crystallizer influenced nucleation. *Chemical Engineering Progress*, **69**, 86.
Berry, D.A. and Ng, K.M., 1996. Separation of quaternary conjugate salt systems by fractional crystallization. *American Institute of Chemical Engineers Journal*, **42**, 2162.
Berry, D.A. and Ng, K.M., 1997. Synthesis of reactive crystallization processes. *American Institute of Chemical Engineers Journal*, **43**, 1737.
Berry, D.A., Dye, S.R. and Ng, K.M., 1997. Synthesis of drowning-out based crystallization separations. *American Institute of Chemical Engineers Journal*, **43**, 91.

Berthoud, A., 1912. Théorie de la formation des faces d'un crystal. *Journal de Chimie Physique*, **10**, 624–653.

Biegler, L.T., Nocedal, J. and Schmid, C., 1995. A reduced Hessian method for large scale constrained optimization. *SIAM Journal of Optimization*, **5**(2), 314.

Bierwagen, G.P. and Saunders, T.E., 1974. Studies of the effects of particle size distribution on the packing efficiency of particles. *Powder Technology*, **10**, 111–119.

Bisio, A. and Kabel, R.L., 1985. *Scale-up of Chemical Processes: Conversion from Laboratory Scale Tests to Successful Commercial Size Design*. New York: Wiley.

Bloor, M.I.G. and Ingham, D.B., 1987. Flow in Industrial Cyclones. *Journal of Fluid Mechanics*, **178**, 507.

Bohlin, M. and Rasmuson, Å.C., 1992. Application of controlled cooling and seeding in batch crystallization. *Canadian Journal of Chemical Engineering*, **70**, 120–126.

Bond, F.C., 1952. Third Theory of Comminution. *American Institute of Mining Engineers, Trans.*, **193**, 484.

Bond, F.C., 1961. Crushing and Grinding Calculations. Part I, *British Chemical Engineering*, **6**(6), 378–385; Part II, *ibid.* **6**(8), 543–548.

Botsaris, G.D., 1976. Secondary nucleation: A review. In *Industrial Crystallization*, Ed. J.W. Mullin. Plenum Press: New York, pp. 3–22.

Bourne, J.R., 1985. Micromixing revisited. *Institution of Chemical Engineers Symposium Series*, **87**(ISCRE 8), 797–813.

Bourne, J.R. and Dell'Ava, P., 1987. Micro- and macromixing in stirred tank reactors of different sizes. *Chemical Engineering Research and Design*, **65**, 180–186.

Bourne, J.R. and Hungerbuehler, K., 1980. *Transactions of the Institution of Chemical Engineers*, **58**, 51.

Bourne, J.R. and Yu, S., 1994. Investigation of micromixing in stirred tank reactors using parallel reactions. *Industrial and Engineering Chemistry Research*, **33**, 41–55.

Bramley, A.S., Hounslow, M.J. and Ryall, R.L., 1996a. Aggregation during precipitation from solution: A method for extracting rates from experimental data. *Journal of Colloid and Interfacial Science*, **183**, 155–165.

Bramley, A.S., Hounslow, M.J. and Ryall, R.L., 1996b. Aggregation during precipitation from solution. Kinetics for calcium oxalate monohydrate. *Chemical Engineering Science*, **52**, 747–757.

Bransom, S.H. and Dunning, W.J., 1949. Kinetics of crystallization in solution. Pt II. *Discussions of the Faraday Society*, **5**, 96–103.

Bransom, S.H., Dunning, W.J. and Millard, B., 1949. Kinetics of crystallization in solution. Pt I. *Discussions of the Faraday Society*, **5**, 83–95.

Bravais, A., 1866. *Etudes Cristallographiques*. Paris: Gauthier-Villars.

Brayshaw, M.D., 1990. Numerical model for the inviscid flow of a fluid in a hydrocyclone to demonstrate the effects of changes in the vorticity function of the flow field on particle classification. *International Journal of Mineral Processing*, **29**, 51.

Brečević, Lj. and Kralj, D., 1989. Factors influencing the distribution of hydrates in calcium oxalate precipitation. *Journal of Crystal Growth*, **97**, 460–468.

Brečević, Lj., Skrtic, D. and Garside, J., 1986. Transformation of calcium oxalate hydrates. *Journal of Crystal Growth*, **74**, 399–408.

Breiman, L., Friedman, J.H., Olshen, R.A. and Stone, C.J., 1984. *Classification and Regression Trees*. Wadsworth and Brooks.

Bristol, E.H., 1966. On a new measure of interactions for multivariable process control, *IEEE Transactions of Automatic Control*, **AC-11**, 133.

Brodley, C.E. and Utgoff, P.E., 1995. Multivariate Decision Trees. *Machine Learning*, **19**, 45.

Broul, M., Nývlt, J. and Söhnel, O., 1981. *Solubility in inorganic two-component systems.* Prague: Academia.

Brown, C.M., Ackermann, D.K., Purich, D.L. and Finlayson, B., 1991. Nucleation of calcium oxalate monohydrate: use of turbidity measurements and computer-assisted simulations in characterising early events in crystal formation. *Journal of Crystal Growth*, **108**, 455–464.

Brown, D.J. and Boyson F., 1987. Modelling of fluid flow in a batch crystallizer. In *Industrial Crystallization '87*. Eds. J. Nývet, S. Záček, Bechyne, Czechoslovakia, September 1987. Academia Prague and Elsevier, 1989, pp. 547–550.

BS2955 Glossary of terms relating to powders.

Buckingham, E., 1914. On physically similar systems; illustrations of the use of dimensional equations. *Physics Review*, **4**, 335–376.

Budz, J., Jones, A.G. and Mullin, J.W., 1987a. On the size-shape dependence of potassium sulphate crystals. *Industrial and Engineering Chemistry Research*, **26**, 820–824.

Budz, J., Jones, A.G. and Mullin, J.W., 1987b. Agglomeration of potassium sulphate in an MSMPR crystallizer. In *Fundamental aspects of crystallization and precipitation processes, American Institute of Chemical Engineers. Symposium Series, No. 253, 83*, New York: American Institute of Chemical Engineers, pp. 78–84.

Bujac, P.D.B., 1976. Attrition and secondary nucleation in agitated crystal slurries. In *Industrial Crystallization*. Ed. J.W. Mullin, New York: Plenum Press, p. 23.

Bujac, P.B. and Mullin, J.W., 1969. A rapid method for the measurement of crystal growth rates in a fluidised bed crystallizer. *Symposium on Industrial Crystallization*. London, 1969. Rugby: Institution of Chemical Engineers, pp. 121–129.

Burke, S.P. and Plummer, W.B., 1928. Gas flow through packed columns. *Industrial and Engineering Chemistry*, **20**, 1196–1200.

Burton, W.K., Cabrera, N. and Frank, F.C., 1951. The growth of crystals and the equilibrium structure of their surfaces. *Philosophical Transactions*, **A243**, 299–358.

Cains, P.W., 1999. Chiral Crystallization. In SPS Crystallization Manual, Volume 5, Part 5.1 *Novel Crystallization Techniques*, Harwell, UK.

Camp, T.R. and Stein, P.C., 1943. Velocity gradients and internal work in fluid motion. *Journal of the Boston Society of Civil Engineers*, **30**, 219.

Capes, C.E. (ed.), 1961, 1977, 1977, 1981, 1985. *International symposia on agglomeration.* Iron and steel society inc. (Canada), Chelsea, MI, USA: Bookcrafters.

Carman, P.C., 1937. Fluid flow through granular beds. *Transactions of the Institution of Chemical Engineers*, **15**, 150–166.

Carman, P.C., 1956. *Flow of gases through porous media.* Oxford: Butterworth-Heinemann.

Cate, A. ten, Derksen, J.J., Kramer, H.J.M., van Rosmalen, G.M. and Van den Akker, H.E.A., 2001. The microscopic modelling of hydrodynamics in industrial crystallizers. *Chemical Engineering Science*, **56**, 2495–2509.

Chakraborty, D. and Bhatia, S.K., 1996a. Formation and aggregation of polymorphs in continuous precipitation. 1. Mathematical modelling. *Industrial and Engineering Chemistry Research*, **35**, 1985–1994.

Chakraborty, D. and Bhatia, S.K., 1996b. Formation and aggregation of polymorphs in continuous precipitation. 2. Kinetics of $CaCO_3$ precipitation. *Industrial and Engineering Chemistry Research*, **35**, 1995–2006.

Chang, C.-T. and Epstein, M.A.F., 1982. Identification of batch crystallization control strategies using characteristic curves. *American Institute of Chemical Engineers Symposium Series*, **78**(215), 68–75.

Chang, C.-T. and Epstein, M.A.F. 1987. Simulation studies of feedback control strategy for batch crystallizers. *American Institute of Chemical Engineers Symposium Series*, **83**(253), 110–119.

Chang, C.J. and Randolph, A.D., 1989. Precipitation of microsized organic particles from supercritical fluids. *American Institution of Chemical Engineers Journal*, **35**, 1876–1882.

Chang, L.-J., Mehta, R.V. and Tarbell, J.M., 1986. An evaluation of models of mixing and chemical reaction with a turbulence analogy. *Chemical Engineering Communications*, **42**, 139–155.

Chang, W.-C. and Ng, K.M., 1998. Synthesis of processing system around a crystallizer. *American Institute of Chemical Engineers Journal*, **44**, 2240.

Chen, J., Zheng, C. and Chen, G., 1996. Interaction of macro- and micromixing on particle size distribution in reactive precipitation. *Chemical Engineering Science*, **51**, 1957–1966.

Cheremisinoff, N.P., 1998. *Liquid filtration*. Oxford: Butterworth-Heinemann.

Chianese, A., Di Berardino, F. and Jones, A.G., 1993. On the effect of crystal breakage on the fine crystal distribution from a seeded batch crystallizer. *Chemical Engineering Science*, **48**, 551–560.

Chianese, A., Mazzarotta, B., Huber, S. and Jones, A.G., 1993. On the Effect of Secondary Nucleation on the Size Distribution of Potassium Sulphate Fine Crystals from Seeded Batch Crystallization. *Chemical Engineering Science*, **48**, 551–560.

Choplin, L. and Villermaux, J., 1994. Viscous mixing in polymer reactors. *American Institute of Chemical Engineers Symposium Series*, **299**, 123–129.

Chung, S.H., Ma, D.L. and Braatz, R.D., 1999. Optimal seeding in batch crystallization. *Canadian Journal of Chemical Engineering*, **77**, 590–596.

Cisternas, L.A., 1999. Optimal design of crystallization-based separation systems. *American Institute of Chemical Engineers Journal*, **45**, 1477–1487.

Cisternas, L.A. and Rudd, D.F., 1993. Process designs for fractional crystallization from solution. *Industrial and Engineering Chemistry Research*, **32**, 1993.

Clark, N.N., 1986. Three techniques for implementing digital fractal analysis of particle shape. *Powder Technology*, **46**, 45–52.

Clark, W.E., 1967. Fluid bed drying. *Chemical Engineering Albany*, **74**(6), 177.

Coe, H.S. and Clevenger, G.H., 1916. Methods for determining the capacities of slime thickening tanks. *Transactions of the American Institute of Mining Engineers*, **55**, 356–384.

Collier, A.P. and Hounslow, M.J., 1999. Growth and aggregation rates for calcite and calcium oxalate monohydrate. *American Institute of Chemical Engineers Journal*, **45**, 2298–2305.

Conti, R. and Nienow, A.W., 1980. Particle Abrasion at High Solids Concentration in Stirred Vessels – II. *Chemical Engineering Science*, **35**, 543–547.

Coombes, D.S., Nagi, G.K. and Price, S.L., 1997. *Chemical Physics Letters*, **265**, 532–537.

Coombes, D.S., Price, S.L., Willock, D.J. and Leslie, M., 1996. *Journal of Physical Chemistry A*, **100**, 7352–7360.

Coulson, J.M. and Richardson, J.F., 1991. *Chemical Engineering. Volume 2: Particle Technology and Separation Processes*, 4th edition. Butterworth-Heinemann.

Cumberland, D.J. and Crawford, R.J., 1987. *The packing of particles*. Amsterdam: Elsevier.

Curie, P., 1885. *Bull. Soc. Franc. Mineral.*, **34**, 145.

Curl, R.L., 1963. Dispersed phase mixing-theory and effects in simple reactors. *American Institution of Chemical Engineers Journal*, **9**, 175.

Cussler, E.L. and Moggridge, G.D., 2001. *Chemical product design*. Cambridge University Press.

Cuthrell, J.E. and Biegler, L.T., 1987. On the optimization of Differential-Algebraic process systems. *American Institution of Chemical Engineers Journal*, **33**(8), 1257.

Da Vinci, Leonardo, ca. 1500 A.D. *Notebooks*.
Davey, R. J. and Rutti, A., 1976. Agglomeration in the crystallization of hexamethylene from aqueous solution. *Journal of Crystal Growth*, **32**, 221–226.
Dallavalle, J.M., 1948. *Micromeritics*, 2nd edition. Pitman.
Damen, A.A.H., Hajdasinski, A.K., 1982. In *Proc. 6th IFAC Symp. Identification and System Parameter Estimation*, Washington, D.C., 903.
Danckwerts, P.V., 1958. The effect of incomplete mixing on homogeneous reactions. *Chemical Engineering Science*, **8**, 93–99.
Danckwerts, P.V., 1970. *Gas–liquid reactions*. New York: McGraw-Hill.
Darcy, H., 1856. *Les fontaines publiques de la ville de Dijon*. Paris: Victor Dalmont.
Daudey, P.J. and de Jong, E.J., 1984. The dynamic behaviour of NaCl crystallization in a 91 L crystallizer. In *Industrial Crystallization 84*. Eds. S.J. Jančić and E.J. de Jong, Amsterdam: Elsevier, pp. 191–194.
Davey, R.J. and Garside, J., 2000. *From molecules to crystallizers*. Oxford: Oxford University Press.
David, R. and Marcant, B., 1994. Prediction of micromixing effects in precipitation: Case of double-jet precipitators. *American Institution of Chemical Engineers Journal*, **40**, 424–432.
Davies, C.W., 1962. *Ion Association*. London: Butterworths.
Davies, O.L., 1979. *The Design and Analysis of Industrial Experiments*, 2nd edition. London: Longman Group Limited.
Davis, R.H. and Gecol, H., 1994. Hindered settling function with no empirical parameters for disperse suspensions. *American Institution of Chemical Engineers Journal*, **40**, 570–575.
Davis, R.F., Palmour, H. and Porter, R.L., 1984. *Emergent processes for high temperature ceramics*. New York: Plenum Press.
Deckwer, W.D., 1992. *Bubble column reactors*. John Wiley and Sons.
Denk, E.G. and Botsaris, G.D., 1972. *Journal of Crystal Growth*, **13/14**, 493.
Derjaguin, B.V. and Landau, L., 1941. The stability of strongly charged lyophobic sols and the adhesion of strongly charged particles in solutions of electrolytes. *Acta Physicochim, URSS*, **14**, 633–662.
Dickinson, E., 1986. Brownian motion and aggregation: from hard spheres to proteins. *Chemistry and Industry*, 158–163.
Di Felice, R., 1994. The voidage function for fluid-particle interaction systems. *International Journal of Multiphase Flow*, **20**, 153–159.
Di Felice, R., 1995. Hydrodynamics of liquid fluidisation. *Chemical Engineering Science*, **50**, 1213–1245.
Dickey, D.S., 1993. Dimensional analysis, similarity and scale-up. *American Institute of Chemical Engineers Symposium Series*, **293**, 143–150.
Dietz, P.W., 1981. Collection efficiency of cyclone separators. *American Institution of Chemical Engineers Journal*, **27**(6), 888.
Docherty, R., Roberts, K.J. and Dowty, E., 1988. MORANG – a computer programme designed to aid in the determination of crystal morphology. *Computer Physics Communications*, **51**, 423–430.
Docherty, R., Clydesdale, G., Roberts, K.J. and Bennema, P., 1991. Application of the Bravais-Freidel-Donnay-Harker attachment energy and Ising models to predicting and understanding the morphology of molecular crystals. *Journal of Physics D: Applied Physics*, **24**, 89–99.
Doki, N., Kubota, N., Sato, A., Yokota, M., Hamada, O. and Masumi, F., 1999. Scaleup experiments on seeded batch cooling crystallization of potassium alum. *American Institution of Chemical Engineers Journal*, **45**(12), 2527–2533.

Donnay, J.D.H. and Harker, D. 1937. *Am. Mineral*, **22**, 463.

Donnet, M., Jongen, N., Lemaître, J. and Bowen, P., 2000. New morphology of calcium oxalate trihydrate precipitated in a segmented flow tubular reactor. *Journal of Materials Science Letters*, **19**, 749–750.

Donnet, M., Jongen, N., Lemaître, J., Bowen, P. and Hofmann, H., 1999. Better control of nucleation and phase purity using a new segmented flow tubular reactor: Model system: Precipitation of calcium oxalate. In *14th International Symposium on Industrial Crystallization*. Cambridge, U.K., September 12–16, Institution of Chemical Engineers, CD ROM, pp. 1–13.

Doucet, J. and Giddey, C., 1966. Automatic control of sucrose crystallization. *International Sugar Journal*, **68**, 131–136.

Douglas, J.M., 1988. *Conceptual design of chemical processes*. New York: McGraw Hill.

Drake, R.L., 1972. A general mathematical survey of the coagulation equation. In *Topics in current aerosol research*, Pt. 2. Eds. G.M. Tidy and J.R. Brocks, New York: Pergamon.

Dunuwila, D. and Berglund, K.A. 1997. ATR FTIR spectroscopy for in situ measurement of supersaturation. *Journal of Crystal Growth*, **179**, 185–193.

Duncan, A.G. and Phillips, V.R., 1979. The dependence of heat exchanger fouling on solution undercooling. *Journal of separation process technology*, **1**, 29–35.

Dye, S.R. and Ng, K.M., 1995. Bypassing eutectics with extractive crystallization: design alternatives and tradeoffs. *American Institute of Chemical Engineers Journal*, **41**, 1456.

Eaton, J.W. and Rawlings, J.B., 1990. Feedback control of chemical processes using on-line optimisation techniques. *Computers and chemical engineering*, **14**, 469.

Ejaz, T. 1997. *Hydrothermal precipitation of zeolite A crystals*. PhD Thesis, University of London, UK.

Englezos, P., Kalogerakis, N., Dholababhai, P.D. and Bishnoi, P.R., 1987a. Kinetics of formation of methane and ethane gas hydrates. *Chemical Engineering Science*, **42**(11), 2647–2658.

Englezos, P., Kalogerakis, N., Dholababhai, P.D. and Bishnoi, P.R., 1987b. Kinetics of gas hydrate formation from mixtures of methane and ethane. *Chemical Engineering Science*, **42**(11), 2659–2666.

Ergun, S., 1952. Fluid flow through packed colomns. *Chemical Engineering Progress*, **48**, 89.

Erlrebacher, G., Hussaini, M.Y., Jameson, L.M., 1996. *Wavelets: theory and applications*. Oxford University Press.

Evans, L.B., 1989. Simulation with respect to solid-fluid systems. *Computers and Chemical Engineering*, **13**, 343.

Evans, T.W., Margolis, G. and Sarofim, A.F., 1974. *American Institute of Chemical Engineers Journal*, **20**, 959.

Falope, G.O., Jones, A.G. and Zauner, R., 2001. On modelling continuous agglomerative crystal precipitation via Monte Carlo simulation. *Chemical Engineering Science*, **56**, 2567–2574.

Falk, L. and Schaer, E., 2001. A PDF modelling of precipitation reactors. *Chemical Engineering Science*, **56**, 2445–2458.

Family, F. and Landau, D.P. (eds.), 1984. *Kinetics of aggregation and gelation*. Amsterdam: Elsevier.

Fan, L.S., 1989. *Gas-liquid-solid fluidization engineering*, Oxford: Butterworth-Heinemann.

Farrell, R.J. and Yen-Cheng Tsai, 1994. Nonlinear controller for batch crystallization: Development and experimental demonstration. In *American Institute of Chemical Engineers National meeting*. Atlanta, Paper 89e.

Farrell, R.J. and Tsai, Y.-C., 1999. Nonlinear controller for batch crystallization: Development and experimental demonstration. *American Institute of Chemical Engineers Journal*, **41**, 2318–2321.

Fayyad, U.M. and Irani, K.B., 1992. On the handling of Continuous-Valued Attributes in Decision Tree Generation. *Machine Learning*, **8**, 87.

Fisher, R.A., 1990. *Statistical Methods, Experimental Design and Scientific Inference*. Oxford University Press.

Font, R. and Laveda, M.L., 1996. Design method of continuous thickeners from semi-batch tests of sedimentation. *Chemical Engineering Science*, **51**, 5007–5015.

Foscolo, P.U., Gibilaro, L.G. and Waldram, S.P., 1983. A unified model for particulate expansion of fluidised beds and flow in porous media. *Chemical Engineering Science*, **38**, 1251–1260.

Fournier, M.-C., Falk, L. and Villermaux, J., 1996. A new parallel competing reaction system for assessing micromixing efficiency – experimental approach. *Chemical Engineering Science*, **51**, 5053–5064.

Franck, R., David, R., Villermaux, J. and Klein, J.P., 1988. Crystallization and precipitation engineering – II. A chemical reaction engineering approach to salicylic acid precipitation: Modelling of batch kinetics and application to continuous operation. *Chemical Engineering Science*, **43**, 69–77.

Frank, F.C., 1949. The influence of dislocations on crystal growth. *Discussions of the Faraday Society*. **5**, 48–54.

Franke, J. and Mersmann, A., 1995. The influence of the operational conditions on the precipitation process. *Chemical Engineering Science*, **50**, 1737–1753.

Friedel, G., 1907. *Bull. Soc. Franc. Mineral*, **30**, 326.

Furth, R. (ed.), 1956. *Albert Einstein, investigations on the theory of Brownian movement*. New York: Dover.

Gahn, C. and Mersman, A., 1999a. Brittle fracture in crystallization processes Part A. Attrition and abrasion of brittle solids. *Chemical Engineering Science*, **54**, 1273–1282.

Gahn, C. and Mersman, A., 1999b. Brittle fracture in crystallization processes Part B. Growth of fragments and scale-up of suspension crystallizers. *Ibid*. pp. 1283–1292.

Gahn, C. and Mersmann, A., 1999c. Brittle fracture in crystallization processes, pp. 1273–1292.

Galilei, Galileo, 1638, German in Ostwalds Klassiker, *Discorsi*. **11**, 106–109.

Gardner, G.L. and Nancollas, G.H., 1975. Kinetics of dissolution of calcium oxalate monohydrate. *Journal of Physical Chemistry*, **79**, 2597–2600.

Garside, J. and Davey, R.J., 1980. Secondary contact nucleation: Kinetics, growth and scale-up. *Chemical Engineering Communications*, **4**, 393.

Garside, J. and Jančić, S.J., 1976. Growth and dissolution of potash alum crystals in the sub-sieve size range. *American Institution of Chemical Engineers Journal*, **22**, 887.

Garside, J. and Jančić, S.J., 1978. Prediction and measurement of crystal size distribution for size-dependent growth. *Chemical Engineering Science*, 4331.

Garside, J. and Jančić, S.J., 1979. Measurement and scale-up of secondary nucleation kinetics for the potash alum-water system. *American Institute of Chemical Engineers Journal*, **25**, 948.

Garside, J. and Larson, M.A., 1978. Direct observation of secondary nuclei production. *Journal of Crystal Growth*, **43**, 694.

Garside, J. and Mullin, J.W., 1966. Continuous measurement of solution concentration in a crystallizer. *Chemistry and Industry*, November 26, pp. 2007–2008.

Garside, J. and Shah, M.B., 1980. Crystallization kinetics from MSMPR crystallizers. *Industrial and Engineering Chemistry Process Design and Development*, **19**, 509–514.

Garside, J. and Tavare, N.S., 1985. Mixing, reaction and precipitation: limits of micromixing in an MSMPR crystallizer. *Chemical Engineering Science*, **40**, 1485–1493.

Garside, J., Rusli, I.T. and Larson, M.A., 1979. *American Institute of Chemical Engineers Journal*, **25**, 57.

Garside, J., Brecevic, Lj. and Mullin, J.W., 1982. The effect of temperature on the precipitation of calcium oxalate. *Journal of Crystal Growth*, **57**, 233–240.

Garside, J., Davey, R.J. and Jones, A.G. (eds.), 1991. *Advances in Industrial Crystallization*. Oxford: Butterworth-Heinemann, ix + 244pp.

Garside, J., Gibilaro, L.G. and Tavare, N.S., 1982. Evaluation of crystal growth kinetics from a desupersaturation curve using initial derivatives. *Chemical Engineering Science*, **37**, 1625–1628.

Geisler, R., Mersmann, A. and Voit, H., 1991. Macro- and micromixing in stirred tanks. *International Chemical Engineering*, **31**, 642–653.

Gelbard, F. and Seinfeld, J.H., 1978. Numerical solution of the dynamic equation for particulate systems. *Journal of Computational Physics*, **28**, 357.

Gelbard, F., Tambour, Y. and Seinfeld, J.H., 1980. Sectional representations for simulating aerosol dynamics. *Journal of Colloid and Interfacial Science*, **76**, 541.

Gertlauer, A., Mitrovic, A., Motz, S. and Gilles, E.-D., 2001. A population balance model for crystallization processes using two independent particles properties. *Chemical Engineering Science*, **56**(7), 2553–2565.

Gibbs, J.W., 1948. *Collected works*, Vol. 1. New York: Longmans Green.

Gibilaro, L.G., 2001. *Fluidisation Dynamics*. Oxford: Butterworth-Heinemann.

Giorgio, S. and Kern, R., 1983. Filtrability, crystal morphology and texture; paraffins and de-waxing aids. *Journal of Crystal Growth*, **62**, 360–374.

Girolami, M.W. and Rousseau, R.W., 1985. Initial breeding in seeded batch crystallizers. *Industrial and Engineering Chemistry Research*, **25**, 66–70.

Glasgow, L.A. and Luecke, R.H., 1980. *Industrial and Engineering Chemistry Fundamentals*, **19**(2), 148–156.

Glasser, D., Crowe, C.M. and Hildebrandt, D., 1987. A geometric approach to steady flow reactors: The attainable region and optimization in concentration space. *Industrial and Engineering Chemistry Research*, **26**, 1803.

Gooch, J.R. and Hounslow, M.J., 1996. Monte Carlo simulation of size-enlargement mechanisms in crystallization. *American Institute of Chemical Engineers Journal*, **42**(7), 1864–1874.

Gottlieb, D. and Orszag, S.A., 1977. *Numerical Analysis of Spectral Methods: Theory and Applications*. SIAM, Philadelphia.

Grace, H.P., 1953. Resistance and compressibility of filter cakes. Part III: Under conditions of centrifugal filtration. *Chemical Engineering Progress*, **49**, 427–436.

Gray, W.A., 1968. *The packing of solid particles*. London: Chapman and Hall.

Green, D.A., Kontomaris, K., Grenville, R.K., Etchells, A.W., Kendall, R.E. and Jacobs, G., 1996. Suspension dynamics and mixing in industrial crystallizers. In *Industrial Crystallization 1996*, Ed. B. Biscans, Toulouse; Progep, 16–19 September 1996, pp. 525–530.

Gregory, J. (ed.), 1984a. *Solid–liquid separation*. Chichester: Society of Chemical Industry/Ellis Horwood.

Gregory, J., 1984b. Flocculation and filtration of colloidal particles. In *Emergent process methods for high temperature ceramics*, Eds. R.F. Dvais *et al.*, Plenum Press, London; p. 59.

Griffith, A.A., 1920. The phenomena of rupture and flow in solids. *Phil. Trans. Roy. Soc. Lond.*, **A221**, 163.

Griffith, A.A., 1924. The theory of rupture. In *Proc. First Internat. Congr. Appl. Mech.*, Eds. C.B. Biezeno and J.M. Burgers, J. Waltman Jnr., Delft, p. 55.

Griffiths, H., 1925. Mechanical crystallization. *J. Soc. Chem. Ind.*, **44**, T7–18.

Groen, H., Hammond, R.B., Lai, X., Mougin, P., Roberts, K.J., Savelli, N., Thomas, A., White, G., Williams, H.L., Wilkinson, D., Baker, M., Dale, D., Erk, P., Latham, D., Merrifield, D., Oliver, R., Roberts, D., Wood, W. and Ford, L.J., 1999. A new on-line batch processing facility for examining the crystallization of organic speciality chemical products. In *Pilot plants and scale up*. Ed. W. Hoyle, Royal Society of Chemistry, London, **236**, 40–62.

Grootscholten, P.A.M., de Leer, B.G.M., de Jong, E.J. and Asselbergs, C.J., 1982. *American Institute of Chemical Engineers Journal*, **28**, 728.

Gruhn, G., Rosenkranz, J., Werther, J. and Toebermann, J.C., 1997. Development of an object-oriented simulation system for complex solids processes. *Computers and chemical engineering*, **21**, S187.

Guichardon, P., Falk, L., Fournier, M.C. and Villermaux, J., 1994. Study of micromixing in a liquid-solid suspension in a stirred reactor. *American Institute of Chemical Engineers Symposium Series*, **299**, 123–130.

Gupta, G. and Tim, D.C., 1971. *Chemical Engineering Progress Symposium Series*, **67**(110), 121.

Gutwald, T. and Mersmann, A., 1990. Determination of crystallization kinetics from batch experiments. In *Industrial crystallization 90*. Garmisch-Partenkirchen, September 1990. Ed. A. Mersmann, Düsseldorf: GVC-VDI, p. 331.

Hall, D.G., 1972. Thermodynamic treatment of some factors affecting the interaction between colloidal particles. *Journal of the Chemical Society Faraday Transactions*, **68**(2), 2169–2182.

Han, C.D., 1969. A control study of isothermal mixed crystallizers. *Industrial and Engineering Chemistry Process Design and Development*, **8**, 150–158.

Harada, M., Arima, K., Eguchi, W. and Nagata, S., 1962. Micromixing in a continuous flow reactor. *Memoir of the Faculty of Engineering*, Kyoto University, Japan, **24**, 431.

Harnby, N., Edwards, M.F. and Nienow, A.W. (eds.), 1992. *Mixing in the process industries*. 2nd edition. Oxford: Butterworth-Heinemann.

Hart, M. and Hart, R., 1989. *Quantitative Methods for quality and productivity Improvements*, Quality Press.

Hartel, R.W. and Randolph, A.D., 1986. Mechanisms and kinetic modelling of calcium oxalate crystals in urine-like liquor: Part II kinetic modelling. *American Institution of Chemical Engineers Journal*, **32**, 1186–1195.

Hartel, R.W., Gottung, B.E., Randolph, A.D. and Drach, G.W., 1986. Mechanisms and kinetic modelling of calcium oxalate crystal aggregation in urine-like liquor: Part I mechanisms. *American Institution of Chemical Engineers Journal*, **32**, 1176–1185.

Hartman, P. and Perdok, W.G., 1955. On the relations between structure and morphology of crystals. *Acta Crystallogr.*, **8**, 49.

Heffels, S.K., de Jong, E.J. and Nienoord, M., 1994. Improved operation and control of batch crystallizers. In *Particle design via crystallization, American Institute of Chemical Engineers Symposium Series*, **87**(284), 170–181.

Helt, J.E. and Larson, M.A., 1977. Effects of temperature on the crystallization of potassium nitrate by direct measurement of super-saturation. *American Institution of Chemical Engineers Journal*, **23**(6), 822.

Hess, W. and Schonert, K., 1981. Brittle-Plastic Transition in Small Particles, *In 1981 Powtech Conference on Particle Technology*, Birmingham, EFCE Event No. 241, Rugby: Institution of Chemical Engineers, pp. D2/I/1–D2/I/9.

Heuer, M. and Leschonski, K., 1985. Results obtained with a new instrument for the measurement of particle size distributions from diffraction patterns. *Particle Characterisation*, **2**, 7–15.

Higbie, R., 1935. The rate of absorption of a pure gas into a still liquid during a short time of exposure. *Transactions of American Institute of Chemical Engineers*, **31**, 365–389.

Hikita, H. and Ishikawa, H., 1969. Physical absorption in agitated vessels with a flat gas–liquid interface. *Bulletin of the UniversityOsaka Prefect*, **A18**, 427–437.

Hildebrandt, D. and Biegler, L.T., 1994. Synthesis of chemical reactor networks. In *Foundations of Computer aided process design (FOCAPD '94)*, Eds. L.T. Biegler and M.F. Doherty, Snowmass CO, p. 52.

Hill, P.J. and Ng, K.M., 1995. New discretization procedure for the breakage equation. *American Institution of Chemical Engineers Journal*, **41**(5), 1204–1217.

Hill, P.J. and Ng, K.M., 1996a. New discretization procedure for the agglomeration equation. *American Institution of Chemical Engineers Journal*, **42**(3), 727.

Hill, P.J. and Ng, K.M., 1996b. Statistics of multiple particle breakage. *American Institution of Chemical Engineers Journal*, **42**, 1600–1608.

Hill, P.J. and Ng, K.M., 1997. Simulation of solids processes accounting for particle size distribution. *American Institute of Chemical Engineers Journal*, **43**, 715.

Holland, F.A. and Chapman, F.S., 1966. *Liquid mixing and processing in stirred tanks*. New York: Reinhold.

Hollander, E.D., Dirksen, J.J., Bruinsma, O.S.L., van den Akker, H.E.A. and van Rosmalen, G.M. A numerical study on the coupling of hydrodynamics and orthokinetic aggregation. *Chemical Engineering Science*, **56**, 2531–2542.

Hostomský, J., 1987. Particle size distribution of agglomerated crystal product from a continuous crystallizer. *Collection of Czechoslovakian Chemical Communications*, **52**, 1186–1197.

Hostomský, J. and Jones, A.G., 1991. Calcium carbonate crystallization kinetics, agglomeration and form during continuous precipitation from solution. *Journal of Physics D: Applied Physics*, **24**, 165–170.

Hostomský, J. and Jones, A.G., 1993a. Modelling and analysis of agglomeration during precipitation from solution. In *Industrial Crystallization 93*. Ed. Z. Rojkowski, University of Warsaw, 1993, pp. 2037–2041.

Hostomský, J. and Jones, A.G., 1993b. *Ibid.*, Crystallization and agglomeration kinetics of calcium carbonate and barium sulphate in the MSMPR crystallizer. *Indem.* pp. 2049–2054.

Hostomský, J. and Jones, A.G., 1993c. *Ibid.*, Modelling of calcium carbonate precipitation in the reaction between gaseous carbon dioxide and aqueous solution of calcium hydroxide. *Indem.* pp. 2055–2059.

Hostomský, J. and Jones, A.G., 1995. A penetration model of the gas-liquid reactive precipitation of calcium carbonate crystals. *Transactions of the Institution of Chemical Engineers*, **73A**, 241–245.

Houcine, I., Plasari, E., David, R. and Villermaux, J., 1997. Influence of mixing characteristics on the quality and size of precipitated calcium oxalate in a pilot scale reactor. *Transactions of the Institution of Chemical Engineers*, **75**, 252–256.

Hounslow, M.J., 1989. Solving the population balance for agglomerating systems. In *5th international symposium on agglomeration*. Brighton, September 1989, Institution of Chemical Engineers, Rugby, pp. 585–598.

Hounslow, M.J., 1990a. A discretized population balance for continuous systems at steady state. *American Institution of Chemical Engineers Journal*, **36**, 106–116.

Hounslow, M.J., 1990b. Nucleation, growth and aggregation rates from steady-state experimental data. *American Institution of Chemical Engineers Journal*, **36**, 1748–1753.

Hounslow, M.J. and Wynn, E.J.W., 1992. Modelling particulate processes: Full solutions and short-cut. *Computers and Chemical Engineering*, **16**, S411–S420.
Hounslow, M.J. and Wynn, E.J.W., 1993. Short-cut models for particulate processes. *Computers and Chemical Engineering*, **17**, 505–516.
Hounslow, M.J., Ryall, R.L. and Marshall, V.R., 1988. A discretized population balance for nucleation, growth and aggregation. *American Institution of Chemical Engineers Journal*, **34**, 1821–1832.
Hounslow, M.J., Mumtaz, H.S., Collier, A.P., Barrick, J.P. and Bramley, A.S., 2001. A micro-mechanical model for the rate of aggregation during precipitation from solution. *Chemical Engineering Science*, **56**, 2543–2552.
Hughmark, G.A., 1969. Mass transfer foe suspended solid particles in agitated liquids. *Chemical Engineering Science*, **24**, 287–291.
Hulburt, H.M. and Katz, S., 1964. Some problems in particle technology – a statistical mechanical formulation. *Chemical Engineering Science*, **19**, 555–574.
Hull, D. and Bacon, D.J., 2001. *Introduction to Dislocations*, 4th edition. Oxford: Butterworth-Heinemann.
Hurley, M.A., Jones, A.G. and Drummond, J.N., 1995. Crystallization kinetics of cyanazine precipitated from aqueous ethanol solutions. *Chemical Engineering Research and Design*, **73B**, 52–57.
Institution of Chemical Engineers, 1975. In *Comminution, Institution of Chemical Engineers Working Party concerned with the theory and practice of the size reduction of solid materials*. Ed. V.C. Marshall, Institution of Chemical Engineers, Rugby, 83pp.
IUPAC, 1980–1991. *Solubility data series*. Ed. J.W. Lorimer, **1–48**, Oxford: Pergamon Press.
Jacques, J., Collet, A. and Wilen, S.H., 1981. *Enantiomers, Racemates and Resolutions*. New York: John Wiley & Sons, Reprinted, 1991.
Jagadesh, D., Chivate, M.R. and Tavare, N.S., 1992. Batch crystallization of Potassium chloride by an ammoniation process. *Industrial and Engineering Chemistry Research*, **31**, 561–568.
Jagannathan, R., Sung, C.V., Youngquist, G.R. and Estrin, J., 1980. *American Institute of Chemical Engineers Symposium Series No. 193*, **76**, 90.
Jager, J.H.J., deWolf, S., Kramer, H.J.M. and deJong, E.J., 1991. Estimation off nucleation kinetics from crystal size distribution transients of a continuous crystallizer. *Chemical Engineering Science*, **46**, 807–818.
Jager, J.H.J., Kramer, M., deJong, E.J., deWolf, S., Bosgra, O.H., Boxman, A., Merkus H.G. and Scarlett, B., 1992. Control of industrial crystallizers. *Powder Technology*, **69**, 11–20.
Jančić, S.B. and Grootscholten, P.A.M., 1984. *Industrial crystallization*. Delft: D. Reidel Publishing.
Janse, A.H. and de Jong, E.J., 1978. On the width of the metastable zone. *Transactions of the Institution of Chemical Engineers*, **56**, 187–193.
Jazaszek, P. and Larson, M.A., 1977. Influence of fines dissolving on crystal size distribution in an MSMPR crystallizer. *American Institution of Chemical Engineers Journal*, **23**, 460–468.
Jézéquel, P.-H., 2001. The concept of 'scaleable reactor in the precipitation of silver halide photographic microcrystals. *Chemical Engineering Science*, **56**(7), 2399–2408.
Johnstone, R.E. and Thring, M.W., 1957. *Pilot Plants, Models and Scale-up Methods in Chemical Engineering*. New York: McGraw Hill.
Jones, A.G., 1972. Programmed cooling crystallization. Ph.D. Thesis, University of London.

Jones, A.G., 1974. Optimal operation of a batch cooling crystallizer. *Chemical Engineering Science*, **29**, 1075–1087.
Jones, A.G., 1984. The Design of Well-Mixed Batch Crystallizers. SPS DR17. (Harwell/ Warren Spring: Separation Processes Service), 40pp.
Jones, A.G., 1985a. Liquid circulation in a draft-tube bubble column. *Chemical Engineering Science*, **40**, 449–462.
Jones, A.G., 1985b. Crystallization and downstream processing interactions. In *POW-TECH 85, Institution of Chemical Engineers. Sympos. Ser. No. 91*. Rugby: Institution of Chemical Engineers, pp. 1–11.
Jones, A.G., 1988. Agglomeration during crystallization and precipitation. In *Crystallization Manual*: (Harwell: Separation Processes Service, AEA Technology), Volume CR II, Part 2, 60pp.
Jones, A.G., 1991. Design and performance of crystallization systems. In *Advances in Industrial Crystallization*. Eds. J. Garside, R.J. Davey and A.G. Jones, Oxford: Butterworth-Heinemann, pp. 213–228.
Jones, A.G., 1994. Particle formation during agglomerative precipitation processes. In *Controlled particle, bubble and droplet formation*. Ed. D.J. Wedlock, Oxford: Butterworth-Heinemann.
Jones, A.G., 1997. Attrition and Breakage. In *Crystallization Manual*: (Harwell: Separation Processes Service, AEA Technology), Volume VI, Part III, 100 pp.
Jones, A.G., Budz, J. and Mullin, J.W., 1987. Batch crystallization and solid–liquid separation of potassium sulphate. *Chemical Engineering Science*, **42**, 619–629.
Jones, A.G., Chianese, A., 1988. Fines destruction during batch crystallization. *Chemical Engineering Communications*, **62**, 5–16.
Jones, A.G. and Mullin, J.W., 1973. Crystallization kinetics of potassium sulphate in a draft-tube agitated vessel. *Transactions of the Institution of Chemical Engineers*, **51**, 362–368.
Jones, A.G. and Mullin, J.W., 1974. Programmed cooling crystallization of potassium sulphate solutions. *Chemical Engineering Science*, **29**, 105–118.
Jones, A.G. and Mydlarz, J., 1989. Crystallization and subsequent solid-liquid separation of potassium sulphate. Part I: MSMPR kinetics. *Chemical Engineering Research and Design*, **67**, 283–293.
Jones, A.G. and Mydlarz, J., 1990a. Continuous crystallization of potash alum. MSMPR kinetics. *Canadian Journal of Chemical Engineering*, **68**, 250–259.
Jones, A.G. and Mydlarz, J., 1990b. Slurry filtrability of potash alum crystals. *Canadian Journal of Chemical Engineering*, **68**, 513–518.
Jones, A.G. and Teodossiev, N.M., 1988. Microcomputer programming of dosage rate during batch precipitation. *Crystal Research and Technology*, **23**, 957–966.
Jones, A.G., Akers, S.R.G. and Budz, J., 1986. Microcomputer programming of temperature in a batch cooling crystallizer. *Crystal Research and Technology*, **21**, 1383–1390.
Jones, A.G., Budz, J. and Mullin, J.W., 1986. Crystallization kinetics of potassium sulphate in an MSMPR agitated vessel. *American Institution of Chemical Engineers Journal*, **32**, 2002.
Jones, A.G., Chianese, A. and Mullin, J.W., 1984. Effect of fines destruction on batch cooling crystallization of potassium sulphate solutions. In *Industrial Crystallization 84*. Eds. S.J. Jančić and E.J. de Jong, Amsterdam: Elsevier, pp. 191–194.
Jones, A.G., Ejaz, T. and Graham, P., 1999. Direct dynamic observation of phase transformations during zeolite crystal synthesis. In *Industrial Crystallization '99*. (Rugby: Institution of Chemical Engineers). Cambridge, 13–15 September 1999. Paper 95, p. 98.

Jones, A.G., Ewing, C. and Melvin, M.V., 1981. Biotechnology of solar saltfields. *Hydrobiologia*, **82**, 391–406.

Jones, A.G., Hostomský, J. and Wachi, S., 1996. Modelling and analysis of particle formation during agglomerative crystal precipitation processes. *Chemical Engineering Communications*, **146**, 105–130.

Jones, A.G., Hostomský, J. and Zhou Li, 1992a. On the effect of liquid mixing rate on primary crystal size during the gas–liquid precipitation of calcium carbonate. *Chemical Engineering Science*, **47**, 3817–3824.

Jones, A.G., Wachi, S. and Delannoy, C.C., 1992b. Precipitation of calcium carbonate in a fluidised bed reactor. In *Fluidization VII*. Eds. O.E. Potter and D.J. Nicklin, New York: Engineering Foundation, pp. 407–414.

Jones, G.L., 1984. Simulating the effects of changing particle characteristics in solids processing. *Computers and Chemical Engineering*, **8**, 329.

Jongen, N., Lemaître, J., Bowen, P. and Hofmann, H., 1996. Oxalate precipitation using a new tubular plug flow reactor. In *Proc. 5th World Congress of Chemical Engineering*. San Diego (California), July 14–18 (New York: American Institute of Chemical Engineers), Vol. V, pp. 2109–2111.

Jongen, N., Lemaître, J., Bowen, P. and Hofmann, H., 1999. Aqueous synthesis of mixed yttrium-barium oxalates. *Chemistry of Materials*, **11**, pp. 712–718.

Juvekar, V.A. and Sharma, M.M., 1973. Absorption of CO_2 in a suspension of lime. *Chemical Engineering Science*, **28**, 825–837.

Kashiki, I. and Suzuki, A., 1986. Flocculation system as a particulate assemblage: A necessary condition for flocculants to be effective. *Industrial and Engineering Chemistry Fundamentals*, **25**, 444–449.

Kavanagh, J.P., 1992. Methods for the study of calcium oxalate crystallization and their application to urolithiasis research. *Scanning Microscopy*, **6**, 685–705.

Kaye, B.H., 1981. *Direct characterisation of fine particles*. New York: Wiley.

Kaye, B.H., 1986. The description of two-dimensional rugged boundaries in fine particle science by means of fractal dimensions. *Powder Technology*, **46**, 245–254.

Keey, R.B., 1972. *Drying principles and practice*. Oxford: Pergamon Press.

Keey, R.B., 1978. *Introduction to industrial drying operations*. Oxford: Pergamon Press.

Keey, R.B., 1991. *Drying of loose and particulate materials*. London and Washington, D.C.: Hemisphere Publishing Corporation.

Keey, R.B., 1992. *Drying of loose particulate materials*. New York: Hemisphere.

Khan, A.R and Richardson, J.F., 1990. Pressure gradient and friction factor for sedimentation and fluidisation of uniform spheres in liquids. *Chemical Engineering Science*, **45**, 255–265.

Kick, F., 1885. *Das Gesetz der Proportionalen Widerstande und Seine Anwendung*, Leipzig.

Kim, K.-J. and Mersmann, A., 2001. Estimation of metastable zone width in different nucleation processes. *Chemical Engineering Science*, **56**(7), 2315–2324.

Kim, W.-S. and Tarbell, J.M., 1996. Micromixing effects on barium sulphate precipitation in an MSMPR reactor. *Chemical Engineering Communications*, **146**, 33–56.

Kirk-Othmer, 1993. *Encyclopaedia of Chemical Technology*. Ed. J.I. Kroschwitz.

Klimpel, R.R. and Austin, L.G., 1984. The back-calculation of specific rates of breakage from continuous mill data. *Powder Technology*, **38**, 77–91.

Kokossis, A.C., Floudas, C.A., 1990. Optimization of complex reactor networks – I: Isothermal operation. *Chemical Engineering Science*, **45**(3), 595.

Kolař, V., 1958. *Collections of the Czechoslovak Communications*, **23**, 1680.

Kolař, V., 1959. *Collections of the Czechoslovak Communications*, **24**, 301.

Kolmogorov, A.N., 1941. *C.R. Academy of Science URSS*, **30**, 301.

Kossel, W., 1934. Zur energetic von Oberflächenvorgängen. *Annalen der Physik*, **21**, 457–480.
Kotaki, Y. and Tsuge, H., 1990. Reactive crystallization of calcium carbonate by gas–liquid and liquid–liquid reactions. *Canadian Journal of Chemical Engineering*, **68**, 435–442.
Kozeny, J., 1927. *Sitzungsber. Akad. Wiss. Wien, Math.-Naturwiss. Kl., Abt. 2A*, **136**, 271.
Kramer, H.J.M., O'Meadhra, R.S., Neumann, A.M. and van Rosmalen, G.M., 1996. Scale-up of ammonium sulphate crystallization in a DTB crystallizer. *Industrial Crystallization 1996*, Toulouse, pp. 619–625.
Krei, G.A. and Von Buschmann, E., 1998. der Laborsynthese zum Produktionsverfahren. *Spektrum der Wissenschaft, Spezial 6: Pharmaforschung*, pp. 38–47.
Kuboi, R., Nienow, A.W. and Conti, R., 1984. Mechanical Attrition of Crystals in Stirred Vessels. In *Industrial Crystallization '84*. Eds. S.J. Jančić and E.J. de Jong, Amsterdam: Elsevier Science Publishers B.V.
Kubota, N., Akazawa, K. and Shimizu, K., 1990. Kinetics of $BaCO_3$ precipitation in an MSMPR crystallizer. *Industrial Crystallization '90*. Garmisch-Partenkirchen, September 1990. Ed. A. Mersmann, Düsseldorf: GVC.VDI, pp. 199–204.
Kuipers, J.A.M. and van Swaaij, W.P.M., 1997. Application of computational fluid dynamics to chemical reaction engineering. *Reviews in Chemical Engineering*, **13**, 1–110.
Kumar, S. and Ramkrishna, D., 1996a. On the solution of population balance by discretization I. A fixed pivot technique. *Chemical Engineering Science*, **51**, 1311–1332.
Kumar, S. and Ramkrishna, D., 1996b. On the solution of population balance by discretization II. A moving pivot technique. *Chemical Engineering Science*, **51**, 1333–1342.
Kynch, G.J., 1952. *Transactions of the Faraday Society 48*, **166**.
Lahav, M. and Leiserowitz, L., 2001. The effect of solvent on crystal growth and morphology. *Chemical Engineering Science*, **56**(7), 2245–2254.
Lapple, C.E., 1950. Dust and Mist Collection, In: *Chemical Engineers Handbook*, 3rd Edn. New York: McGraw Hill, pp. 1013–1050.
Larson, M.A. and Wolff, P.R., 1971. Crystal size distributions from multistage crystallizers. *Chemical Engineering Progress Symposium Series*, **67**(110), 97.
Larson, M.A. and Garside, J., 1973. Crystallizer design techniques using the population balance. *Chemical Engineer*, London, June, p. 318.
Larson, M.A. and Klekar, S.A., 1973. In-situ measurement of supersaturation in crystallization from solution. Presented at *American Institute of Chemical Engineers 66th Annual Meeting*, Philadelphia, November.
Laufhütte, H. and Mersmann, A., 1987. Local Energy Dissipation in Agitated Turbulent Fluids and its Significance for the Design of Stirring Equipment. *Chemical Engineering Technology*, **10**, 56–65.
Lawn, B.R., 1983. *Fracture of Brittle Solids*, 2nd edition. Cambridge University Press.
Lawn, B.R. and Wilshaw, T.R., 1975. *Fracture of Brittle Solids*. Cambridge University Press, p. 8.
Lee, Y.-M. and Lee, L.J., 1987. Effect of mixing and reaction on a fast step growth polymerization. *International Polymer Processing*, **1**, 144–152.
Lee, P.L. and Sullivan, G.R., 1988. Generic model control. *Computers and Chemical Engineering*, **12**(6), 573–580.
Lei, S., Shinnar, R. and Katz, S., 1971a. The stability and dynamic behaviour of a continuous crystallizer with a fines trap. *American Institute of Chemical Engineers Journal*, **17**, 1459–1470.

Lei, S., Shinnar, R. and Katz, S., 1971b. The regulation of a continuous crystallizer with fines trap. *Chemical Engineering Progress Symposium Series*, **67**(110), 129–144.

Lemaître, J., Jongen, N., Vacassy R. and Bowen, P., 1996/97. Production of powders. Swiss Patent Application No. 1752/96, PCT application EP97/03817, (July15, 1997).

Leusen, F.J.J., Noordik, J.H. and Karfunkel, H.R., 1993. *Tetrahedron*, **49**(24), 5377.

Levenspiel, O., 1972. *Chemical Reaction Engineering*, 2nd edition. New York: John Wiley.

Levich, V.G., 1962. *Physicochemical Hydrodynamics*. New York: Prentice-Hall Inc.

Linhoff, B. and Hindmarsh, E., 1983. The pinch design method for heat exchanger networks. *Chemical Engineering Science*, **38**, 745.

Litster, J.D., Smit, J.D. and Hounslow, M.J., 1995. Adjustable discretized population balance for growth and aggregation. *American Institution of Chemical Engineers Journal*, **41**(3), 591–603.

Ljung, L., 1987. *System Identification: Theory for the user*. New York: Prentice-Hall.

Lommerse, J.P.M., Motherwell, W.D.S., Ammon, H.L., Dunitz, J.D., Gavezzotti, A., Hofmann, D.W.M., Leusen, F.J.J., Mooij, W.T.M., Price, S.L., Schweizer, B., Schmidt, M.U., van Eijck, B.P., Verwer, P. and Williams, D.E., 2000. A test of crystal structure prediction of small organic molecules. *Acta Crystallographica Section B-Structural Science*, **56**, 697.

Low, G.C., 1975. *Agglomeration effects in aluminium trihydroxide precipitation*. Ph.D. Dissertation, University of Queensland, Australia.

Low, G.C. and White, E.T., 1975. Agglomeration effects in alumina precipitation. In *Symposium on extractive metallurgy*. Melbourne, pp. III.5.1–10.

Luss, R. and Jakola, T.H., 1973. Optimisation by direct search and systematic reduction of the size of search region, *American Institute of Chemical Engineers Journal*, **19**, 760.

Lyklema, J., 1985. The colloidal background of agglomeration. In *4th International symposium on agglomeration*. Ed. C.E. Capes, Iron and Steel Co. Inc. American Institute of Mining Engineers, Bookcrafters.

Ma, D.L., Chung, S.H. and Braantz, R.D., 1999. Worst-case performance analysis of optimal batch control trajectories. *American Institution of Chemical Engineers Journal*, **45**(7), 1469–1476.

Mahajan, A.J. and Kirwan, D.J., 1996. Micromixing effects in a two-impinging-jets precipitator. *American Institution of Chemical Engineers Journal*, **42**, 1801–1814.

Mandlebroot, B.B., 1977. *Fractals; form, chance and dimension*. San Francisco: Freeman.

Mandlebroot, B.B., 1982. *The fractal geometry of nature*. San Francisco: Freeman.

Manninen, M. and Syrjänen, J., 1998. Modelling turbulent flow in stirred tanks. *CFX Update*, **16**, 10–11.

Mantovani, G., Vaccari, G., Accorsi, C.A., Aquilano, D. and Rubbo, M., 1983. Twin growth of sucrose crystals. *Journal of Crystal Growth*, **62**, 595–602.

Marcant, B., 1996. Prediction of mixing effects in precipitation from laser sheet visualisation. *Industrial Crystallization 1996*, Toulouse (Rugby: Institution of Chemical Engineers), pp. 531–538.

Marchal, P., David, R., Klein, J.P. and Villermaux, J., 1988. Crystallization and Precipitation engineering – I: An efficient method for solving population balance in crystallization with agglomeration. *Chemical Engineering Science*, **43**(1), 59.

Marmo, L., Manna, L., Chiampo, F., Sicardi, S. and Bersano, G., 1996. Influence of mixing on the particle size distribution of an organic precipitate. *Journal of Crystal Growth*, **166**, 1027–1034.

Masters, K., 1985. *Spray Drying*. London: George Godwin.

Mathews, H.B. and Rawlings, J.B., 1998. Batch crystallization of a photochemical: Modelling, control and filtration. *American Institution of Chemical Engineers Journal*, **44**(5), 1119–1127.

Mathews, H.B., Miller, S.J. and Rawlings, J.B., 1996. Model identification for crystallization: theory and experimental verification. *Powder Technology*, **88**, 227–235.
Matteson, M.J. and Orr, C. (eds.), 1987. *Filtration: principles and practice*, 2nd edition. New York: Marcel Dekker.
Mayrhofer, B. and Nývlt, J., 1988. Programmed cooling of batch crystallizers. *Chemical Engineering Processing*, **24**, 217–220.
Mazzarotta, B., 1992. Abrasion and Breakage Phenomena in Agitated Crystal Suspensions. *Chemical Engineering Science*, **47**, 3105–3111.
Mazzarotta, B., Di Cave, S. and Bonifazi, G., 1996. Influence of time on crystal attrition in a stirred vessel. *American Institution of Chemical Engineers Journal*, **42**, 3554–3558.
McCabe, W.L., 1929a,b. Crystal growth in aqueous solutions. *Industrial and Engineering Chemistry*, **21**, 30, 112.
McNeil, T.J., Weed, D.R. and Estrin, J., 1978. A note on modelling laboratory batch crystallizers. *American Institution of Chemical Engineers Journal*, **24**(4), 728–731.
Mehta, R.V. and Tarbell, J.M., 1983. A four environment model of mixing and chemical reaction. Part I – model development. *American Institution of Chemical Engineers Journal*, **29**, 320.
Meklenburgh, J.C. and Hartland, S., 1975. *Theory of backmixing*. Wiley-Interscience.
Mersmann, A. (ed.), 2001. *Crystallization Technology Handbook*, 2nd edition. New York: Dekker.
Mersmann, A. and Braun, B., 2001. Agglomeration. In *Crystallization Technology Handbook*. Ed. A. Mersmann, 2nd edition. New York: Dekker.
Mersmann, A. and Laufhütte, H.D., 1985. Scale-up of agitated vessels for different mixing processes. *5th European Conference on Mixing 1985*, Würzburg, pp. 273–284.
Mersmann, A., Angerhöfer, M. and Franke, J., 1994. Controlled precipitation. *Chemical Engineering Technology*, **17**, 1–9.
Mersmann and Geisler, R., 1991. Determination of the local turbulent energy dissipation rates in stirred vessels and its significance for different mixing tasks. In *4th World Congress of Chemical Engineering*. Karlsruhe, Germany.
Mersmann, A., Angerhöfer, M., Gutwald, T., Sangl, R. and Wang, S., 1990. General Prediction of Mean Crystal Sizes. In *11th International Symposium on Industrial Crystallization*. Ed. A. Mersmann, Garmisch-Partenkirchen, Germany.
Middleman, S., 1965. Mass transfer from particles in agitated systems: Application of the Kolmogorov theory. *American Institute of Chemical Engineers Journal*, **11**, 750–752.
Miers, H.A. and Isaac, F., 1907. The spontaneous crystallization of binary mixtures. *Proceedings of the Royal Society*, **A79**, 322–351.
Mingers, J., 1989. An Empirical Comparison of Pruning Methods for Decision Tree Induction. *Machine Learning*, **4**, 227.
Misra, C. and White, E.T., 1971. Kinetics of aluminium trihydroxide from seeded caustic aluminate solutions. *American Institute of Chemical Engineers Symposium Series*, **67**(110), 53–65.
Momonaga, M., Yazawa, H. and Kagara, K., 1992. Reactive crystallization of methyl α-methoxyimino acetoacetate. *Journal of Chemical Engineering of Japan*, **25**, 237–242.
Monnier, O., Fevotte, G., Hoff, C. and Klein, J.P., 1997. Model identification of batch cooling crystallizations through calorimetry and image analysis. *Chemical Engineering Science*, **52**, 1125–1139.
Moroyama, K. and Fan, L.S., 1985. *American Institution of Chemical Engineers Journal*, **31**, 1.

Mullin, J.W., 2001. *Crystallization*, 4th edition. Oxford: Butterworth-Heinemann.
Mullin, J.W. and Raven, K.D., 1961a. Nucleation in agitated solutions. *Nature*, **190**, 251.
Mullin, J.W. and Raven, K.D., 1961b. Influence of mechanical agitation on the nucleation of some aqueous salt solutions. *Nature*, **195**, 35–38.
Mullin, J.W. and Garside, J., 1967. Crystallization of aluminium potassium sulphate: a study in the assessment of crystallizer design data: I: Single crystal growth rates, II Growth in a fluidised bed. *Transactions of the Institution of Chemical Engineers*, **45**, 285–295.
Mullin, J.W. and Nývlt, J., 1971. *Chemical Engineering Science*, **26**, 369.
Mumtaz, H.S., Hounslow, M.J., Seaton, N.A. and Paterson, W.R., 1997. Orthokinetic aggregation during precipitation: A computational model for calcium oxalate monohydrate. *Transactions of the Institution of Chemical Engineers*, **75**, 152–159.
Muralidar, R. and Ramkrishna, D., 1986. An inverse problem in agglomeration kinetics. *Journal of Colloid and Interfacial Science*, **112**, 348–361.
Mydlarz, J. and Jones, A.G., 1989. Crystallization and subsequent solid–liquid separation of potassium sulphate. Part II: Slurry filtrability. *Chemical Engineering Research and Design*, **67**, 294–300.
Mydlarz, J. and Jones, A.G., 1989. On modelling the size-dependent growth rate of potassium sulphate in an MSMPR crystallizer. *Chemical Engineering Communications*, **90**, 47–56.
Mydlarz, J. and Jones, A.G., 1990. On the estimation of size-dependent crystal growth rate functions in MSMPR crystallizers. *Chemical Engineering Journal*, **53**, 125–135.
Myerson, A.S. (ed.), 1999. *Molecular Modelling Applications in Crystallization*. Cambridge University Press.
Myerson, A.S. (ed.), 2001. *Handbook of Industrial Crystallization*, 2nd edition. Oxford: Butterworth-Heinemann.
Nancollas, G.H., 1966. *Interactions in Electrolyte Solutions*. Elsevier Publishing Company.
Nancollas, G.H. and Gardner, G.L., 1974. Kinetics of crystal growth of calcium oxalate monohydrate. *Journal of Crystal Growth*, **21**, 267–276.
Nancollas, G.H. and Reddy, M.M., 1971. The crystallization of calcium caronate. II Calcite growth mechanism. *Journal of Colloid and Interfacial Science*, **37**, 824–833.
Natarajan, R. and Schechter, R.S., 1987. Electrokinetic behaviour of colloidal particles with thin ionic double layers. *American Institution of Chemical Engineers Journal*, **33**, 1110–1123.
Naumova, T.N., Efimof, V.M., Goltsova, K.N. and Stroganova, L.A., 1990. Investigation of low soluble compound crystallization process. *Industrial Crystallization '90*. Garmisch-Partenkirchen, September 1990. Ed. A. Mersmann, Düsseldorf: GVC-VDI, pp. 217–222.
Nicmanis, N. and Hounslow, M.J., 1998. Finite-element methods for steady-state population balance equations. *American Institution of Chemical Engineers Journal*, **44**(10), 2258–2272.
Ness, J.N. and White, E.T., 1976. Collision in an agitated crystallizer. *American Institute of Chemical Engineers Symposium Series*, **72**(153), 64–73.
Neville, J.M. and Seider, W.D., 1980. Coal pre-treatment – extensions of *FLOWTRAN* to model solids-handling equipment. *Computers and Chemical Engineering*, **4**, 49.
Newton, I., 1687. *Principia*.
Ng, K.M., 1991. Systematic separation of a multicomponent mixture of solids based on selective crystallization and dissolution. *Separations Technology*, **1**, 108.

Nielsen, A.E., 1969. Nucleation and growth of crystals at high supersaturation. *Kristall und Technik*, **4**, 17–38.
Nielsen, A.E. and Toft, J.M., 1984. Electrolyte crystal growth kinetics. *Journal of Crystal Growth*, **67**, 278–288.
Nielsen, A.E. and Söhnel, O., 1971. Interfacial tensions, electrolyte crystal aqueous solution, from nucleation data. *Journal of Crystal Growth*. **11**, 233–242.
Nienow, A.W., 1985. The mixer as a reactor: Liquid/solid systems. In *Mixing in the process industries*. Eds. N. Harnby, M.F. Edwards and A.W. Nienow, 1992, 2nd edition. Oxford: Butterworth-Heinemann.
Nienow, A.W. and Conti, R., 1978. Particle abrasion at high solids concentration in stirred vessels. *Chemical Engineering Science*, **33**, 1077–1086.
Nonhebel, G. and Moss, A.A.H., 1971. *Drying of solids in the chemical industry*. London: Butterworths.
Noor, P. and Mersmann, A., 1993. Batch precipitation of calcium carbonate. *Chemical Engineering Science*, **48**, 3083–3088.
Nývlt J., 1970. *Industrial crystallisation from solution*. London: Butterworth Group.
Nývlt, J., 1982. *Industrial Crystallization-The State of The Art*, 2nd edition. Weinheim: Verlag Chemie.
Nývlt, J., 1989. Calculation of crystallization kinetics based on a single batch experiment. *Collection of Czechoslovakian Chemical Communications*, **54**, 3187–3197.
Nývlt J., 1992. *Design of crystallizers*. Boca Raton: CRC Press.
O'Hara, M. and Reid, R.C., 1973. *Modelling Crystal Growth Rates from Solution*. Englewood Cliffs: Prentice-Hall.
Ohsol, E.O., 1973. What does it cost to pilot a process? *Chemical Engineering Progress*, **69**, 17–20.
Ohtaki, H., 1998. *Crystallization Processes*, Wiley Series in Solution Chemistry, Vol. 3. West Sussex: John Wiley & Sons.
Oldshue, J.Y., 1983. *Fluid Mixing Technology*. New York: McGraw-Hill.
Oldshue, J.Y., 1985. Scale-up of unique industrial fluid mixing processes. *5th European Conference on Mixing*, Würzburg, pp. 35–51.
Osborne, D.G., 1990. Gravity thickening. In *Solid–liquid separation*, 3rd edition. Ed. L. Svarovsky. Oxford: Butterworth-Heinemann.
Ostwald, W., 1896. *Lehrbuch der Allgemeinen Chemie*. Leipzig: Engelmann, **2**, 444.
Ostwald, W., 1897. Studien über de bildung und umwandlung fester körper. *Zietschrift für physikalishe chemie*, **22**, 289–330.
Ottens, E.P.K and de Jong, E.J., 1973. A Model for Secondary Nucleation in a Stirred Vessel Cooling Crystallizer. *Industrial and Engineering Chemistry Fundamentals*, **12**, 179–184.
Ottens, E.P.K., Janse, A.H. and de Jong, E.J., 1972. *Journal of Crystal Growth*, **13/14**, 500.
Ouchyama, N. and Tanaka, T., 1981. Porosity of a mass of particles having a range of sizes. *Industrial and Engineering Chemistry Fundamentals*, **20**, 66–71.
Overbeek, J. Th., 1949. Reversible systems. In *Colloid Science*, Vol. 2. Ed. H.R. Kruyt. Amsterdam: Elsevier.
Oyanader, M.A., Guerrero, C.J. and Cisternas, L.A., 1997. SSS, salt separation system by fractional crystallization. *Information Technology*, **8**, 11.
Parfitt, G.D. (ed.), 1981. *Dispersion of powders in liquids*, 3rd edition. Elsevier.
Parks, R.M. and Rousseau, R.W., 1979. *JACC Proceedings*, 819.
Patwardhan, V.S. and Chi Tien, 1985. Sedimentation and liquid fluidisation of solid particles of different sizes and densities. *Chemical Engineering Science*, **40**, 1051–1060.

Pepin, X., Simons, S.J.R., Blanchon, S., Rosstti, D. and Couarraze, G., 2001. Hardness of moist agglomerates in relation to interparticle friction, granule liquid content and nature. *Powder Technology*, **117**, 123–138.

Petanate, A.M. and Glatz, C.E., 1983. Isoelectric precipitation of soy protein. I. Factors affecting particle size distributions. II. Kinetics of protein aggregate growth and breakage. *Biotechnology and Bioengineering*, **25**, 3049.

Pethica, B.A., 1986. A development chemists guide to colloidal stability. *Colloids and Surfaces*, **20**, 151–170.

Pietsch, W., 1991. *Size enlargement by agglomeration*. Chichester: John Wiley.

Ploß, R. and Mersmann, A., 1989. A New Model of the Effect of Stirring Intensity on the Rate of Secondary Nucleation. *Chemical Engineering Technology*, **12**, 137–146.

Ploß, R., Tengler, T. and Mersmann, A., 1986. Scale-up of MSMPR-Crystallizers. *German Chemical Engineering*, **1**, 42–48.

Plummer, L.N. and Busenberg, E., 1982. *Geochim. Cosmochim. Acta*, **46**, 1011–1040.

Pohorecki, R. and Baldyga, J., 1983. The use of new model of micromixing for determination of crystal size in precipitation. *Chemical Engineering Science*, **38**, 79–83.

Pohorecki, R. and Baldyga, J., 1988. The effects of micromixing and the manner of reactor feeding on precipitation in stirred tank reactors. *Chemical Engineering Science*, **43**, 1949–1954.

Polisch, J. and Mersmann, A., 1988. The Influence of Stress and Attrition on Crystal Size Distribution. *Chemical Engineering Technology*, **11**, 40–49.

Pontryagin, L.S., Boltyanski, V.G., Gankredlize, R.V. and Mischenko, E.F., 1962. *The mathematical theory of optimal processes*, English Edition. New York: Interscience.

Pope, S.B., 1979. Probability distribution in turbulent shear flows. In *Turbulent shear flows 2*. Berlin: Springer, pp. 7–16.

Pope, S.B., 1985. PDF methods for turbulent reactive flows. *Progress in Energy and Combustion Science*, **11**, 119–192.

Pope, S.B., 2000. *Turbulent flows*. Cambridge University Press.

Potter, B.S., Palmer, R.A., Withnall, R., Chowdhry, B.Z. and Price, S.L., 1999. *Journal of Molecular Structure*, **349**, 485–486.

Powers, H.E.C., 1963. Nucleation and early crystal growth. *Industrial Chemist*, **39**, 351–355.

Prasad, P.B.V., 1985. Twinning in palmitic acid crystals. *Journal of Crystal Growth*, **72**, 663–669.

Pratola, F., Simons, S.J.R. and Jones, A.G., 2000. Micro-Mechanics of Agglomerative Crystallization Processes. *Proceeding of Advances in particle formation*, American Institute of Chemical Engineers National Meeting, November 2000, Paper 22 g.

Price, S.L. and Wibley, K.S., 1997. *Journal of Physical Chemistry A*, **101**, 2198–2206.

Prior, M.H., Prem, H. and Rhodes, M.J., 1990. Size reduction. In *Principles of powder technology*. Ed. M. Rhodes. Chichester: John Wiley.

Purchas, D.B., 1981. *Solid/Liquid Separation Technology*. London: Uplands Press.

Purchas, D.B. and Wakeman, R.J. (eds.), 1986. *Solid/Liquid Separation Equipment Scale-up*, 2nd edition. London: Uplands Press.

Qian, R., Chen, Z., Ni, H., Fan, Z. and Cai, F., 1987. Crystallization kinetics of potassium chloride from brine and scale-up criterion. *American Institution of Chemical Engineers Journal*, **33**, 1690–1697.

Qui, Yangeng and Rasmusen, Å.C., 1991. Nucleation and growth of succinic acid crystals in a batch stirred crystallizer. *American Institute of Chemical Engineers Journal*, **36**, 665–676.

Qui, Yangeng and Rasmuson, Å.C., 1994. Estimation of crystallization kinetics from batch cooling experiments. *American Institute of Chemical Engineers Journal*, **40**, 799–812.

Quinlan, J.R., 1990. Decision Trees and Decision making. *IEEE Trans Systems Man Cybernetics*, **20**(2), 339.
Quinlan, J.R., 1993. *C4.5: Programs for Machine Learning*, Morgan Kaufmann.
Rajagopal, S., Ng, K.M. and Douglas, J.M., 1988. Design of solids processes: Production of potash. *Industrial and Engineering Chemistry Research*, **27**, 2071.
Rajagopal, S., Ng, K.M. and Douglas, J.M., 1992. A hierachical decision procedure for the conceptual design of solids processes. *Computers and Chemical Engineering*, **16**, 675.
Raleigh, Lord, 1915. The principle of similitude. *Nature*, **95**, 66–68.
Ramkrishna, D., 1981. Analysis of population balance – IV. The precise connection between Monte Carlo simulation and population balances. *Chemical Engineering Science*, **36**, 1203–1209.
Ramkrishna, D., 1985. The status of population balances. *Reviews in Chemical Engineering*, **3**, 49–95.
Ramkrishna. D., 2000. *Population balances. Theory and applications to particulate systems in engineering*. New York: Academic Press.
Ranade, V.V., 1997. An efficient computational model for simulating flow in stirred vessels: a case of Rushton turbine. *Chemical Engineering Science*, **52**, 4473–4484.
Randolph, A.D., 1969. Effect of crystal breakage on crystal size distribution from a mixed-suspension crystallizer. *Industrial and Engineering Chemistry Fundamentals*, **8**, 58.
Randolph, A.D., 1980. *International Symposium of Fine Chemical Processing*, S.M.E. of A.I.M.E. New York, pp. 3–14.
Randolph, A.D. and Cise, M.D., 1972. Nucleation kinetics of potassium sulphate-water system. *American Institute of Chemical Engineers Journal*, pp. 4181, 798.
Randolph, A.D. and Larson, M.A., 1962. Transient and steady state size distribution in continuous mixed suspension crystallizers. *American Institution of Chemical Engineers Journal*, **8**, 639–649.
Randolph, A.D. and Larson, M.A., 1988. *Theory of Particulate Processes*, 2nd edition. New York: Academic Press.
Randolph, A.D. and Sikdar, S.K., 1976. Creation and survival of secondary crystal nuclei. The potassium sulphate-water system. *Industrial and Engineering Chemistry Fundamentals*, **15**, 64.
Randolph, A.D. and Tan, C., 1978. Numerical design techniques for staged classified recycle crystallizers: Examples of continuous alumina and sucrose crystallizers. *Industrial and Engineering Chemistry Process Design and Development*, **17**(2), 189.
Randolph, A.D. and White, E.T., 1977. *Chemical Engineering Science*, **32**, 1067–1076.
Randolph, A.D., Beckman, J.R. and Kraljevich, Z., 1977. Crystal size distribution dynamics in a classified crystallizer. Pts I and II. *American Institute of Chemical Engineers Journal*, **23**, 500–520.
Randolph, A.D., Chen, L. and Tavana, A., 1987. Feedback control of CSD in a KCl crystallizer with a fines dissolver. *American Institution of Chemical Engineers Journal*, **33**(4), 583–591.
Rawlings, J.B., Miller, S.M. and Witkowski, W.R., 1993. Model identification and control of solution crystallization processes: A review. *Industrial and Engineering Chemistry Research*, **32**, 1275–1296.
Rawlings, J.B., Sink, C.W. and Miller, S.M., 2001. Control of crystallization processes. In *Handbook of Industrial Crystallization*. Ed. A.S. Myerson, 2nd edition. Oxford: Butterworth-Heinemann.
Rawlings, J.B., Witkowski, W.R. and Eaton, J.W., 1992. Modelling and control of crystallizers. *Powder Technology*, **69**, 3.

Redman, T. and Rohani, S., 1992. Control of crystal purity in a continuous KCl crystallizer. In *First separations division topical conference on separations technologies: New developments and opportunities*, 2–6 November 1992, Miami Beach Fl. (New York: American Institute of Chemical Engineers), p. 685.

Redman, T., Rohani, S. and Strathdee, G., 1997. Control of the crystal mean size in a pilot plant potash crystallizer. *Transactions of the Institution of Chemical Engineers*, **75A**, 183–192.

Rekoske, J.E., 2001. *American Institute of Chemical Engineers Journal*, **47**(1), 2.

Rhodes, M., 1990. *Particle Technology*. Chichester: John Wiley.

Rice, R.G. and Do, D.D., 1995. *Applied Mathematics and modelling for chemical engineers*. New York: Wiley.

Rice, R.W. and Baud, R.E., 1990. The role of micromixing in the scale-up of geometrically similar batch reactors. *American Institution of Chemical Engineers Journal*, **36**, 293–298.

Richardson, J.F. and Zaki, W.N., 1954. *Chemical Engineering Science*, **3**, 65.

Richardson, J.F. and Zaki, W.N., 1954. Sedimentation and fluidisation. *Transactions of the Institution of Chemical Engineers*, **32**, 35.

Rielly, C.D. and Marquis, A.J., 2001. A particle's eye view of crystallizer fluid mechanics. *Chemical Engineering Science*, **56**, 2475–2493.

Rigopoulos, Stelios and Alan G. Jones, 2001. Dynamic Modelling of a Bubble Column for Particle Formation via a Gas–Liquid Reaction. *Chemical Engineering Science* (in press).

Ristic, R.I., Sherwood, J.N. and Shripathi, T., 1991. The role of dislocations and mechanical deformation in growth rate dispersion in potash alum crystals. In *Advances in Industrial Crystallization*. Eds. J. Garside, R.J. Davey, and A.G. Jones. Oxford: Butterworth-Heinemann, pp. 77–91.

Ristic, R.I., Sherwood, J.N. and Wojciechowski, K., 1988. *Journal of Crystal Growth*, **91**, 163.

Ristic, R.I. and Sherwood, J.N., 2001. The influence of mechanical stress on the growth and dissolution of crystals. *Chemical Engineering Science*, **56**, 2267–2280.

Ritchie, B.W. and Togby, A.H., 1979. A three-environment micromixing model for chemical reactors with arbitrary separate feed streams. *Chemical Engineering Journal*, **17**, 173.

Rohani, S. and Paine, K., 1987. Measurement of solids concentration of a soluble compound in a saturated slurry. *Canadian Journal of Chemical Engineering*, **65**, 163.

Rohani, S. and Paine, K., 1991. Feedback control of CSD in a continuous cooling crystallizer. *Canadian Journal of Chemical Engineering*, **69**, 165.

Rohani, S. and Bourne, J., 1990a. A simplified approach to the operation of a batch crystallizer. *Canadian Journal of Chemical Engineering*, **45**, 3457–3466.

Rohani, S. and Bourne, J., 1990b. Self-tuning control of crystal size distribution in a batch cooling crystallizer. *Chemical Engineering Science*, **45**, 3457–3466.

Rohani, S., 2001. Control of crystallizers. In *Crystallization Technology Handbook*, 2nd edn. Ed. A. Mersmann. New York: Marcel Dekker.

Rohani, S., Tavare, N.S. and Garside, J., 1990. Control of crystal size distribution in a batch cooling crystallizer. *Canadian Journal of Chemical Engineering*, **68**, 260–267.

Rossitter, A.P., 1986. Design and optimisation of solids processes: Part III. *Chemical Engineering Research and Design*, **64**, 191–196.

Rossiter, A.P. and Douglas, J.M., 1986. Design and optimisation of solids processes: Parts I&II. *Chemical Engineering Research and Design*, **64**, 175–190.

Roth, H.J., Kleeman, A. and Beisswenger, T., 1988. *Pharmaceutical Chemistry*, **1**, Chichester: Ellis Horwood.

Rousseau, R.W. and Howell, T.R., 1982. Comparison of simulated crystal size distribution control systems based on nuclei density and super-saturation. *Industrial and Engineering Chemistry Process Design and Development*, **21**, 606.

Rovang, R.D. and Randolph, A.D., 1980. On-line particle size analysis in the fines loop of a KCl crystallizer. *American Institute of Chemical Engineers Symposium Series*, **76**(193), 18.

Rowe, P.N., 1987. A convenient empirical expression for the estimation of the Richardson-Zaki exponent. *Chemical Engineering Science*, **42**, 2795–2796.

Rudd, D.F. and Watson, C.C., 1968. *Strategy of process engineering*. New York: John Wiley.

Rudd, D.F., Powers, G.J. and Sirola, J.J., 1973. *Process synthesis*. Englewood Cliffs, New Jersey: Prentice Hall.

Rumpf, H., 1990. *Particle Technology*. Translated by F.A. Bull. London: Chapman and Hall, 199pp.

Russel, W.B., 1987. *The dynamics of colloidal systems*. Madison: University of Wisconsin Press.

Sada, E., Kumazawa, H. and Butt, M.A., 1977. Single gas absorption with reaction in a slurry containing fine particles. *Chemical Engineering Science*, **32**, 1165–1170.

Sada, E., Kumazawa, H. and Aoyama, M., 1988. Reaction kinetics and controls of size and shape of geothite fine particles in the production of ferrous hydroxide. *Chemical Engineering Fundamentals*, **71**, 73–82.

Sada, E., Kumazawa, H. and Lee, C.H., 1983. Chemical absorption in a bubble column loading concentrated slurry. *Chemical Engineering Science*, **38**, 2047–2051.

Sada, E., Kumazawa, H., Lee, C. and Fujiwara, N., 1985. Gas–liquid mass transfer characteristics in a bubble column with suspended sparingly soluble fine particles. *Industrial and Engineering Chemistry Process Design and Development*, **24**, 255–261.

Saeman, W.C., 1956. Crystal size distribution in mixed suspensions. *American Institute of Chemical Engineers Journal*, **2**, 107–112.

Saraiva, P.M., 1995. Inductive and analogical learning: Data-driven improvement of Process operation. In *Intelligent Systems in Process Engineering*. Eds. G. Stephanopoulos and C. Han, Advances in Chemical Engineering, **22**, 377.

Saraiva, P.M. and Stephanopoulos, G., 1992a. Continuous process improvement through inductive and analogical learning. *American Institution of Chemical Engineers Journal*, **38**(2), 161.

Saraiva, P.M. and Stephanopoulos, G., 1992b. An exploratory data analysis robust optimization approach to continuous process improvement. *Working Paper*, Dept. Chem. Eng. MIT, Cambridge MA.

Shah B.C., McCabe, W.L. and Rousseau, R.W., 1973. Polyethylene vs. stainless steel impellers for crystallization processes, *American Institute of Chemical Engineers Journal*, **19**, 194.

Schenk, R., Donnet, M., Hessel, V., Hofmann, H., Jongen, N. and Löwe, H., 2001. Suitability of various types of micromixers for the forced precipitation of calcium carbonate. In *5th International Conference on Microreaction Technology* (IMRET 5), Strasbourg, France 27–30, May 2001.

Schierholtz, P.M. and Stevens, J.D., 1975. Determination of the kinetics of precipitation in a dilute system. *American Institute of Chemical Engineers Symposium Series*, **151**, 248–256.

Shippey, T.A. and Garside, J., 1982. Presented at *Institution of Chemical Engineers, Research Meeting*, London.

Schonert, K., 1972. Role of fracture physics in understanding comminution phenomena, *Transactions of the Society of Mining Engineers, American Institute of Mining Engineers*, **252**, 21–26.

Schreiner, A., Jones, A.G., Donnet, M. and Jongen, N., 2001. Precipitation of $CaCO_3$ in the Segmented Flow Tubular Reactor (SFTR). In *BIWIC-8 International Workshop on Industrial Crystallization*, Delft 19–20 September 2001.
Schubert, H., 1981. Principles of agglomeration. *International Chemical Engineering*, **21**, 363–377.
Schuler, H., 1996. *Prozeßsimulation*. Weinheim: VCH-Verlag.
Scott, K.J., 1968a. *Trans.IMM*, **77**, 185.
Scott, K.J., 1968b. Experimental study of continuous thickening of a flocculated silica slurry. *Industrial and Engineering Chemistry*, **7**, 582.
Scrutton, A., Grootscholten, P.A.M. and de Jong, E.J., 1982. *Transactions of the Institution of Chemical Engineers*, **60**, 345.
Seckler, M.M., Brinsma, O.S.L. and van Rosmalen, G.M. 1995. Influence of hydrodynamics on precipitation: a computational study. *Chemical Engineering Communications*, **135**, 113–131.
Seidell, A., 1958. *Solubilities of inorganic and metalorganic compounds*. Ed. W.F. Linke, 4th edition, **1**. Washington: American Chemical Society.
Sengupta, B. and Dutta, T.K., 1990. Effect of dispersions on CSD in continuous MSMPR crystallizers. *Chemical Engineering and Technology*, **13**(6), 426–431.
Shah, B.H., Borwanker, J.D. and Ramkrishna, D., 1977. Simulation of particle systems using the concept of the interval of quiescence. *American Institute of Chemical Engineers Journal*, **23**, 897–904.
Shah, Y.T., Kelkar, B.G., Godbole, S.P. and Deckwer, W.D., 1982. Design parameter estimations for bubble column reactors. *American Institute of Chemical Engineers Journal*, **28**, 353.
Sharratt, P.N., 1996. Crystallization: Meeting the environmental challenge. *Chemical Engineering Research and Design*, **74**, 732–738.
Shaw, D.J., 1980. *Introduction to colloid and surface chemistry*, 3rd edition. Oxford: Butterworth-Heinemann.
Sheikh, A.Y., 1997. *Synthesis, optimisation and control of crystallization systems*. PhD Thesis, University of London, UK.
Sheikh, A.Y. and Jones, A.G., 1996. Dynamic flow sheet model for an MSMPR crystalliser. In *Industrial Crystallization '96*. Ed. B. Biscans, Toulouse, Progep, 16–19 September 1996, pp. 583–588.
Sheikh, A.Y. and Jones, A.G., 1997. Crystallization process optimisation through a revised machine learning methodology. *American Institution of Chemical Engineers Journal*, **43**, 1448–1457.
Sheikh, A.Y. and Jones, A.G., 1998. Optimal synthesis of stage wise continuous crystallization networks. *American Institution of Chemical Engineers Journal*, **44**, 1637–1645.
Shekunov, B. Yu., Baldyga, J. and York, P., 2001. Particle formation by mixing with supercritical antisolvent at high Reynolds numbers. *Chemical Engineering Science*, **56**(7), 2421–2433.
Sherrington, P.J. and Oliver, R., 1981. *Granulation*. London: Heyden.
Sherwin, M.B., Shinner, R. and Katz, S., 1967. Dynamic behavior of the well-mixed isothermal crystallizer. *Chemical Engineering Progress Symposium Series*, **65**(95), 59–90.
Shippey, T.A. and Garside, J., 1982. Presented at *Institution of Chemical Engineers, Research Meeting*.
Shock, R.A.W., 1983. Encrustation of crystallizers. *Journal of separation process technology*, **4**, 1–13.
Sie, S.T. and Krishna, R., 1998. Process development and scale-up: 1. Process development strategy and methodology. *Reviews in Chemical Engineering*, **14**, 46–87.

Sikdar, S.K., 1977. Size-dependent growth rate from curved log n(L) vs. L steady-state data. *Industrial and Engineering Chemistry Fundamentals*, **16**, 390.
Sikdar, S.K. and Randolph, A.D., 1976. *American Institution of Chemical Engineers Journal*, **22**, 110.
Simons, S.J.R., 1996. Modelling of agglomerating systems: from spheres to fractals. *Powder Technology*, **87**, 29–41.
Simons, S.J.R. and Fairbrother, R.J., 2000. Direct observation of liquid binder-particle interactions: the role of wetting behaviour in agglomerate growth. *Powder Technology*, **110**, 44–58.
Sinnott, R.K., 1999. *Chemical Engineering, Volume 6. Chemical Engineering Designs*, 3rd edn. Eds, J.M. Coulson and J.F. Richardson. Oxford: Butterworth-Heinemann.
Skovborg, P. and Rasmussen, P., 1994. A mass transport limited model for the growth of methane and ethane gas hydrates. *Chemical Engineering Science*, **49**(8), 1131–1143.
Smith, G.D., 1985. *Numerical Solution of Partial Differential Equations: Finite Difference Methods*, 3rd edition. Clarendon Press.
Smoluchowski, M.V., 1916. Drei Vorträge über Diffusion, Brownsche Molekularbewegung und Koagulation von Kolloidteilchen. *Physik Zeitung*, **XVII**, 557–599.
Smoluchowski, M.V., 1916. Three lectures on diffusion, Brownian movement and coagulation of colloidal systems. *Physik Zeitung*, **17**, 557.
Smoluchowski, M.V., 1917. Mathematical theory of the kinetics of coagulation of colloidal systems. *Zeitschrift für Physikalische Chemie*, **92**, 129–168.
Söhnel, O. and Matejcková, E., 1981. Batch precipitation of alkaline earth carbonates. Effect of reaction conditions on filterability of resulting suspensions. *Industrial and Engineering Chemistry Process Design and Development*, **20**, 525–528.
Söhnel, O. and Garside, J., 1992. *Precipitation*. Oxford: Butterworth-Heinemann.
Söhnel, O. and Mullin, J.W., 1978. A method for the determination of crystallization induction periods. *Journal of Crystal Growth*, **44**, 377–382.
Söhnel, O. and Mullin, J.W., 1988. The role of time in metastable zone width determinations. *Chemical Engineering Research and Design*, **66**, 537–540.
Söhnel, O., Chianese, A. and Jones, A.G., 1991. Theoretical approaches to the design of precipitation systems: The present state of the art. *Chemical Engineering Communications*, **106**, 151–175.
Söhnel, O., Mullin, J.W. and Jones, A.G., 1988. Nucleation, growth and agglomeration during the batch precipitation of strontium molybdate. *Industrial and Engineering Chemistry Research*, **27**, 1721–1728.
Sowul, L. and Epstein, M.A.F., 1981 Crystallization kinetics of sucrose in a CMSMPR evaporative crystallizer. *Industrial and Engineering Chemistry Process Design and Development*, **29**(2), 197–203.
Spanos, N. and Koutsoukos, P.G., 1998. The transformation of vaterite to calcite: effect of the conditions of the solutions in contact with the mineral phase. *Journal of Crystal Growth*, **191**, 783–790.
Stanley-Wood, N.G., 1990. Size enlargement. In *Principles of powder technology*. Ed. M. Rhodes. Chichester: John Wiley.
Stephen, H. and Stephen, T., 1963. *Solubility of inorganic and organic compounds* (5 vols). London: Pergamon Press.
Šterbáček, Z. and Tausk, P., 1965. *Mixing in the chemical industry*. Oxford: Pergamon Press.
Stokes, G.G., 1851. On the effect of the internal friction of fluids on the motion of pendulums. *Transactions of the Cambridge Philosophical Society*, **9**, 8.
Strickland-Constable, R.F., 1979. *Journal of the Chemical Society Faraday Transaction*, I, **75**, 921.

Sung, C.Y., Estrin, J. and Youngquist, G.R., 1973. *American Institute of Chemical Engineers Symposium Journal*, **19**, 957.

Suzuki, M. 1997. Description of particle assemblies. In *Powder technology handbook*. Eds. K. Gotoh, H. Masuda and K. Higashanti, 2nd edition. New York: Marcel Dekker.

Svarovsky, L., 1990. Characterisation of powders. In *Principles of powder technology*. Ed. M. Rhodes. Chichester: John Wiley.

Svarovsky, L., 2000. *Solid–liquid separation*, 4th edition. Oxford: Butterworth-Heinemann.

Swinney, L.D., Stevens, J.D. and Peters, R.W., 1982. Calcium Carbonate Crystallization Kinetics. *Industrial and Engineering Chemistry Fundamentals*, **21**, 31.

Synowiec, P., Jones, A.G. and Shamlou, P.A., 1993. Crystal break-up in dilute turbulently agitated suspensions. *Chemical Engineering Science*, **48**, 3485–3495.

Tai, C.Y. and Chen, F.-B., 1998. Polymorphism of $CaCO_3$ precipitated in a constant-composition environment. *American Institution of Chemical Engineers Journal*, **44**, 1790–1798.

Talmage, W.P. and Fitch, E.B., 1955. Determining thickener unit areas. *Industrial Chemist*, **47**, 38–41.

Tanimoto, A.K., Kobayashi, K. and Fujita, S., 1964. Overall crystallization rate of copper sulfate pentahydrate in an agitated vessel. *International Chemical Engineering*, **4**(1), 153.

Tavare, N.S., 1986. Mixing in continuous crystallizers. *American Institution of Chemical Engineers Journal*, **32**, 705–732.

Tavare, N.S., 1986. Crystallization kinetics from transients of an MSMPR crystallizer. *The Canadian Journal of Chemical Engineering*, **64**, 752–758.

Tavare, N.S., 1987. Batch crystallizers: A review. *Chemical Engineering Communications*, **61**, 259–318.

Tavare, N.S., 1995. *Industrial crystallization: Process simulation, analysis and design*. New York: Plenum.

Tavare, N.S. and Garside, J., 1982. *Chemical Engineering Journal*, **25**, 229.

Tavare, N.S. and Garside, J., 1986. Simultaneous estimation of crystal nucleation and growth kinetics from batch experiments. *Chemical Engineering Research and Design*, **64**, 109.

Tavare, N.S., Garside, J. and Chivate, M.P., 1980. *Industrial and Engineering Chemistry Process Design and Development*, **19**, 653–665.

Thompson, P.D., 1968. A transformation of the stochastic equation for droplet coalescence. In *Proceedings of the international conference on cloud physics*, Toronto, Canada, pp. 1115–1126.

Timm, D.C. and Cooper, T.R., 1971. *American Institution of Chemical Engineers Journal*, **17**, 285.

Timm, D.C. and Larson, M.A., 1968. Effect of nucleation kinetics on the dynamic behaviour of a continuous crystallizer. *American Institute of Chemical Engineers Journal*, **14**(3), 452–457.

Tipnis, S.K., Penney, W.R. and Fasano, J.B., 1994. An experimental investigation to determine a scale-up method for fast competitive parallel reactions in agitated vessels. *American Institute of Chemical Engineers Symposium Series*, **299**, 78–91.

Togkalidou, T., Braatz, R.D., Johnson, B.K., Davidson, O. and Andrews, A., 2001. Experimental design and inferential modelling in pharmaceutical crystallization. *American Institution of Chemical Engineers Journal*, **47**(1), 160–168.

Tomazic, B. and Nancollas, G.H., 1979. The kinetics of dissolution of calcium oxalate hydrates. *Journal of Crystal Growth*, **46**, 355–361.

Torbacke, M. and Rasmuson, Å.C., 2001. Influence of different scales of mixing in reaction crystallization. *Chemical Engineering Science*, **56**, 2549–2474.

Tosun, G., 1988. An experimental study of the effect of mixing on the particle size distribution in $BaSO_4$ precipitation reaction. *6th European Conference on Mixing*, Pavia, pp. 161–170.

Tosun, G., 1992. A mathematical model of mixing and polymerization in a semibatch stirred tank reactor. *American Institution of Chemical Engineers Journal*, **38**, 425–437.

Tsutsumi, A., Nieh, J.Y. and Fan, L.S., 1991. Role of the bubble wake in fine particle production of calcium carbonate in bubble column systems. *Industrial and Engineering Chemistry Research*, **30**, 2328–2333.

Ueyama, K. (ed.), 1993. *Handbook of bubble columns and three phase reactors*. IPC, Tokyo.

Uhl, V.W. and Gray, J.B., 1986. *Mixing: Theory and practice* (3 vols). New York: Academic Press.

Utgoff, P.E., 1989. Incremental Induction of Decision Trees. *Machine Learning*, **4**, 161.

Vacassy, R., Jongen, N., Lemaître, J., Bowen, P. and Hofmann, H., 1998. Development of the new segmented flow tubular reactor for powder technology. In *Word Congress on Particle Technology 3*, Brighton, UK, July 6–9, 1998. (Rugby: Institution of Chemical Engineers). CD ROM version, paper N° 376.

van Der Heijden, A.E.D.M., Van der Eerden, J.P. and Van Rosmalen, G.M., 1994. The secondary nucleation rate: A physical model. *Chemical Engineering Science*, **49**(18), 3103–3113.

van Hook, A., 1961. *Crystallization: Theory and practice*. New York: Reinhold.

van Leeuwen, M.L.J., 1998. *Precipitation and mixing*. PhD Thesis, University of Delft, The Netherlands.

van Leeuwen, M.L.J., Bruinsma, O.S.L. and van Rosmalen, G.M., 1996a. Influence of mixing on the product quality in precipitation. *Chemical Engineering Science*, **51**, 2595–2600.

van Leeuwen, M.L.J., Bruinsma, O.S.L. and van Rosmalen, G.M., 1996b. Three-zone approach for precipitation of barium sulphate. *Journal of Crystal Growth*, **166**, 1004–1008.

van't Land, C.M. and Wienk, B.G., 1976. In *Industrial Crystallization*. Ed. J.W. Mullin. New York: Plenum Press, p. 51.

Vega, A., Diez, F. and Alvarez, J.M., 1995. Programmed cooling control of a batch crystallizer. *Computers and Chemical Engineering*, **9**, 471–476.

Versteeg, H.K. and Malalasekera, W., 1995. *An introduction to Computational Fluid Dynamics*. Longman.

Vervey, E.J.W. and Overbeek, J.T.G., 1948. *The stability of lyophobic colloids*. Amsterdam: Elsevier.

Villermaux, J., 1989. A simple model for partial segregation in a semibatch reactor. *American Institute of Chemical Engineers Annual Meeting*, San Francisco, Paper 114a.

Villermaux, J. and Devillon, J.C., 1975. Représentation de la coalescence et de la redispersion des domains de ségrégation dans un fluide par un modèle d'interaction phénoménologique. In *Proceedings of the second international conference of chemical reaction engineering*. Amsterdam, pp. B1–13.

Villermaux, J. and Falk, L., 1994. A generalised mixing model for initial contacting of reactive fluids. *Chemical Engineering Science*, **49**, 5127–5140.

Virtanen, J. 1984. Automatic control of batch evaporative crystallization. In *Industrial Crystallization 84*. The Hague September 1984. Eds. S.J. Jančić and E.J. de Jong. Elsevier Science.

Volmer, M., 1939. *Kinetic der Phasenbildung*. Steinkopf, Leipzig.

von Rittinger, P.R., 1867. *Lehrbuch der Aufbereitungskunde*. Berlin.

Wachi, S. and Jones, A.G., 1990. Model calculations of precipitation with gas–liquid rapid chemical reaction. In *Industrial Crystallization '90*. Garmisch-Partenkirchen, September 1990. Ed. A. Mersmann, Düsseldorf: GVC.VDI, pp. 229–235.

Wachi, S. and Jones, A.G., 1991a. Mass transfer with chemical reaction and precipitation. *Chemical Engineering Science*, **46**, 1027–1033.

Wachi, S. and Jones, A.G., 1991b. Effect of gas–liquid mass transfer on crystal size distribution during the batch precipitation of calcium carbonate. *Chemical Engineering Science*, **46**, 3289–3293.

Wachi, S. and Jones, A.G., 1992. Dynamic modelling of particle size distribution and degree of agglomeration during precipitation. *Chemical Engineering Science*, **47**, 3145–3148.

Wachi, S. and Jones, A.G., 1995. Aspects of gas–liquid precipitation systems for precipitate particle formation. *Reviews in Chemical Engineering*, **11**, 1–51.

Wachi, S., Morikawa, H. and Ueyama, K., 1987. Gas hold-up and axial dispersion in gas–liquid concurrent bubble column. *Journal of Chemical Engineering Japan*, **20**, 309–316.

Wadell, H., 1932. Volume, shape and roundness of rock particles. *Journal of Geology*, **40**, 443–451.

Wakeman, R.J., 1975. Packing densities of particles with log-normal size distributions. *Powder Technology*, **11**, 297–299.

Wakeman, R.J., 1990a. Pressure filters. In *Solid-liquid separation*, 3rd edition. Ed. L. Svarovsky. Oxford: Butterworth-Heinemann.

Wakeman, R.J., 1990b. Filter cake washing. In *Solid–liquid separation*, 3rd edition. Ed. L. Svarovsky. Oxford: Butterworth-Heinemann.

Wallas, 1988. *Chemical Process Equipment Selection*, Butterworths, p. 329.

Wallis, G.B., 1969. *One-dimensional two-phase flow*. New York: McGraw Hill.

Walton, A.G., 1967. *The formation and properties of precipitates*. New York: Wiley, Interscience.

Wang, Y.-D. and Mann, R., 1992. Partial segregation in stirred batch reactors: effect of scale-up on the yield of a pair of competing reactions. *Transactions of the Institution of Chemical Engineers*, **70**, 282–290.

Wedlock, D.J. (ed.), 1994. *Controlled particle, droplet and bubble formation*. Oxford: Butterworth-Heinemann.

Wei, H.Y. and Garside, J., 1997. Application of CFD modelling to precipitation systems. *Transactions of the Institution of Chemical Engineers*, **75**, 219–227.

Westerterp, K.R., van Swaaij, W.P.M. and Beenackers, A.A.C.M., 1995. *Chemical Reactor Design and Operation*. New York: John Wiley & Sons.

Wibowo, C. and Ng, K.M., 2000. Unified approach for synthesising crystallization-based separation processes. *American Institution of Chemical Engineers Journal*, **46**, 1400–1421.

Williams, R.A., 1994. *Colloid and surface engineering*. Oxford: Butterworth-Heinemann.

Willock, D.J., Price, S.L., Leslie, M. and Catlow, C.R.A., 1995. *Journal of Computational Chemistry*, 16, 628–647.

Wisniak, J. and Herskowitz, M., 1984. *Solubility of gasses and solids – A literature source*. Amsterdam: Elsevier.

Witkowski, W.R., Miller, S.M. and Rawlings, J.B., 1990. Light scattering measurements to estimate kinetic parameters of crystallization. In *Crystallization as a separation process, ACS Symposium Series*, **438**, 102.

Woinaroschy, A., Isopescu, R. and Fillipescu, L., 1994. Crystallisation process optimisation using artificial neural networks, *Chemical Engineering Technik*, **17**, 269–272.

Wójcik, J. and Jones, A.G., 1997. Experimental investigation into dynamics and stability of continuous MSMPR agglomerative precipitation of $CaCO_3$ crystals. *Transactions of the Intitution of Chemical Engineers*, **75**, 113–118.

Wójcik, J. and Jones, A.G., 1998a. Dynamics and stability of continuous MSMPR agglomerative precipitation: numerical analysis of the dual particle coordinate model. *Computers and Chemical Engineering*, **22**, 535–545.

Wójcik, J. and Jones, A.G., 1998b. Particle disruption of precipitated $CaCO_3$ crystal agglomerates in turbulently agitated suspensions. *Chemical Engineering Science*, **53**, 1097–1101.

Wood, W.M.L., 1997. *Crystal science techniques in the manufacture of chiral compounds.* Ch. 7 in *Chirality in Industry II*. Eds. N.A. Collins, G.N. Sheldrake and Crosby. New York: J. Wiley & Sons.

Wright, H. and Ramkrishna, D., 1992. Solutions of inverse problems in population balance aggregation kinetics. *Computers and Chemical Engineering*, **16**(2), 1019–1030.

Wuklow, M., Gerstlauer, A. and Nieken, U., 2001. Modeling and simulation of crystallization processes using parsival. *Chemical Engineering Science*, **56**(7), 2575–2588.

Wulff, G., 1901. *Z. Kristallogr.*, **34**, 449.

Wyn-Jones, E. and Gormally J. (eds.), 1983. Aggregation processes in solution. In *Studies in physical and theoretical chemistry*, **26**. Amsterdam: Elsevier.

Xu, Y. and McGrath, G., 1996. CFD predictions of stirred tank flows. *Transactions of the Institution of Chemical Engineers*, **74**, 471–475.

Yagi, H., 1986. Kinetics of solid production accompanying gas-liquid reaction. *Proceedings of World Congress III Chemical Engineering, Tokyo*, **4**, 20–23.

Yagi, H., 1988. Semi-batch precipitation accompanying gas–liquid precipitation. *Chemical Engineering Communications*, **65**, 109–119.

Yagi, H., Iwazawa, A., Sonobe, R., Matsubara, T. and Hikita, H., 1984. Crystallization of calcium carbonate accompanying chemical absorption. *Industrial and Engineering Chemistry Fundamentals*, **23**, 153–158.

Yagi, H., Nagashima, S. and Hikita, H., 1988. Semibatch precipitation accompanying gas–liquid reaction. *Chemical Engineering Fundamentals*, **65**, 109–119.

Yates, F., 1937. *Design and Analysis of Factorial Experiments.* Imperial Bureau of Soil Science, London.

Yates, B., 1981. Systems engineering in powder processing. In *POWTECH 81, Institution of Chemical Engineers. Symposium Series No. 63*, Rugby: Institution of Chemical Engineers, D4/Z/1-D4/Z/16.

Yu, A.B. and Standish, N., 1987. Porosity calculations of multi-component mixtures of spherical particles. *Powder Technology*, **52**, 233–241.

Zauner, R., 1999. *Scale-up of precipitation processes*, PhD thesis, University College London, UK.

Zauner, R. and Jones, A.G., 2000a. Determination of nucleation, growth, agglomeration and disruption kinetics from experimental precipitation data: The calcium oxalate system. *Chemical Engineering Science*, **55**, 4219–4232.

Zauner, R. and Jones, A.G., 2000b. Scale-up of continuous and semibatch precipitation processes. *Industrial and Engineering Chemistry Research*, **39**, 2392–2403.

Zauner, R. and Jones, A.G., 2000c. Mixing effects on product characteristics from semibatch precipitation. *Transactions of the Institution of Chemical Engineers*, **78(A)**, 894–901.

Zeitsch, K., 2000. Centrifugal Filtration. In *Solid–liquid separation*, 4th edition. Ed. L. Svarovsky. Oxford: Butterworth-Heinemann.

Zienkiewicz, O.C. and Taylor, R.L., 1991. *The Finite Element Method – Solid and Fluid Mechanics*. McGraw-Hill.

Zlokarnik, M., 1985. Modellübertragung bei partieller Ähnlichkeit. *Chemie-Ingenieur-Technik*, **57**, 410–416.

Zlokarnik, M., 1991. *Dimensional Analysis and Scale-up in Chemical Engineering*. Springer-Verlag.

Zweitering, T.N., 1958. Suspension of solid particles in liquids by agitators. *Chemical Engineering Science*, **8**, 244.

Author index

Abegg, C.F., 153, 299
Accorsi, C.A., 314
Ackermann, D.K., 302
Adler, P.M., 179, 299
Ajinkya, M.B., 287, 299
Akazawa, K., 313
Akers, S.R.G., 311
Allen, T., 7, 18, 26, 299
Allenby, B.R., 297, 299
Al-Rashed, M.H., 241, 250, 251, 252, 253, 254, 299
Alvarez, J.M., 325
Ammon, H.L., 314
Andrews, A., 324
Angerhöfer, M., 315
Aoki, Y., 160, 299
Aoun, M., 136, 299
Aoyama, M., 321
Aquilano, D., 160, 299
Arima, K., 308
Ash, S.G., 164, 299
Åslund, B.L., 226, 299
Asselbergs, C.J., 308
Astarita, G., 236, 299
Attarian, C., 291, 299
Austin, L.G., 176, 312

Backhurst, J.R., 106, 299
Bacon, D.J., 6, 310
Baker, C.G., 299
Baker, M., 308
Bakker, A., 48, 220, 299
Balakrishna S., 281, 282, 299, 300
Baldyga, J., 49, 50, 51, 56, 215, 216, 218, 229, 230, 231, 236, 300, 318, 322
Bamforth, A.W., 58, 190, 300
Barrick, J.P., 310
Barton, G.W., 261, 300
Baud, R.E., 320
Beckman, J.R., 319
Beckmann, J.R., 291, 300
Beddow, J.K., 7, 300
Beenackers, A.C.M., 326
Beisswenger, T., 320
Bemrose, C.R., 137, 300
Bennema, P., 304
Bennett, R.C., 149, 150, 270, 300
Berglund, K.A., 289, 305

Berry, D.A., 275, 300
Bersano, G., 314
Berthoud, A., 127, 301
Bhatia, S.K., 259, 302
Biegler, L.T., 279, 281, 282, 299, 300, 301, 303, 309
Bierwagen, G.P., 17, 301
Biscans, B., 299, 307, 322
Bishnoi, P.R., 305
Bisio, A., 301
Blanchon, S., 318
Bloor, M.I.G., 115, 301
Bohlin, M., 288, 301
Boltyanski, V.G., 318
Bond, F.C., 139, 301
Bonifazi, G., 315
Borwanker, J.D., 322
Bosgra, O.H., 310
Botsaris, G.D., 148, 149, 301, 304
Bourne, J.R., 49, 50, 51, 150, 215, 216, 231, 288, 300, 301, 320
Bowen, P., 305, 312, 314, 325
Boxman, A., 310
Boyson, F., 241, 302
Braantz, R.D., 314
Bramley, A. S., 301, 310
Bransom, S.H., 53, 135, 136, 301
Braun, B., 173, 233, 315
Bravais, A., 4, 301
Brayshaw, M.D., 301
Brečević, L.J., 307
Breiman, L., 301
Bridgwater, J., 137, 300
Bruinsma, O.S.L., 309, 325
Bristol, E.H., 293, 301
Brodley, C.E., 301
Broul, M., 61, 302
Brown, C.M., 47, 182, 183, 302
Buckingham, E., 302
Budz, J., 160, 302, 311
Bujac, P.D.B., 135, 151, 302
Bull, F.A., 321
Burke, S.P., 41, 42, 302
Burton, W.K., 182, 302
Busenberg, E., 233, 318
Butt, M.A., 302, 321

Cabrera, N., 302
Cai, F., 318

Calabrese, R.V., 299
Camp, T.R., 186, 302
Capes, C.E., 302, 314
Carman, P.C., 39, 42, 302
Cate, A.ten, 47, 302
Catlow, A., 326
Chakraborty, D., 259, 302
Chang, C.J., 60, 303
Chang, C.-T., 288, 302
Chang, L.-J., 51, 303
Chang, W.-C., 273, 303
Chapman, F.S., 26, 44, 309
Chen, F.-B., 324
Chen, G., 303
Chen, J., 135, 303
Chen, L., 327
Chen, Z., 318
Cheremisinoff, N.P., 80, 303
Chiampo, F., 314
Chianese, A., 152, 201, 303
Chivate, M.P., 310, 324
Choplin, L., 303
Chowdhry, B.Z., 318
Chung, S.H., 200, 303, 316
Cise, M.D., 319
Cisternas, L.A., 275, 276, 303, 317
Clark, N.N., 12, 303
Clark, W.E., 121, 303
Clevenger, G.H., 84, 303
Clydesdale, G., 304
Coe, H.S., 303
Collet, A., 310
Collier, A.P., 187, 303, 310
Collins, N.A., 327
Conti, R., 142, 143, 144, 303, 313, 317
Coombes, D.S., 4, 303
Cooper, T.R., 147, 324
Couarraze, G., 318
Coulson, J.M., 7, 26, 37, 40, 80, 87, 106, 116, 121, 303, 323
Crawford, R.J., 17, 303
Crowe, C.M., 307
Cumberland, D.J., 303
Curie, P., 4, 303
Curl, R.L., 51, 303
Cussler, E.L., 276, 277, 303
Cuthrell, J.E., 303

Da Vinci, Leonardo, 304
Dale, D., 308
Dallavalle, J.M., 31, 304
Damen, A.A.H., 304
Danckwerts, P.V., 49, 236, 304
Darcy, H., 304
Daudey, P.J., 136, 304

Davey, R.J., 4, 146, 150, 157, 304, 306, 307, 311, 320
David, R., 304, 306, 309, 314
Davidson, O., 324
Davies, C.W., 304
Davies, O.L., 304
Davis, R., 299
Davis, R.F., 157, 304
Davis, R.H., 34, 304
de Jong, E.J., 132, 136, 142, 143, 149, 150, 153, 304, 308, 310, 311, 313, 317, 322, 325
de Leer, B.G.M., 308
Deckwer, W.D., 235, 304, 322
Delannoy, C.C., 312
Dell'Ava, P., 301
Denk, E.G., 148, 304
Derjaguin, B.V., 304
Derksen, J.J., 302
Devillon, J.C., 51, 325
deWolf, S., 310
Dholababhai, P.D., 305
Di Berardino, F., 303
Di Cave, S., 315
Di Felice, R., 27, 33, 34, 304
Dickey, D.S., 304
Dickinson, E., 304
Dietz, P.W., 304
Diez, F., 325
Dirksen, J.J, 309
Do, D.D., 284, 320
Docherty, R., 4, 304
Doherty, M.F., 309
Doki, N., 200, 304
Donnay, J.D.H., 4, 305
Donnet, M., 258, 305, 321, 322
Doucet, J., 288, 305
Douglas, J.M., 261, 271, 272, 279, 305, 319, 320
Dowty, E., 304
Drach, G.W., 308
Drake, R.L., 305
Drummond, J.N., 310
Duncan, A.G., 296, 305
Dunitz, J.D., 314
Dunning, W.J., 53, 135, 194, 301
Dunuwila, D., 289, 305
Dutta, T.K., 248, 322
Dye, S.R., 275, 300, 305

Eaton, J.W., 288, 305, 319
Edwards, M.F., 308, 317
Efimof, V.M., 316
Eguchi, W., 308
Einstein, A., 161
Ejaz, T., 305, 311

Englezos, P., 305
Epstein, M.A.F., 136, 287, 288, 302, 323
Ergun, S., 41, 43, 305
Erk, P., 308
Erlrebacher, G., 305
Estrin, J., 310, 315, 324
Etchells, A.W., 307
Evans, L.B., 146, 149, 305
Evans, T.W., 261, 277, 305
Everett, D.H., 299
Ewing, C., 312

Fairbrother, R.J., 174, 323
Falk, L., 50, 231, 305, 306, 308, 325
Falope, G.O., 249, 251, 305
Family, F., 157, 305
Fan, L.S., 235, 305, 315, 318, 325
Fan, Z., 318
Farrell, R.J., 288, 305, 306
Fasano, J.B., 324
Fayyad, U.M., 306
Fevotte, G., 315
Fiedelman, H., 300
Fillipescu, L., 326
Finlayson, B., 302
Fisher, R.A., 306
Fitch, E.B., 87, 324
Floudas, C.A., 281, 312
Font, R., 87, 306
Ford, L.J., 308
Foscolo, P.U., 35, 306
Fournier, M.-C., 306
Franchini-Angela, M., 160, 299
Franck, R., 51, 306
Frank, F.C., 128, 302, 306
Franke, J., 306, 315
Friedel, G., 4, 306
Friedman, J.H., 301
Fujita, S., 324
Fujiwara, N., 321
Furth, R., 306

Gahn, C., 153, 306
Galilei, Galileo, 306
Gankredlize, R.V., 318
Gardner, G.L., 182, 306, 316
Garside, J., 4, 51, 58, 62, 76, 123, 125, 127, 131, 136, 146, 147, 148, 149, 150, 151, 182, 190, 197, 206, 211, 215, 216, 236, 288, 301, 304, 306, 307, 311, 313, 316, 320, 321, 322, 323, 324, 326
Gavezzotti, A., 314
Gecol, H., 34, 304
Geisler, R., 45, 307, 315
Gelbard, F., 56, 168, 307
Gerstlauer, A., 327

Gertlauer, A., 153, 307
Gibbs, J.W., 4, 58, 61, 307
Gibilaro, L.G., 26, 306, 307
Giddey, C., 288, 305
Gilles, E.-D., 307
Giorgio, S., 261, 307
Girolami, M.W., 151, 307
Glasgow, L.A., 142, 307
Glasser, D., 307
Glatz, C.E., 141, 318
Godbole, S.P., 322
Goltsova, K.N., 316
Gooch, J.R., 249, 307
Gormally, J., 157, 327
Gotoh, K., 324
Gottlieb, D., 48, 307
Gottung, B.E., 308
Grace, H.P., 113, 307
Graham, P., 311
Gray, J.B., 26, 44, 325
Gray, W.A., 17, 307
Green, D.A., 307
Gregory, J., 157, 163, 307
Grenville, R.K., 307
Griffith, A.A., 138, 307, 308
Griffiths, H., 195, 196, 308
Groen, 289, 308
Grootscholten, P.A.M., 58, 150, 190, 308, 310, 322
Gruhn, G., 261, 308
Guerrero, C.J., 317
Guichardon, P., 49, 308
Gupta, G., 291, 308
Gutwald, T., 136, 308, 315

Hajdasinski, A.K., 294, 304
Hall, D.G., 164, 308
Hamada, O., 304
Hammond, R.B., 308
Han, C.D., 136, 291, 308, 321
Harada, M., 308
Harker, D., 4, 305
Harker, J.H., 106, 299
Harnby, N., 26, 44, 236, 308, 317
Hart, M., 279, 308
Hart, R., 279, 308
Hartel, R.W., 171, 179, 184, 308
Hartland, S., 252, 315
Hartman, P., 4, 308
Hearn, S.J., 300
Heffels, S.K., 289, 308
Helt, J.E., 147, 291, 308
Herskowitz, M., 61, 326
Hess, W., 308
Hessel, V., 321
Heuer, M., 309

Higashanti, K., 324
Higbie, R., 309
Hikita, H., 309, 327
Hildebrandt, D., 307, 309
Hill, P.J., 56, 176, 261, 309
Hindmarsh, E., 281, 314
Hoff, C., 315
Hofmann, D.W.M., 314
Hofmann, H., 305, 312, 321, 325
Holland, F.A., 26, 44, 309
Hollander, E.D., 173, 309
Hostomský, J., 169, 251, 309, 312
Houcine, I., 216, 309
Hounslow, M.J., 56, 157, 168, 173, 187, 249, 277, 301, 303, 307, 309, 310, 314, 316
Howell, T.R., 291, 321
Huber, S., 303
Hughmark, G.A., 44, 310
Hulburt, H.M., 52, 310
Hull, D., 6, 310
Hungerbuehler, K., 150, 301
Hurley, M.A., 135, 310
Hussaini, M.Y., 305

Ingham, D.B., 115, 301
Irani, K.B., 306
Isaac, F., 58, 315
Ishikawa, H., 235, 309
Isopescu, R., 326
Iwazawa, A., 327

Jacobs, G., 307
Jacques, J., 310
Jagadesh, D., 310
Jagannathan, R., 142, 310
Jager, J., 136, 292, 294, 295, 310
Jakola, T.H, 314
Jameson, L.M., 305
Jančić, S.J., 147, 149, 150, 206, 306, 310
Janse, A.H., 310, 317
Jazaszek, P., 210, 310
Jézéquel, P.-H., 229, 310
Jindal, V., 299
Johnson, B.K., 324
Johnstone, R.E., 310
Jones, A.G., 56, 60, 63, 76, 78, 98, 135, 136, 137, 151, 153, 157, 159, 169, 174, 175, 177, 180, 181, 182, 183, 184, 185, 186, 187, 191, 195, 197, 198, 199, 200, 201, 202, 216, 217, 219, 220, 221, 222, 223, 224, 225, 226, 228, 229, 231, 233, 235, 237, 238, 239, 242, 244, 245, 246, 247, 248, 250, 251, 252, 253, 254, 255, 256, 259, 261, 267, 268, 278, 279, 280, 282, 285, 286, 287

Jones, G.L., 240, 312
Jongen, N., 305, 312, 314, 321, 322, 325
Juvekar, V.A., 312

Kabel, R.L., 301
Kagara, K., 315
Kalogerakis, N., 305
Karfunkel, H.R., 314
Kashiki, I., 163, 312
Katz, S., 52, 252, 310, 313, 314, 322
Kavanagh, J.P., 181, 312
Kaye, B.H., 7, 12, 312
Keey, R.B., 80, 312
Kelkar, B.G., 322
Kendall, R.E., 307
Kern, R., 261, 307
Khan, A.R., 312
Kick, F., 139, 140, 312
Kim, K.-J., 63, 312
Kim, W.-S., 223, 312
Kimpel, R., 299
Kirk-Othmer, 232, 312
Kirwan, D.J., 215, 314
Kleeman, A., 320
Klein, J.P., 306, 314, 315
Klekar, S.A., 147, 313
Klimpel, R.R., 176, 312
Kobayashi, K., 324
Kokossis, A.C., 312
Kolar, V., 312
Kolmogorov, A.N., 45, 312, 315
Kontomaris, K., 307
Kossel, W., 313
Kotaki, Y., 313
Koutsoukos, P.G., 323
Kozeny, J., 42, 313
Kralj, D., 301
Kraljevich, Z., 319
Kramer, H.J.M., 302, 310, 313
Krei, G.A., 313
Krishna, R., 322
Kruyt, H.R., 317
Kuboi, R., 142, 313
Kubota, N., 304, 313
Kuipers, J.A.M., 313
Kumar, S., 56, 175, 176, 221, 313
Kumazawa, H., 321
Kynch, G.J., 87, 313

Lahav, M., 4, 130, 313
Lai, X., 308
Landau, D.P., 157, 305
Landau, L., 165, 304
Lapple, C.E., 313
Laroche, R.D., 299

Larson, M.A., 7, 17, 26, 52, 54, 55, 58, 70, 75, 136, 141, 147, 148, 166, 175, 177, 190, 194, 197, 210, 211, 212, 213, 281, 291, 299, 306, 307, 308, 310, 313, 319, 324
Latham, D., 308
Laufhütte, H., 142, 313
Laufhütte, H.D., 315
Laveda, M.L., 87, 306
Lawn, B.R., 313
Lee, C., 321
Lee, P.L., 288, 313
Lee, Y.-M., 313
Lei, S., 291, 313, 314
Leiserowitz, L., 4, 130, 313
Lemaître, J., 257, 305, 312, 314, 325
Leschonski, K., 309
Leslie, M., 303, 326
Leusen, F.J.J., 6, 314
Levenspiel, O., 314
Levich, V.G., 314
Li, Zhou, 312
Linhoff, B., 281, 314
Litster, J.D., 314
Ljung, L., 314
Lommerse, J.P.M., 4, 314
Lorimer, J.W., 310
Low, G.C., 156, 314
Löwe, H., 321
Luecke, R.H., 142, 307
Luss, R., 275, 314
Lyklema, J., 161, 314

M. Leslie, C.R., 303, 326
Ma, D.L., 289, 303, 314
McCabe, W.L., 127, 130, 193, 194, 315, 321
McGrath, G., 48, 220, 327
McNeil, T.J., 135, 315
Mahajan, A.J., 215, 314
Malalasekera, W., 48, 325
Mandlebroot, B.B., 12, 314
Mann, R., 326
Manna, L., 314
Manninen, M., 47, 220, 314
Mantovani, G., 160, 314
Marcant, B., 216, 304, 314
Marchal, P., 314
Margolis, G., 305
Marmo, L., 314
Marquis, A.J., 45, 186, 320
Marshall, V.R., 310
Masters, K., 80, 314
Masuda, H., 324
Masumi, F., 304
Matejcková, E., 261, 323
Mathews, H.B., 269, 289, 314, 315

Matsubara, T., 327
Matteson, M.J., 80, 315
Mayrhofer, B., 197, 315
Mazzarotta, B., 142, 151, 303, 315
Mehta, R.V., 51, 303, 315
Meklenburgh, J.C., 252, 315
Melvin, M.V., 312
Merkus, H.G., 310
Merrifield, D., 308
Mersmann, A., 45, 58, 63, 123, 135, 136, 142, 150, 153, 173, 190, 215, 259, 306, 307, 308, 312, 313, 315, 316, 317, 318, 320, 326
Middleman, S., 44, 315
Miers, H.A., 315
Millard, B., 301
Miller, S.M., 315, 319, 326
Mingers, J., 315
Mischenko, E.F., 318
Misra, C., 135, 315
Mitrovic, A., 307
Moggridge, G.D., 261, 276, 277, 303
Momonaga, M., 315
Monnier, O., 289, 315
Mooij, W.T.M., 314
Morikawa, H., 326
Moroyama, K., 235, 315
Moss, A.A.H., 80, 317
Motherwell, W.D.S., 314
Mougin, P., 308
Mullin, J.W., 1, 3, 58, 61, 62, 63, 116, 127, 131, 132, 133, 134, 135, 190, 194, 195, 197, 199, 200, 201, 215, 288, 301, 302, 306, 307, 311, 316, 323, 325
Mumtaz, H.S., 173, 186, 187, 310, 316
Muralidar, R., 177, 316
Mydlarz, J., 76, 153, 159, 267, 268, 311, 316
Myerson, A.S., 1, 4, 58, 123, 190, 215, 316, 319

Nagashima, S., 327
Nagata, S., 308
Nagi, G.K., 303
Nakamuto, Y., 160, 299
Nancollas, G.H., 182, 306, 316, 324
Natarajan, R., 162, 316
Naumova, T.N., 316
Ness, J.N., 135, 316
Neumann, A.M., 313
Neville, J.M., 316
Newton, I., 316
Ng, K.M., 56, 142, 175, 180, 261, 273, 275, 277, 278, 300, 303, 305, 309, 316, 319, 326
Ni, H., 318
Nicmanis, N., 56, 316

Author index

Nieh, J.Y., 325
Nieken, U., 327
Nielsen, A. E., 134, 182, 317
Nienoord, M., 308
Nienow, A.W., 44, 142, 143, 144, 303, 308, 313, 317
Nocedal, J., 301
Nonhebel, G., 80, 317
Noor, P., 135, 317
Noordik, J.H., 314
Nývlt, J., 58, 136, 190, 195, 196, 197, 281, 287, 302, 315, 316, 317

O'Hara, M., 127, 129, 317
Ohsol, E.O., 317
Ohtaki, H., 317
Oldshue, J.Y., 26, 44, 227, 317
Oliver, R., 123, 155, 308, 322
Olshen, R.A., 301
O'Meadhra, R.S., 313
Orciuch, W., 56, 229, 230, 300
Orr, C., 80, 315
Orszag, S.A., 48, 307
Osborne, D.G., 87, 317
Ostwald, W., 58, 77, 232, 317
Ottens, E.P.K., 142, 143, 149, 150, 317
Ouchyama, N., 17, 317
Overbeek, J.Th., 165, 317, 325
Oyanader, M.A., 317

Paine, K., 293, 320
Palmer, R.A., 318
Palmour, H., 304
Parfitt, G.D., 161, 317
Parks, R.M., 291, 317
Paterson, W.R., 316
Patwardhan, V.S., 34, 317
Penney, W.R., 324
Pepin, X., 174, 318
Perdok, W.G., 4, 308
Perkins, J.D., 261, 300
Petanate, A.M., 141, 318
Peters, R.W., 324
Pethica, B.A., 164, 318
Phillips, V.R., 296, 305
Pietsch, W., 123, 155, 318
Plasari, E., 299, 309
Ploû, R., 142, 150, 318
Plummer, L.N., 41, 42, 233, 302, 318
Plummer, W.B., 41, 42, 233, 302, 318
Podgorska, W., 300
Pohorecki, R., 51, 236, 300, 318
Polisch, J., 318
Pontryagin, L.S., 198, 318
Pope, S.B., 56, 318
Porter, R.L., 304

Potter, B.S., 4, 312, 318
Powers, G.J., 321
Powers, H.E.C., 142, 318
Prasad, P.B.V., 160, 318
Pratola, F., 174, 318
Prem, H., 318
Price, S.L., 4, 299, 303, 314, 318, 326
Prior, M.H., 79, 122, 123, 137, 192, 201, 255, 264, 298, 318
Purchas, D.B., 80, 87, 97, 106, 318
Purich, D.L., 302

Qian, R., 318
Qui, Yangeng, 136, 318
Quinlan, J.R., 319

Radke, C., 299
Rajagopal, S., 261, 273, 274, 319
Raleigh, Lord, 319
Ramkrishna, D., 26, 56, 175, 176, 177, 221, 248, 313, 316, 319, 322, 327
Ranade, V.V., 47, 220, 319
Randolph, A.D., 7, 17, 26, 52, 54, 55, 58, 60, 65, 70, 75, 136, 141, 147, 153, 166, 171, 172, 175, 177, 179, 184, 190, 194, 210, 211, 212, 213, 281, 291, 292, 300, 303, 308, 319, 321, 323
Rasmuson, Å.C., 318
Rasmussen, P., 237, 323
Raven, K.D., 316
Rawlings, J.B., 269, 287, 288, 289, 292, 305, 314, 315, 319, 326
Ray, W. H., 287, 299
Reddy, M.M., 233, 316
Redman, T., 293, 294, 295, 320
Reid, R.C., 127, 129, 317
Rekoske, J.E., 320
Rhodes, M., 320, 323, 324
Rhodes, M.J., 318
Rice, R.G., 284, 320
Rice, R.W., 320
Richardson, J.F., 7, 26, 32, 34, 37, 40, 80, 84, 87, 106, 116, 121, 303, 312, 320, 323
Rielly, C.D., 45, 186, 320
Rigopoulos, S., 252, 255, 256, 320
Ristic, R.I., 153, 320
Ritchie, B.W., 51, 320
Roberts, D., 308
Roberts, K.J., 304, 308
Rohani, S., 287, 288, 290, 293, 294, 320
Rosenkranz, J., 308
Rossiter, A.P., 320
Rosstti, D., 318
Roth, H.J., 5, 320
Rousseau, R.W., 151, 291, 307, 317, 321
Rovang, R.D., 291, 321

Rowe, P.N., 33, 321
Rubbo, M., 314
Rudd, D.F., 261, 275, 303, 321
Rumpf, H., 7, 321
Rusli, I.T., 307
Russel, W.B., 161, 321
Rutti, A., 157, 304
Ryall, R.L., 301, 310

Sada, E., 233, 321
Saeman, W.C., 321
Sangl, R., 315
Saraiva, P.M., 279, 321
Sarofim, A.F., 305
Sato, A., 304
Saunders, T.E., 17, 301
Savage, K., 299
Savelli, N., 308
Scarlett, B., 310
Schechter, R.S., 162, 316
Schenk, R., 321
Schierholtz, P.M., 135, 321
Schmid, C., 301
Schmidt, M.U., 131, 314
Schonert, K., 138, 308, 321
Schreiner, A., 258, 322
Schubert, H., 174, 322
Schuler, H., 322
Schweizer, B., 314
Scott, K.J., 17, 18, 87, 322
Scrutton, A., 322
Seaton, N.A., 316
Seidell, A., 61, 322
Seider, W.D., 261, 316
Seinfeld, J.H., 56, 168, 307
Sengupta, B., 248, 322
Shah B. C., 145
Shah, B.H., 322
Shah, M.B., 147, 148, 306
Shah, Y.T., 235, 322
Sharma, M.M., 233, 312
Sharratt, P.N., 297, 322
Shaw, D.J., 241, 322
Sheikh, A.Y., 278, 279, 285, 291, 322
Shekunov, B.Yu., 61, 322
Sheldrake, G.N., 327
Sherrington, P.J., 155, 322
Sherwin, M.B., 295, 322
Sherwood, J.N., 131, 153, 320
Shimizu, K., 313
Shinnar, R., 322
Shippey, T.A., 151, 152, 321, 322
Shock, R.A.W, 296, 322
Shoji, V., 299
Shripathi, T., 320
Sicardi, S., 314

Sie, S.T., 322
Sikdar, S.K., 75, 147, 319, 323
Simons, S.J.R., 12, 174, 318, 323
Sink, C.W., 319
Sirola, J.J., 321
Skovborg, P., 237, 323
Skrtic, D., 301
Smit, J.D., 314
Smith, G.D., 48, 323
Smoluchowski, M.V., 170, 173, 178, 185, 186, 187, 189, 323
Söhnel, O., 58, 62, 76, 123, 125, 127, 132, 133, 134, 172, 173, 183, 215, 261, 266, 302, 317, 323
Sonobe, R., 327
Sowul, L., 136, 323
Spanos, N., 323
Standish, N., 17, 327
Stanley-Wood, N.G., 123, 155, 323
Stein, P.C., 186, 302
Stephanopoulos, G., 278, 279, 321
Stephen, H., 61, 323
Stephen, T., 61, 323
Šterbácek, Z., 26, 44, 323
Stevens, J.D., 135, 299, 321, 324
Stokes, G.G., 323
Stone, C.J., 301
Strathdee, G., 320
Strickland-Constable, R.F., 323
Stroganova, L.A., 316
Sullivan, G.R, 288, 313
Sung, C.V., 310
Sung, C.Y., 142, 324
Suzuki, A., 163, 312
Suzuki, M., 324
Svarovsky, L., 80, 115, 317, 324, 326, 327
Svrcek, W.Y., 299
Swinney L.D., 259, 324
Synowiec, P., 142, 143, 144, 145, 146, 149, 150, 179, 324
Syrjänen, J., 47, 220, 314

Tai, C.Y., 324
Talmage, W.P., 87, 324
Tambour, Y., 307
Tan, C., 281, 319
Tanaka, T., 17, 317
Tanimoto, A.K., 135, 324
Tarbell, J.M., 51, 223, 303, 312, 315
Tausk, P., 26, 44, 323
Tavana, A., 319
Tavare, N.S., 51, 58, 135, 136, 153, 190, 191, 197, 236, 307, 310, 320, 324
Taylor, R.L., 48, 327
Tengler, T., 318

Teodossiev, N.M., 201, 202, 311
Thomas, A., 308
Thompson, P.D, 179, 246, 324
Thring, M.W., 310
Tien, Chi, 317
Timm, D.C., 136, 147, 291, 324
Tipnis, S.K., 324
Toft, J.M., 181, 182, 317
Togby, A.H., 51, 320
Togkalidou, T., 269, 324
Tomazic, B., 324
Torbacke, M., 229, 324
Tosun, G., 325
Tsai, Y.-C., 288, 305, 306
Tsai, Yen-Cheng, 288, 305, 306
Tsuge, H., 233, 313
Tsutsumi, A., 240, 325

Ueyama, K., 235, 325, 326
Uhl, V.W., 26, 44, 325
Utgoff, P.E., 301, 325

Vacassy, R., 258, 314, 325
Vaccari, G., 314
van den Akker, H.E.A., 302, 309
van der Eerden, J.P., 325
van Der Heijden, A.E.D.M., 153, 325
van Eijck, B.P., 314
van Hook, A., 190, 325
van Leeuwen, M.L.J., 52, 216, 325
van Rosmalen, G.M., 302, 309, 313, 322, 325
van Swaaij, W.P.M., 47, 313, 326
van't Land, C.M., 151, 325
Vega, A., 288, 325
Versteeg, H.K., 325
Vervey, E.J.W., 165, 325
Verwer, P., 314
Villermaux, J., 51, 216, 299, 303, 306, 308, 309, 314, 325
Virtanen, J., 288, 325
Voit, H., 307
Volmer, M., 325
Von Buschmann, E., 313
von Rittinger, P.R., 139, 325

Wachi, S., 231, 233, 235, 237, 239, 250, 253, 256, 259, 312, 326
Wadell, H., 326
Wakeman, R.J., 17, 80, 87, 116, 318, 326
Waldram, S.P., 306
Wallas, 114, 326
Wallis, G.B, 28, 326
Walton, A.G, 326
Wang, M.H., 299
Wang, S., 315
Wang, Y.-D., 326
Watson, C.C., 261, 321
Wedlock, D.J., 161, 311, 326
Weed, D.R., 315
Wei, H.Y., 216, 326
Werther, J., 308
Westerterp, K.R., 326
White, E.T., 135, 153, 156, 314, 315, 316, 319
White, G., 308
Wibley, K.S., 318
Wibowo, C., 277, 326
Wienk, B.G., 151, 325
Wilen, S.H., 310
Wilkinson, D., 308
Williams, D.E, 314
Williams, H.L., 308
Williams, R.A., 161, 326
Willock, D.J., 4, 303, 326
Wilshaw, T.R., 138, 313
Wisniak, J., 326
Withnall, R., 318
Witkowski, W.R., 136, 319, 326
Woinaroschy, A., 273, 326
Wojciechowski, K., 320
Wójcik, J., 56, 174, 184, 185, 187, 326, 327
Wolff, P.R., 281, 286, 313
Wood, W.M.L, 5, 308, 327
Wright, H., 248, 327
Wuklow, M., 327
Wulff, G., 4, 327
Wynn, E.J.W., 278, 281, 310

Xu, Y., 48, 220, 327

Yagi, H., 233, 327
Yates, B., 261, 327
Yates, F., 327
Yazawa, H., 315
Yokota, M., 304
Youngquist, G.R., 310, 324
Yu, A.B., 322, 327
Yu, S., 17, 301

Zaki, W.N., 32, 34, 84, 320
Zauner, R., 47, 175, 177, 180–7, 216, 217, 219–26, 228, 229, 305, 327
Zeitsch, K., 106, 327
Zheng, C., 303
Zienkiewicz, O.C., 48, 327
Zlokarnik, M., 327, 328
Zweitering, T.N, 43, 328

Subject index

Acetic acid, 234
Adducts, 6
Adipic acid, 273, 275
Adsorption layer, 126
Ageing of precipitate, 77, 232
Agglomerate particle form, 155–61
　agglomerative crystallization, 155–7
Agglomeration:
　disruption during precipitation, 171–4
　collection efficiency, 173–4
　strength, 174
Algebraic stress model (ASM), 47
Alum, 159
Ammonium:
　phosphate, 232
　sulphate, 143, 147, 232
Analytical approximations, population balance, 168–71
　continuous MSMPR, 168
　nucleation and agglomeration, 169
　nucleation and crystal growth, 168
　orthokinetic aggregation, 170–1
　perikinetic aggregation, 170
　well-mixed batch, 170
Anti-solvents, 60–1
Aragonite, 4
Artificial neural networks (ANN), 273
Attenuated total reflection (ATR), 289

Barium sulphate, 52, 135, 217, 223, 229, 230, 232
Batch crystallizer design and operation, 190–2, 192–3
　heat and mass balances, 192–3
　operating policy, 192
　performance measures, 192
　vessel sizing, 192
Benzoic acid, 229
BFDH theory, 4
Bond's law, 139
Buchner funnel, 88

Calcite, 4, 224, 225, 233
Calcium
　chloride, 221
　dichloride, 233
　hydroxide, 233
　nitrate, 233
　oxalate trihydrate (COT), 160

Carman–Kozeny equation, 40, 42, 91, 97, 265
Centrifugal filtration, 112–14
Centrifugation, theory of, 107
　separating effect, 107
Centrifuge types, classification of, 106–7
　construction, 106
　mechanical construction, 107
　operation, 106
　particle size, 106
Chlorobenzoic acids, 277
Coalescence-redispersion (CRD) model, 51
Colloid and surface science 161–6
　adsorption, 164
　charge, 162–3
　particle forces, 162
　particle interactions, 163
　stability, 164–6
　steric interactions, 163
Computational fluid dynamics (CFD), 47, 57, 220, 251
Computer aided design (CAD), 48
Continuous crystallizers, design and performance of 203–14
　mass balance, 203
　MSMPR interactions, 203
Continuous rotary vacuum filtration, 103–6
　effect of drum speed, 105–6
　operating equations, 104–5
　operating principle, 103–4
　submergence fraction, 104
Copper sulphate, 143
Coulter Counter, 19, 181, 225, 291
Crystal agglomeration and disruption, 155–89
Crystal drying, 116–120
　time, 120
　types of drier, 120–2
Crystal growth and nucleation kinetics, determination of, 135
　batch methods, 135–6
　continuous method, 136
Crystal number, area and mass, 71
　anomalous crystal growth, 75–6
　coefficient of variation, 75
　distributions, 71–2

Crystal number, area and mass (*Continued*)
  mean size, 75
  mean sizes of the CSD, 74–5
  median size, 72
  modal size, 73–4
Crystal size distribution (CSD), 54, 64, 65, 66, 67, 69, 76, 79, 135, 136, 145, 152, 153, 167, 195, 201, 203, 204, 211, 214, 230, 248, 260, 264, 265, 266, 268, 284, 285, 287, 289
Crystal:
  characteristics, 1–7
  defects, 6
  formation and breakage, 123–54
  growth, 125–31
  habit, 4
  particle, 7
  polymorphs and enantiomorphs, 46
  structure, 1–2
  symmetry, 2
  systems, 3
  washing, 116
Crystal, equilibrium moisture content, 116–120
  drier design equations, 119–20
  drying periods, 119
  drying rates, 118
  drying tests, 117–18
  total drying time, 119
  wet-bulb protection, 120
Crystallization practice, 295–7
  crystal caking, 296
  crystal encrustation, 296
  scale-up 295–6
  sustainability, 297
Crystallization process instrumentation and control, 287–95
  batch crystallizers, 287–9
  black box modelling for dynamics, 293–5
  continuous crystallizers, 289–91
  physical model based control, 291–3
Crystallization process synthesis, 271–7
  biotechnological systems, 276–7
  costs, 271–2
  design level, 272
  multi-component systems, 275–6
  operational level, 272–5
  procedure, 277
Crystallization process systems simulation, 277–87
  machine learning methodology, 278–9
  network synthesis methodology, 281–2
  problem formulation of crystallizer networks, 282–7
  stage-wise crystallization process synthesis, 279–81

Crystallization:
  design data, 263–4
  equipment, 262–3
  precipitation kinetics, 123
  principles and techniques, 58–79
  systems design, 261–98
Crystallizer:
  design and performance, 190–214
  mixing and scale-up, 215–60
CSD, prediction of 193–7
  control of supersaturation, 195–7
  population balance, 194–5

Darcy's law, 38
  equation, 90
Derjaguin–Landau and Verwey–Overbeek (DLVO) model, 164
Diamond 4
Diffusion-reaction model, 127
Direct numerical solution (DNS), 46, 47, 57
Down time, 101, 102

Effect of:
  bed voidage and particle size, 39–40
  fluid viscosity, 38
Electrical double layer, surface charge, 162–3
Energy for size reduction, 139–41
Engulfment-deformation-diffusion (EDD) theory, 51
Ergun equation, 43

Filter:
  medium, 88, 90, 95, 97, 102, 103, 112
  and filtration methods 87–8
Filtration time, 101–3, 105, 113–14
Filtration, theory of 90–7
  effect of the filter medium, 95–6
  modified Darcy equation, 90–2
  simplified filtration equation, 96–7
Fines suspension density sensor (FSDS), 293
Fluid flow through porous media, 37–43
  compressible beds, 40
  empirical methods, 41
  laminar flow, 38
  turbulent flow, 41
Fluid–particle transport processes, 26–57
Forced circulation (FC), 64, 65, 288, 290, 293
Fourier transform infrared (FTIR), 289
Friction factors, 41–3
  packed bed friction factor, 41
  packed bed Reynolds number, 42–3

Gas–liquid crystal precipitation, 231–57
  industrial gas–liquid precipitation
    reactions, 232–4
  reactor designs, 255–7
  types of reactor, 234–6
Gas–liquid reactor modelling, 250–2
  bubble column reactor modelling,
    252–5
General model control (GMC), 288
Gibbs–Thompson equation, 61, 125
Graphite, 4
Gypsum, (calcium sulphate) 232, 233, 281

Hagen–Poisseuille equation, 39
Hexamethylene tetramine (HMT), 157
Hindered drag coefficients, Slurries, 33–4
  force balance, 33–4
Hydrocyclones, 114–15
Hydrodynamics determinants of crystal
  growth rates, 130–1

Induction periods, 131–5
  interfacial tension, 135
Interaction by exchange with mean (IEM)
  model, 51
Inter-model control (IMC), 294
Isomorphous substances, 4

Jar test, 35, 36, 84

Kick's law, 139
Kinetic energy (KE), 28, 41, 46

Lamella settler, 82
Large eddy simulation (LES), 47
Laser Doppler anemometry (LDA), 45, 48
Leaf test, 97–100
Liquid–liquid crystallizaton system,
    215–31
  continuous mode of operation, 221–2
  determination of mixing times
    using Computational Fluid
    Dynamics, 220
  effect of feed point position, 226–7
  mixing model: Segregated Feed Model
    (SFM), 216–17
  model solutions, 221
  model validation, 221
  probability density function (PDF)
    methods, 229–31
  scale-up criteria, 227
  semibatch mode of operation, 223–6
  SMF applied to continuous
    precipitation, 217–19
  SMF methodology, 228–9
  volume mean size, L_{43}, 223

Magnesium oxide, 277
  sulphate, 147, 276
Mass deposition rate, 203–4
  classified product crystal
    withdrawal, 212
  crystal size-dependent product removal
    crystallizers, 210
  crystallizer dynamics, 212–14
  design equations for MSMPR
    crystallizers, 205–10
  fines destruction, 210–12
  supersaturation, 204
Mass separating agent
    (MSA), 277
Mean:
  particle size, 14
  sizes of distributions, 14–16
  surface size (d_{ms}), 16
  weight (or volume) size
    (d_{mw}), 15
  surface mean size (d_{sm}), 15–16
  weight (or volume) mean size
    (d_{wm}), 14–15
Microcrystallizers, 257–9
Mixed suspension, mixed product removal
    (MSMPR) crystallizer, 65–8
  continuous MSMPR operation, 68
  crystal yield, 66
  mass balance, 66–7
  modelling, 66
  prediction and analysis of
    CSD, 67–8
Mixing models, 49–51
  macromixing and micromixing, 49
  mesomixing and micromixing, 50
  phenomenological or mechanistic,
    50–1
  physical, 51–2
Models, crystallization process,
    264–71
  crystal size distribution, 264
  crystal slurry filtrability, 268–9
  crystallizer volume, 265
  effect of downstream processing on
    product characteristics, 270–1
  improving solid–liquid separability,
    269–70
  kinetics, 267–8
  median crystal size, 264–5
  particle characteristics, 266–7
  solid–liquid separation, 265–6
  solids hold-up, 265
Modes of operation, 92–5
  constant pressure filtration, 93–4
  constant rate filtration, 93
Monodisperse particles, 32–3

Nickel ammonium sulphate, 143
Nucleation, 124–5
  heterogeneous, 124, 125, 132
  homogeneous, 124, 125, 132, 134
  primary, 125

Optimum cycle time during batch filtration, 101–3
  maximizing filter throughput, 102–3
  pre-coat and bodyfeed, 103
Ordinary differential equation (ODE), 55, 95, 177, 221, 282
Oswald–Freundlich equation, 61
  ripening, 77, 232

Packing characteristics, 17–18
  effect of size distributions, 18
  voidage and internal surface area, 17–18
Palmitic acid, 160
Particle breakage processes, 136–41
  nature of comminution, 137
  single, 137–8
  typical comminution, 138
Particle size distribution, 12–17
  coefficient of variation (CV), 16
  frequency histogram, 12
  mass fractions, 12–13
  specific surface (S_p), 16
Particle–fluid hydrodynamics, 26–7
Particulate crystal characteristics, 1–25
PBC theory, 4
Periodic bond chain (PBC), 4
Phase equilibria, 58–63
  crystallization and precipitation modes, 61–2
  solubility, 58–60
  supersaturation, 62–3
Phosphoric acid, 232, 281
Plug flow reactor (PFR), 49, 286
Population balance, 52–7, 68–76
  moment transformation, 54–5
  MSMPR crystal size distribution (CSD), 69–71
  numerical methods, 56–7
  turbulent, 56
Potassium:
  chloride, 160, 234, 276
  nitrate, 147, 279, 285
  sulphate, 147, 151, 159, 160, 197–9, 249, 267, 268, 269
Precipitated calcium carbonate (PCC), 233
Precipitation kinetics, determination of 175–87
  agglomeration kernel, 178, 185–7
  crystal growth rates, 181–2
  disruption kernel, 179–80, 184–5

  experimental method, 180–1
  growth and nucleation, 177–8
  invariant agglomeration kernel, 179
  nucleation rates, 182–3
  parameter estimation, 177
  Smoluchowski kernel, 178
  Thompson kernel, 179
Precipitation, 76–9
  microscopic analysis, 78–9
  phase transformation, 77–8
Primary agglomeration, 157
  dendrite growth, 157–8
  parallel growth, 158–9
Probability density function (PDF), 229, 230, 231
Product particle size distribution, prediction of 141–6
  hydrodynamic determinants of attrition, 142
  impact attrition, 143
  population balance, 141–2
  turbulent attrition, 143–6
Programmed cooling crystallization, 197–202
  fines destruction during batch crystallization, 201
  optimal operation, 197–9
  programmed cooling crystallizer, 199–201
  programmed precipitation, 201–2

Residence time distribution (RTD), 49, 50
Reynolds stress model (RSM), 47
Richardson and Zaki equation, 34, 84
Rittinger's law, 139
Role of gas–liquid mass transfer, 236–55
  Monte Carlo methodology, 249–50
  Monte Carlo simulation, 248–9
Rotary vaccum filter (RVF), 88
Rotation–reflection axis, 2

Salicylic acid, 51
Secondary nucleation, 146
  effect of supersaturation, 151–4
  growth of secondary nuclei, small crystals and attrition fragments, 151
  kinetics, 148–50
  mechanisms of, 148
  MSMPR crystallizer studies, 147–8
  scale-up of 150–1
Sedimentation, 27–31
  drag on particles, 28
  force balance, 29
  laminar flow: Stokes law ($Re_p < 0.2$), 29–30
  particle Reynolds number, 30

terminal velocity in Stokes' flow ($Re_p < 0.2$),
turbulent flow, 30–1
Segmented flow tubular reactor (SFTR), 257
Segregated feed model (SFM), 49, 216, 229
Settling tanks, 80–2
Shape factors of single particles, 9
  specific surface (F), 10
  specific surface (S_p), 10
  sphericity (Ψ), 10
  surface (f_s), 9
  volume (f_v), 9–10
Single input single output (SISO), 291
Single particle size, 7–12
  cube, 8
  equivalent dimensions, 9
  fractals, 12
  irregular particles, 8–9
  shape, 9
  sphere, 8
Slurries, 31–7
  hindered settling rates, 31–3
  mass flux, 37
  polydisperse suspensions, 34–5
  sedimentation of slurries, 35–6
  typical settling rates, 36–7
Sodium:
  chloride, 181
  sulphate, 233, 276
Solid bowl centrifuge, 107–9
Solid–liquid separation processes, 80–122
Sulphur oxide, 233
Sulphuric acid, 232
Supercritical fluids (CSFs), 60
Surface integration models, 128
  birth and spread (B&S), 129–30
  continuous growth, 128
  screw dislocation, or BCF, 128–9
Suspension of settling particles by agitation, 43–9
  computational fluid dynamics (CFD), 47–8
  conservation equations, 45–7
  energy dissipation, 44–5
  stimulation of fluid dynamics in agitated vessels, 48–9
Sympatec Helos laser scattering analyser, 225

Techniques for particle sizing and characterization, 18–22
  data presentation and analysis from sizing tests, 21–2
  laser diffraction, 19
  mesh number (N), 20
  Sieve test, 19–20
  Sieve test method, 21
  zone sensing, 19
Terephthalic acid, 231, 234
Thermodynamic theory, 166
Thickeners, 82–7
Thin layer solid bowl centrifuge, 109–12
Total annual cost (TAC), 271
Troubleshooting agglomeration, 187–9
Twining, 159–60
  secondary agglomeration, 160–1
Two-impinging-jets (TIJ) precipitator, 215
Types of crystallizer, 63–5
  agitated vessels, 63–4
  draft-tube and baffled (DTB), 64–5
  oslo fluidized-bed, 65
  swenson forced circulation (FC), 65
  unstirred vessels, 63
Types of filter, 88–90
  belt, 90
  press, 88
  rotary vacuum, 88–90

Vaterite, 4, 224, 225, 259

Whatman cellulose nitrate, 181

Zeolite, 78, 79